ENVIRONMENTAL REMEDIATION TECHNOLOGIES, REGULATIONS AND SAFETY

RENEWABLE ENERGY GRID INTEGRATION: TECHNICAL PERFORMANCE AND REQUIREMENTS

ENVIRONMENTAL REMEDIATION TECHNOLOGIES, REGULATIONS AND SAFETY

Additional books in this series can be found on Nova's website under the Series tab.

Additional E-books in this series can be found on Nova's website under the E-books tab.

ENVIRONMENTAL REMEDIATION TECHNOLOGIES, REGULATIONS AND SAFETY

RENEWABLE ENERGY GRID INTEGRATION: TECHNICAL PERFORMANCE AND REQUIREMENTS

MITCHELL B. FERGUSON
EDITOR

Nova Science Publishers, Inc.
New York

Copyright ©2011 by Nova Science Publishers, Inc.

All rights reserved. No part of this book may be reproduced, stored in a retrieval system or transmitted in any form or by any means: electronic, electrostatic, magnetic, tape, mechanical photocopying, recording or otherwise without the written permission of the Publisher.

For permission to use material from this book please contact us:
Telephone 631-231-7269; Fax 631-231-8175
Web Site: http://www.novapublishers.com

NOTICE TO THE READER

The Publisher has taken reasonable care in the preparation of this book, but makes no expressed or implied warranty of any kind and assumes no responsibility for any errors or omissions. No liability is assumed for incidental or consequential damages in connection with or arising out of information contained in this book. The Publisher shall not be liable for any special, consequential, or exemplary damages resulting, in whole or in part, from the readers' use of, or reliance upon, this material. Any parts of this book based on government reports are so indicated and copyright is claimed for those parts to the extent applicable to compilations of such works.

Independent verification should be sought for any data, advice or recommendations contained in this book. In addition, no responsibility is assumed by the publisher for any injury and/or damage to persons or property arising from any methods, products, instructions, ideas or otherwise contained in this publication.

This publication is designed to provide accurate and authoritative information with regard to the subject matter covered herein. It is sold with the clear understanding that the Publisher is not engaged in rendering legal or any other professional services. If legal or any other expert assistance is required, the services of a competent person should be sought. FROM A DECLARATION OF PARTICIPANTS JOINTLY ADOPTED BY A COMMITTEE OF THE AMERICAN BAR ASSOCIATION AND A COMMITTEE OF PUBLISHERS.

Additional color graphics may be available in the e-book version of this book.

Library of Congress Cataloging-in-Publication Data

Renewable energy grid integration. Technical performance and requirements / editor, Mitchell B. Ferguson.
  p. cm.
 Includes bibliographical references and index.
 ISBN 978-1-60741-325-7 (hardcover : alk. paper)
 1. Photovoltaic power systems. 2. Building-integrated photovoltaic systems. 3. Interconnected electric utility systems. 4. Distributed generation of electric power. 5. Renewable energy resources--Technological innovations. I. Ferguson, Mitchell B.
 TK1087.R4643 2009
 621.31--dc22
                          2009038433

Published by Nova Science Publishers, Inc. † New York

CONTENTS

Preface		vii
Chapter 1	Transmission System Performance Analysis for High-Penetration Photovoltaics *S. Achilles, S. Schramm, and J. Bebic*	1
Chapter 2	Distributed Photovoltaic Systems Design and Technology Requirements *Chuck Whitaker, Jeff Newmiller, Michael Ropp and Benn Norris*	57
Chapter 3	Distribution System Voltage Performance Analysis for High-Penetration Photovoltaics *E. Liu and J. Bebic*	107
Chapter 4	Power System Planning: Emerging Practices Suitable for Evaluating the Impact of High-Penetration Photovoltaics *J. Bebic*	141
Chapter 5	Enhanced Reliability of Photovoltaic Systems with Energy Storage and Controls *D. Manz, O. Schelenz, R. Chandra, S. Bose, M. de Rooij, and J. Beb*	171
Chapter 6	Renewable Systems Interconnection Study: Cyber Security Analysis *Annie McIntyre*	211
Chapter Sources		233
Index		235

PREFACE

Now is the time to plan for the integration of significant quantities of distributed renewable energy into the electricity grid. Concerns about climate change, the adoption of state-level renewable portfolio standards and incentives, and accelerated cost reductions are driving steep growth in U.S. renewable energy technologies. The number of distributed solar photovoltaic (PV) installations, in particular, is growing rapidly. As distributed PV and other renewable energy technologies mature, they can provide a significant share of our nation's electricity demand. However, as their market share grows, concerns about potential impacts on the stability and operation of the electricity grid may create barriers to their future expansion. This book describes the Renewable Systems Interconnection (RSI) that the U.S. Department of Energy launched in 2007 to facilitate more extensive adoption of renewable distributed electric generation. The technical and analytical challenges that must be addressed to enable high penetration levels of distributed renewable energy technologies are also addressed in this book. Because integration-related issues at the distribution system are likely to emerge first for PV technology, the RSI study focuses on this area. This is an edited, excerpted and augmented edition of various government publications.

Chapter 1- A combination of incentive programs, technology advancements, and further increases in prime energy costs could result in vast deployment of photovoltaics (PV) in power distribution systems. In such a scenario, power systems could be characterized by large quantities of generation embedded throughout the electric power system. Wind generation has been massively installed in some areas of the world (such as Spain and northern Germany) and has had a substantial impact on power system performance. For example, transmission faults that were previously characterized by short voltage sags can become significant system events with large power imbalance issues between control zones because a considerable amount of wind generation was disconnected during faults. Thus, wind generation technologies provide enhanced performance characteristics such as tolerance to voltage sags. Although PV generation is more distributed in nature than wind generation, many concerns about significant wind penetration are relevant to PV.

Chapter 2- Distributed photovoltaic (PV) systems currently make an insignificant contribution to the power balance on all but a few utility distribution systems. Interest in PV systems is increasing and the installation of large PV systems or large groups of PV systems that are interactive with the utility grid is accelerating, so the compatibility of higher levels of distributed generation needs to be ensured and the grid infrastructure protected. The

\variability and nondispatchability of today's PV systems affect the stability of the utility grid and the economics of the PV and energy distribution systems.

Chapter 3- Currently, electrical distribution systems are designed and operated based on the assumption of centralized generation, with the corollary that the power always flows from the distribution substation to the end-use customers. With the increasing penetration of residential and commercial PV, the PV power generation could not only offset the load, but could also cause reverse power flow through the distribution system. Significant reverse power flow may cause operational issues for the traditional distribution system, including:

- Over-voltage on the distribution feeder (loss of voltage regulation).
- Increased short circuit currents, potentially reaching damaging levels.
- Protection desensitization and potential breach of protection coordination.
- Incorrect operation of control equipment that may lead to an increase in the number of operations and related equipment wear, or to further aggravation of problems that affect more equipment and more customers.

Chapter 4- This report explores the impact of high-penetration renewable generation on electric power system planning methodologies, and outlines how these methodologies are evolving to enable effective integration of variable-output renewable generation sources. All three areas of system planning are considered—generation, transmission, and distribution—and the impact of high penetration of solar PV analyzed relative to each.

Chapter 5- We observed enhanced customer reliability through the management of energy storage systems and photovoltaics (PV) during a utility outage. In an outage, the affected community would intentionally island and meet only its critical load. The timing, duration, and number of customers affected by each outage event were obtained for a single utility in 2005. These data were used to simulate outage events for a community on a distribution feeder. Overall, this technology resulted in a community experiencing fewer outages and outages of shorter duration.

Chapter 6- Integration of renewable energy into the U.S. critical power-generation infrastructure will likely continue to increase throughout the next decade. Incorporating PV energy into the existing bulk power distribution grid requires consideration of potential effects associated with that integration. Cyber security issues are among these potential effects. Addressing cyber security now affords the opportunity to optimize design and build a more secure, cost-efficient solution up front. Secure operations are beneficial in that they protect both the energy consumer and the provider, ensuring continued availability and creating greater economic stability.

Chapter 1

TRANSMISSION SYSTEM PERFORMANCE ANALYSIS FOR HIGH-PENETRATION PHOTOVOLTAICS

S. Achilles, S. Schramm, and J. Bebic

LIST OF ACRONYMS

ACL	active current limit
AVR	automatic voltage regulator
FC	frequency droop control
GE	General Electric
IEEE	Institute of Electrical and Electronics Engineers
LVRT	low-voltage ride-through
MPP	maximum power point tracking
PLL	phase-locked loop
PSLF	positive sequence load flow
PSS	power system stabilizer
PV	photovoltaic

EXECUTIVE SUMMARY

A combination of incentive programs, technology advancements, and further increases in prime energy costs could result in vast deployment of photovoltaics (PV) in power distribution systems. In such a scenario, power systems could be characterized by large quantities of generation embedded throughout the electric power system. Wind generation has been massively installed in some areas of the world (such as Spain and northern Germany) and has had a substantial impact on power system performance. For example, transmission faults that were previously characterized by short voltage sags can become significant system events with large power imbalance issues between control zones because a considerable amount of wind generation was disconnected during faults. Thus, wind generation technologies provide enhanced performance characteristics such as tolerance to voltage sags.

Although PV generation is more distributed in nature than wind generation, many concerns about significant wind penetration are relevant to PV.

This research developed an understanding of the impact of significant PV penetration on transmission system reliability and performance. A simulation database was developed to allow analysis of system behavior and interactions with high levels of PV penetration under realistic system conditions. These explorations showed that for high PV penetration, the performance requirements for the PV units are stricter to keep similar reliability and performance. Also, the criterion to dispatch and commit conventional generation for high penetration of PV has a significant impact on system behavior.

The main observations associated with systems aspects with large penetration PV follow. (See Section 6 for more detail.)

- Unit commitment strategy has a significant impact on system performance at high PV penetration levels:
  - System inertia and frequency regulation capabilities are reduced as conventional generation is de-committed.
  - Thermal units could operate at less efficient load levels
  - Reactive power support in the transmission system is reduced as conventional generation is de-committed.
  - Dynamic stability of the system can be affected.
- Considerable dispatch flexibility of conventional generation is required to accommodate high-penetration PV.
- With substantial PV penetration that is compliant with IEEE 1547,[1] there is considerable reduction in system reliability caused by extensive loss of PV generation during transmission faults.
- PV generation could provide primary frequency control for frequency excursions above nominal without significantly reducing energy production.
- Anti-islanding schemes of PV can affect the oscillatory stability of the bulk power system.

The main observations on PV potential performance are:

- The low-voltage ride-through capability of PV would reduce the negative impact on system reliability of high-penetration PV.
- Even if the PV stays connected during and after a system fault, voltage sags are prone to cause prolonged PV power output reductions.

The simulation work presented in this report focused on system performance after electrical faults and power imbalances caused by generation trips. Other aspects of PV integration that were not analyzed with simulations in this effort, but are considered relevant, follow. (For more detail see Section 6.)

- The commitment of fewer regulating units to accommodate PV generation increases the requirement of load-following reserve in the system and for individual units.

Additionally, the variability of PV generation may also increase load-following requirements.
- The implementation of voltage control on individual PV systems is challenging. There is potential for undesirable interactions between PV systems connected to the same feeder and phase and between PV systems and other voltage-regulating devices.

Based on the analysis and observations, we recommend that future research in this area:

- Develop models that are accurate enough to estimate aggregated behavior of PV systems for system planning. The behavior of aggregated PV during and after faults is most relevant. Converter technology and control can result in considerable differences between systems.
- Develop guidelines for enhancing transmission planning databases to accommodate such models, including aggregated representation of medium- and low-voltage networks.
- Improve understanding and provide guidelines to quantify the performance and economic impact of PV penetration on regulation and load-following requirements.
- Develop methodologies for estimating the required flexibility of the generation assets to meet regulation and load-following requirements; in particular, the requirements
- for generating units to ramp production up or down and to stop and start.
- Develop a unit commitment and dispatch strategy for conventional units in systems with high-penetration PV. The proposed strategy should include reliability requirements, operational costs, regulation, and load-following costs. The use of PV forecasts in unit commitment is also instrumental to increase the value of PV generation for high penetration. The approach may require that part of the unit scheduling be done a few hours in advance (instead of 24 to 48 hours currently required) to improve PV production forecast accuracy. Extending such research to systems with PV and wind generation is also recommended.
- Provide guidance to quantify the value in terms of performance and the economic benefits of potentially mitigating measures of PV power variability (modifications of control zone constraints, flexible conventional generation, centralized and local energy storage, forecast, etc.). These guidelines could be applied to different systems and generation resources. Extending such research to systems with PV and wind generation is also recommended.
- Develop methods to reliably forecast PV generation at regional levels.
- Develop methods to estimate the actual PV generation to help with system operation.
- Develop active anti-islanding schemes or tuning guidelines that do not affect regional system performance.
- Develop strategy and specification of PV voltage control.

Develop potential remuneration mechanisms for ancillary services associated with voltage support of PV.

1.0 INTRODUCTION

A combination of incentive programs, technology advancements, and a further increase in prime energy costs could result in a vast deployment of PV in future power distribution systems. In such a scenario, power systems could be characterized by large quantities of generation embedded throughout the electric power system. Wind generation that has been massively installed in some areas of the world (such as Spain and northern Germany) has had a substantial impact on power system performance and therefore should be examined when planning for high-penetration PV scenarios. For example, transmission faults that were previously characterized by short voltage sags became significant system events with large power imbalance issues between control zones because of the considerable amount of wind generation that was disconnected during the faults. Thus, wind generation technologies provided the enhanced performance required as the penetration increased, such as tolerance to voltage sags. Although PV generation is more distributed in nature than wind generation, many concerns about significant wind penetration are relevant to PV.

This research developed an understanding of the impact of significant PV penetration on transmission system reliability and performance. A simulation database was developed to allow analysis of system behavior and interactions with high levels of PV penetration under realistic system conditions. These explorations showed that for high PV penetration, the performance requirements for the PV units are stricter to keep similar reliability and performance. Also, the criterion to dispatch and commit conventional generation for high penetration of PV has a significant impact on system behavior.

Section 2 presents the simulation database and describes the developed aggregated PV model. It also includes a short description of scenarios and contingencies. Section 3 presents the results of load flow calculations and time simulations. The results are summarized in tables and plots are presented to support the conclusions. Sections 4, 5, and 6 state the needs in the industry to phase in high-penetration PV, the conclusions of this effort, and the recommended research programs in the area of transmission planning and high PV penetration.

2.0 PROJECT APPROACH

The potential impact of high levels of PV penetration was assessed by using stability simulations of a transmission system with different levels of penetration in the simulation environment: General Electric (GE) Positive Sequence Load Flow (PSLF). The transmission system data are based on an Institute of Electrical and Electronic Engineers (IEEE) benchmark system. Section 0 describes the database and its implementation on GE PSLF. This section also presents the modifications performed on the IEEE benchmark system to apply PV generation and adapt the database to the purpose of this effort.

The characteristics of the PV generation had a significant impact on the results. The analysis performed considered different behaviors of the PV systems. An ad-hoc simulation model was developed to represent the dynamic behavior of aggregated PV for technology assumptions such as IEEE 1547 compliance, voltage control, active power control, low-voltage ride-through (LVRT), and sensitivities to voltage sags.

Various starting scenarios and contingencies were considered to evaluate voltage and frequency performance with different PV characteristics and levels of PV penetration and load.

2.1 Transmission System

The transmission system database selected for this analysis is based on the IEEE 39 Bus System [1]. The system was implemented in the simulation environment GE PSLF for load flow and time simulations analysis.

Figure 2.1 presents a one-line diagram of the system [1].

Generator 1 connected to bus 39 represents a neighboring transmission system. The rest of the generators represent generating stations.

Some modifications were made to the system model presented in [1] to adapt the model to the needs of this effort. In particular, the load representation was modified to better represent the impact of PV connected at the distribution level.

2.1.1 Steady-State Database

Data in [1] were used to create a load flow database in GE PSLF. The load flow results of the GE PSLF match well with the reported voltages of the benchmark system in [1]. Appendix A shows the comparison of results.

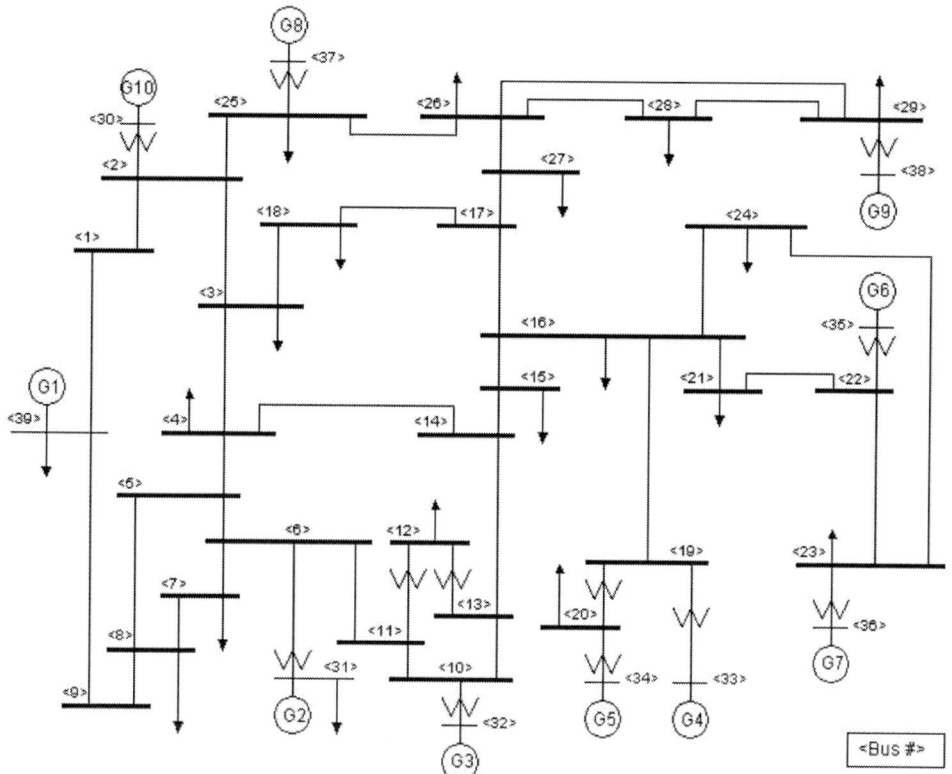

Figure 2.1. IEEE 39 Bus System [1].

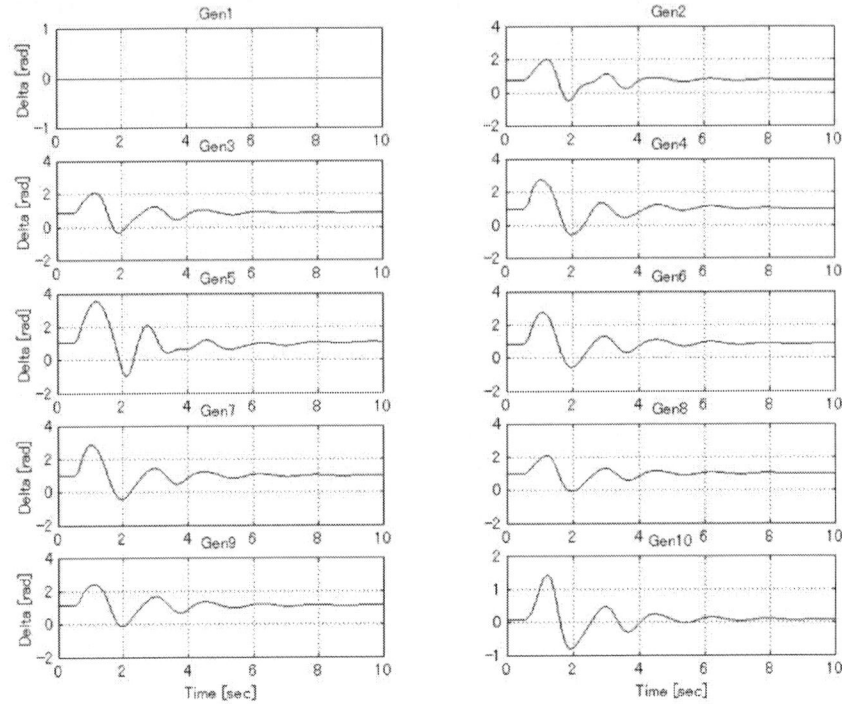

Figure 2.2. Rotor angle plot [deg] of the IEEE 39 benchmark model [1].

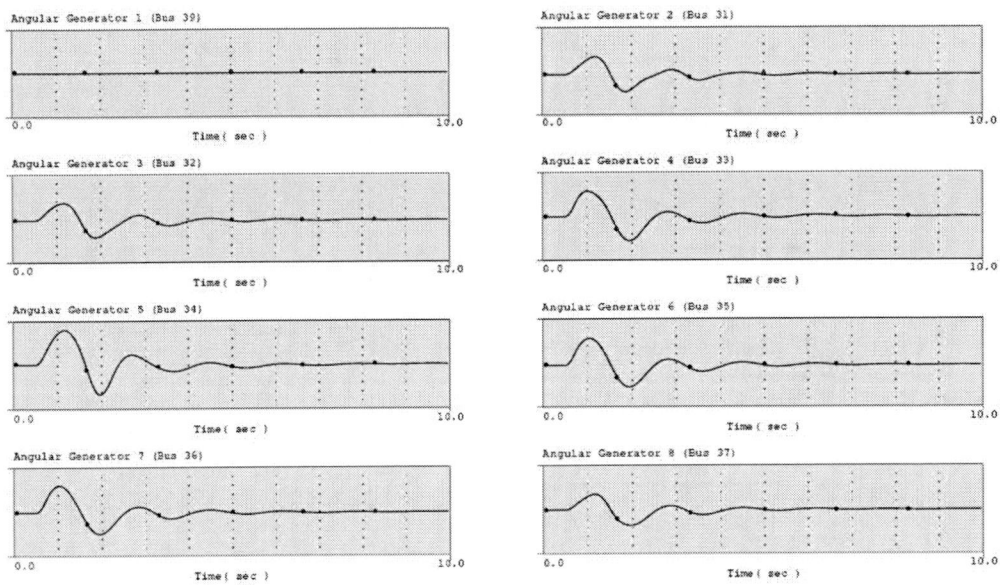

Figure 2.3. Rotor angle deviation to Gen1(39) (acting as swing bus in this example). Gen10 (30), Gen2(31), Gen3(32), Gen4(33), and Gen5(34).

2.1.2 Dynamic Database

Information in [1] was used to set up a dynamic database in PSLF. The selected PSLF models and parameters are presented in Appendix B. Reference [1] provides generator, automatic voltage regulator (AVR), and power system stabilizer (PSS) dynamic data. Governor models were added in PSLF to allow analysis of frequency events. Reference [1] does not indicate generator ratings and all generator parameters are expressed in per unit of the system base. Generator ratings were calculated to achieve physically meaningful transient reactances. Generators in [1] are represented with a fourth-order machine model; PSLF models include subtransient components based on typical values. The benchmark system [1] considers constant mechanical torque. Loads are treated as constant impedance loads.

2.1.3 Validation Runs

A fault at bus 16 was simulated for validation. The solid fault is applied at $t = 0.5$ seconds and cleared 0.2 seconds later. Figure 2.2 and Figure 2.4 present rotor angle and speed results from [1] and Figure 2.3 and 2.5 from PSLF for comparison. Scales of PSLF plots coincide with plots extracted from [1]. Results match well.

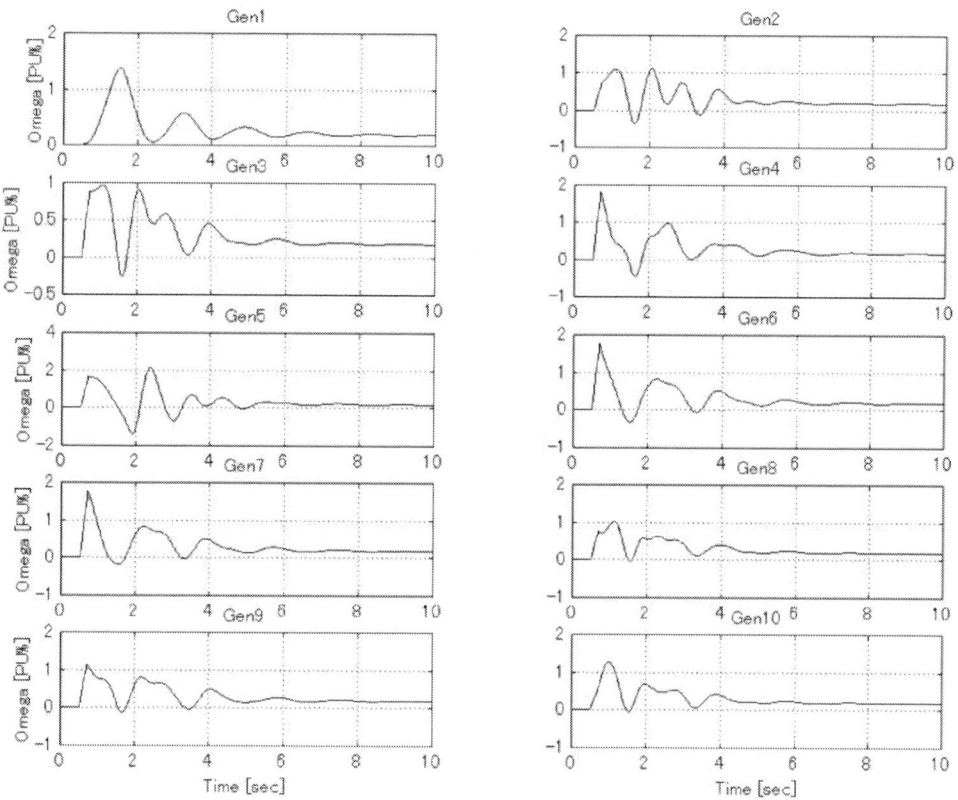

Figure 2.4. Rotor speed plot [PU%], IEEE 39 benchmark model [1].

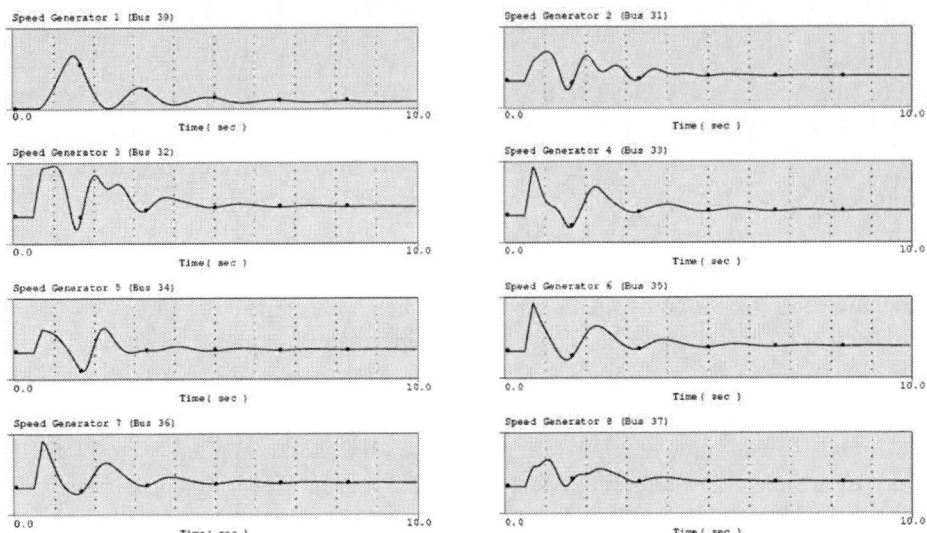

Figure 2.5. Rotor Speed Plot [PU%] - Gen1(39), Gen10 (30), Gen2(31), Gen3(32), Gen4(33), and Gen5(34).

2.1.4 Model Modification

2.1.4.1 Load busses

In the original benchmark system, loads are connected to high-voltage buses. The load representation for this effort was modified to include two levels of voltage transformation, medium-voltage capacitor compensation, and feeder impedances. Each load transformer incorporated represents an aggregate of parallel load transformers.

Figure 2.6 shows the enhanced load model. The load was moved to the low-voltage bus. Medium-voltage capacitors were sized to obtain about the same reactive power in the high-voltage bus (Q in the figure).

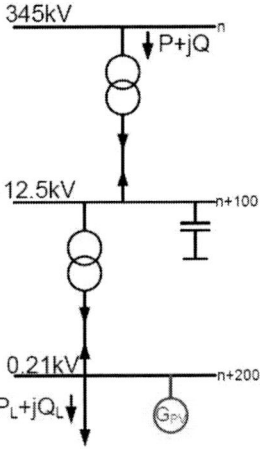

Figure 2.6. Load representation.

Table 2.1. Impedances of Load Model Extension in per Unit of Transformer Rating

Low Voltage				Medium Voltages			
Transformer		Line/Cable		Transformer		Line/Cable	
R	X	R	X	R	X	R	X
0	0.05	0.1	0.1	0	0.07	0.01	0.03

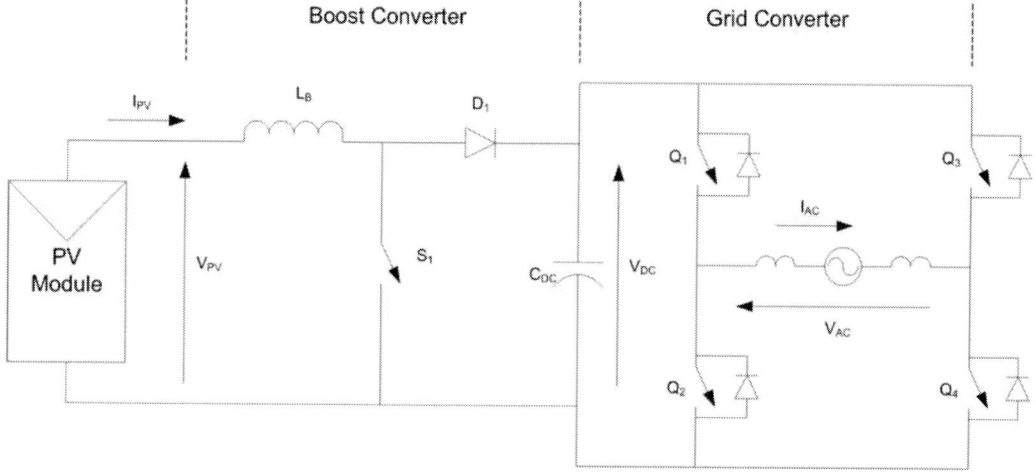

Figure 2.7. Generic PV inverter arrangement considered for dynamic model.

Table 2.1 indicates impedances considered on a transformer base. The transformer impedances in the database are used to represent transformer and cable impedances. All PV generation in the low-voltage system fed from a high-voltage bus is represented with a single generator (G_{pv}) connected to the low-voltage bus.

Load flow results after extending the model are indicated in Appendix C. The results on the high-voltage distribution line are within 0.6% compared with the benchmark model.

2.1.4.2 Governors

To enable meaningful analysis of frequency transients, turbine-governor models were included for all generators. Also, the generator bases were adapted when required to allow reasonable levels of rotating reserve in the simulations. Governor model structure and parameters are presented in Appendix C.

2.2 Aggregated PV Representation

As described in Section 2.6, PV generation is represented at low-voltage load buses. A single generator is used to represent all PV generators connected at a low-voltage bus of the equivalent transformers. A dynamic model for the equivalent PV generator was created as part of this effort. The model includes a number of features of commercial systems and some that are not common.

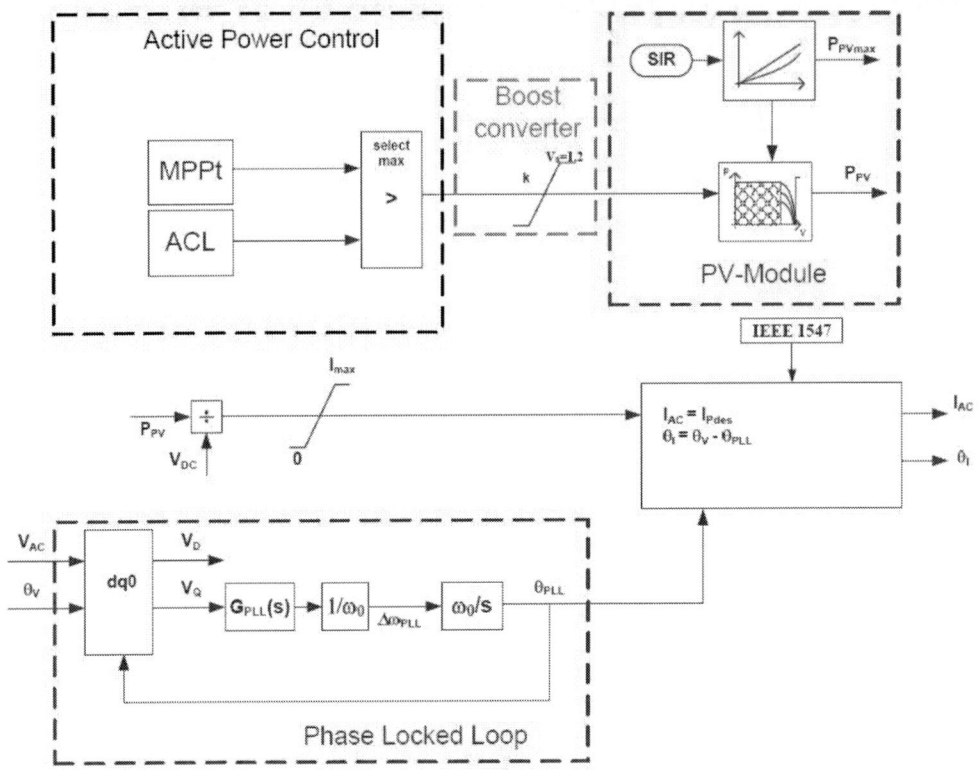

Figure 2.8. Block diagram representation of aggregated PV generation model.

The generic PV inverter arrangement in Figure 2.7 was used to derive the dynamic model. The grid converter is controlled to keep the direct current link (V_{DC}) constant. The reactive power of the converter can be controlled as long as the current capability of the components is not exceeded. In present systems, the reactive power is kept constant, equal to zero, despite the potential to control it. The duty cycle of the boost converter switch S is controlled to apply the desired DC voltage at the PV module (V_{PV}). In present systems, the duty cycle of switch S is controlled to track the point of operation that extracts maximum power from the PV modules. Assumptions associated with the model are:

- Single-phase PV modules are aggregated and represented as a single three-phase source.
- All dynamics faster than a fundamental frequency cycle were neglected (the typical approach for transient stability simulations).
- PV systems periodically perform screenings of the relationship between current and voltage. The simulation model does not incorporate these phenomena.
- All magnitudes are normalized.

The proposed simulation model for transient stability of aggregated standard PV systems is presented in Figure 2.8. The model includes:

- PV panel power as a function of applied dc voltage V_{PV}. Two irradiation levels are considered.
- Simplified representation of the maximum power point (MPP) tracking. The representation is designed to replicate power output fluctuations caused by MPP operation during or after system disturbances, and does not capture implementation details of actual MPP tracking algorithms.
- Over- and undervoltage and frequency disconnection according to IEEE 1547
- Current source representation of inverter
- Fixed power factor (unity)
- Phase-locked loop (PLL).

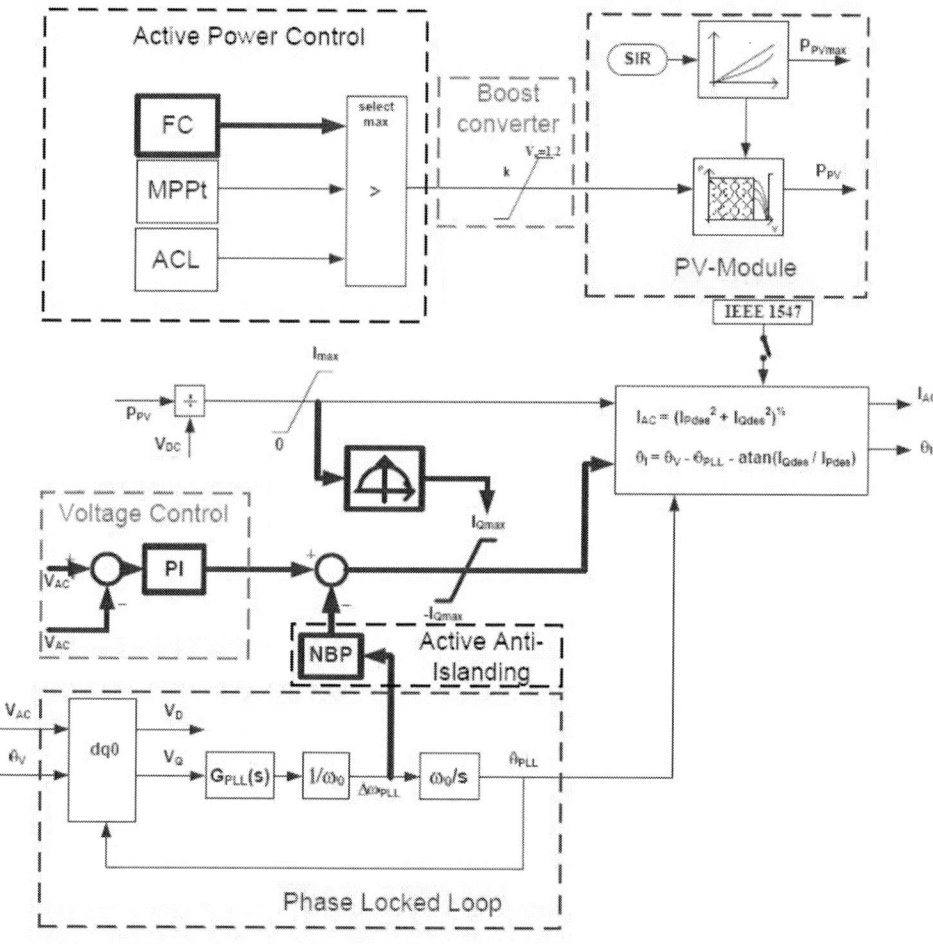

Figure 2.9. Block diagram representation of aggregated PV generation model for enhanced performance.

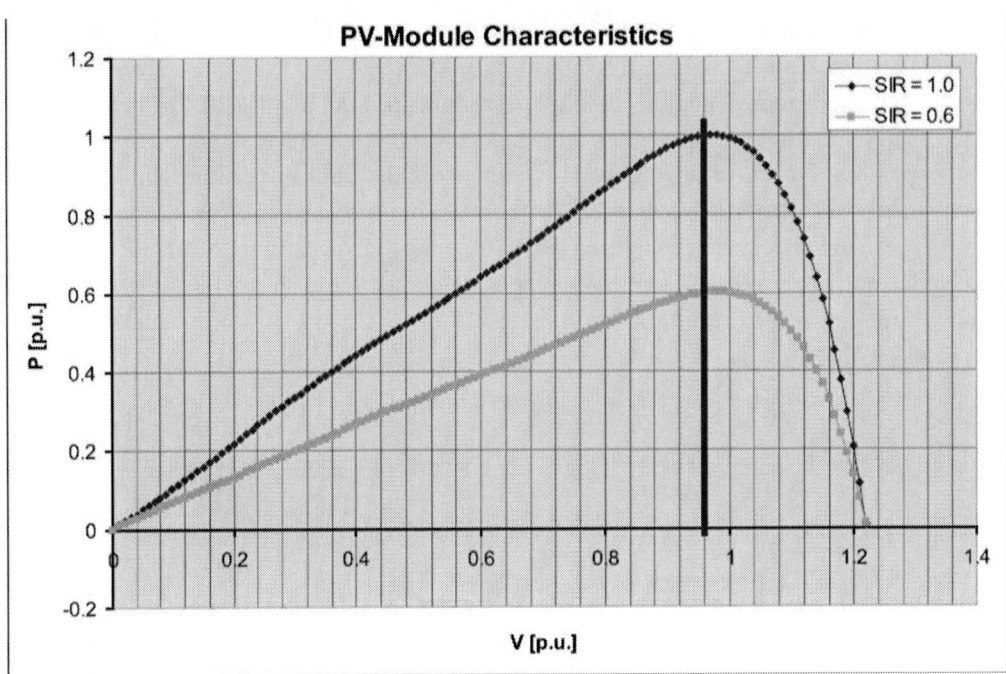

Figure 2.10. Normalized PV-module characteristic implemented in the PV-dynamic model.

Another model, presented in Figure 2.9, was also created to incorporate additional control capabilities to assess their impact for high PV penetration levels. The model also includes:

- Terminal voltage control
- Frequency droop control and active power reserve
- Inverter current limitation control
- Active anti-islanding.

The next sections describe the different blocks of these diagrams.

2.2.1 Photovoltaic Module Representation

The PV module is represented with an algebraic transfer function between DC voltage and power output. The function consists of a sixth-order polynomial equation and was obtained for two solar irradiation levels (see Figure 2.10). For each irradiation level there is an MPP for a DC voltage V_{PV} of about 1 pu.

In present systems, the module is operating close to the MPP voltage. In this effort, different operating conditions are explored to support the transmission system operation. Under normal operating conditions, the module is assumed to operate with DC voltages between the MPP voltage (1 pu) and the open circuit voltage (about 1.2 pu). This region of operation is preferred, to avoid high direct currents in switching operations, e.g., caused by faults.

2.2.2 Boost Converter

The boost converter is assumed to have a limited duty cycle. The limits applied in the model are associated with the maximum ratio between the voltages V_{PV} and V_{DC} in Figure 2.7. Temperature dependence of the module was not considered relevant for this study.

2.2.3 Active Power Control

The active power control includes three different control loops (Figure 2.9) and has as an output the desired ratio between the voltages V_{PV} and V_{DC} (Figure 2.7). A maximum selector is used to select the control loop governing the operation. As indicated in Section 2.2.1, the module will be normally operated in the region above the MPP voltage. Hence, selecting the maximum voltage request results in the lowest produced power. The three control loops and the boost converter limit are implemented to ensure smooth transitions between control loops and avoids wind-ups. The three control loops are:

- MPP tracking, indicated as MPPt in Figure 2.8 and Figure 2.9. This consists of a simplified representation of an actual MPP loop with the objective of capturing the relatively slow power output variations caused by this control.
- Active current limit, indicated as ACL in Figure 2.8 and Figure 2.9. This loop is intended to limit the module active power when the current capability of the inverter is reached (or exceeded). This control loop will only be active during the voltage sags and when the PV generation is not tripped.
- Frequency droop control and active power reserve, indicated as FC in Figure 2.9. This control option was not considered for all simulations. It includes a frequency signal to allow PV generation to perform frequency control and to be dispatched with primary reserve.

2.2.4 Photovoltaic Inverter Model

The PV inverter is represented as a current source. The dynamics of the current control loops are neglected and the control current references are assumed to be equal to the currents. The active current component is set depending on the active power of the PV module. This simple approach is associated with the assumptions that the inverter losses are neglected and that the inverter controls the DC voltage V_{DC} with high bandwidth. The reactive current component is set to zero or by the voltage control. The effect of phase jumps on the inverter operation (for example during faults or fault clearings) is considered using the error between the actual phase of the inverter AC voltage and the PLL angle.

The inverter model also includes the under- and overvoltage and frequency disconnection criteria according to IEEE 1547. This function was disabled in some simulation cases.

2.2.5 Active Anti-Islanding

An active anti-islanding loop was incorporated in the PV model. The scheme considered has the frequency measurement of the PLL as an input and acts on the reactive power signal of the converter. The loop includes a band pass filter to avoid noise injection and permanent reactive power modifications caused by frequency changes of normal system operation.

2.2.6 Phase-Locked-Loop

The phase-locked-loop is represented to allow for RMS voltage input and with a typical structure and tuning. Filtering and angle compensations are neglected.

2.2.7 Voltage Control

The voltage control loop was not considered in all simulations, as it is considered optional. The plug-in controller sets the reactive current component based on the AC voltage measurement. The bandwidth of the control is set relatively slow. No voltage drooping was considered. When this control is not active, the reactive current is set to zero.

2.2.8 Sensitivity of Photovoltaics to Voltage Sags

More economical (but less efficient) PV systems do not have boost converters (see Figure 2.7). In such systems, the DC link voltage is applied to the PV modules and controlled with an MPP tracking to maximize energy capture. The power output of such systems is expected to be more sensitive to terminal voltage fluctuations. Some simulation cases were performed considering the sensitivity to voltage variations of such systems.

2.2.9 Photovoltaics Model Verification

The models in Figure 2.8 and Figure 2.9 were implemented in GE PSLF. A number of simulations were performed in a simple system to verify the behavior of the model under different disturbances and with different control options. The primary results are presented in Appendix E.

2.3 Case Studies

2.3.1 Contingencies

The contingencies simulated are presented in Table 2.2. The bus numbers correspond to numbering presented in the one-line diagram of Figure 2.1.

2.3.2 Study Scenarios

Different starting points were considered for the simulations to account for:

- Different load levels
- Different PV penetration
- Different unit commitment strategy
- Different primary reserve distribution

Table 2.2. Contingencies

Contingency	Description
G1	Disconnection from neighbor transmission system.
L1	Large load disconnection.
F3	Fault at bus 16. Voltage sag to 0%. 100ms clearing of 16-17 line.

3.0 PROJECT RESULTS

3.1 Study Scenarios

Different load flow scenarios were considered as a starting point for time simulations. Table 3.1 describes the scenarios considered. Single line diagrams of the scenarios are presented in Appendix D.

The benchmark system presented in Section 2.1 was assumed to be a peak load operating condition and was used as scenario s1.

Variations of this scenario were produced to account for behavior of the system under the following conditions and operating rules:

- Peak (s1, s3, s5, and s7) and moderate load (s2, s4, s6, and s8)
- With (s3 to s8) and without (s1 and s2) PV generation
- Maintaining dispatch and commitment to accommodate PV generation (s3 and s4)
- Modifying dispatch and commitment to accommodate PV generation (s5 to s8)
- PV generation with and without power reserve for frequency support.

Scenario s2 assumes a 50% load reduction with respect to s1. Generation units 3, 4, 6, and 10 were de-committed. This implies that a good degree of flexibility was assumed in the generation portfolio. The scenarios that include PV (s3 to s8) assume total PV generation rating of 30% of the peak load (1.8 GW). The different combinations associated with load level and assumptions resulted in different penetration levels with respect to system load (Table 3.1).

Consistent with Ye et al. [3], the peak scenario s3 assumed that PV takes the load increase over future years. That is, the load was increased to match the PV generation. Scenario 4 was similarly created using scenario s2 (low load) as a reference. In these two scenarios, the conventional generation remained unchanged.

Table 3.1. Load Flow Scenarios

	Filename	Scenario	PV Generation (% of load)
s1	nrel_lf_s1.sav	Peak load (IEEE benchmark load). No PV generation.	0
s2	nrel_lf_s2.sav	Low load (50% of peak). No PV generation.	0
s3	nrel_lf_s3.sav	Peak (plus 30%). 30% PV generation.	23
s4	nrel_lf_s4.sav	Low load plus 30%. 30% PV generation.	37
s5	nrel_lf_s5.sav	Peak load. 30% PV generation. Conventional generation de-committed.	30
s6	nrel_lf_s6.sav	Low load. 30% PV generation. Conventional generation de-committed.	60
s7	nrel_lf_s7.sav	Peak load. 30% PV generation. Conventional generation not de-committed.	30
s8	nrel_lf_s8.sav	Low load. 30% PV generation. Conventional generation not de-committed.	60

Table 3.2. Active Power Data of Different Scenarios

	P_{gen} (MW)	P_{PV} (MW)	P_{Trans} (MW)	P_{Load} (MW)	P_{loss} (MW)
s1	5543	0	1080	6089	533
s2	3002	0	200	3044	158
s3	5589	1826	1080	7916	580
s4	3033	1826	200	4871	188
s5	4348	1826	200	6089	285
s6	1274	1826	0	3044	56
s7	3768	1826	756	6089	261
s8	1203	1826	40	3044	25

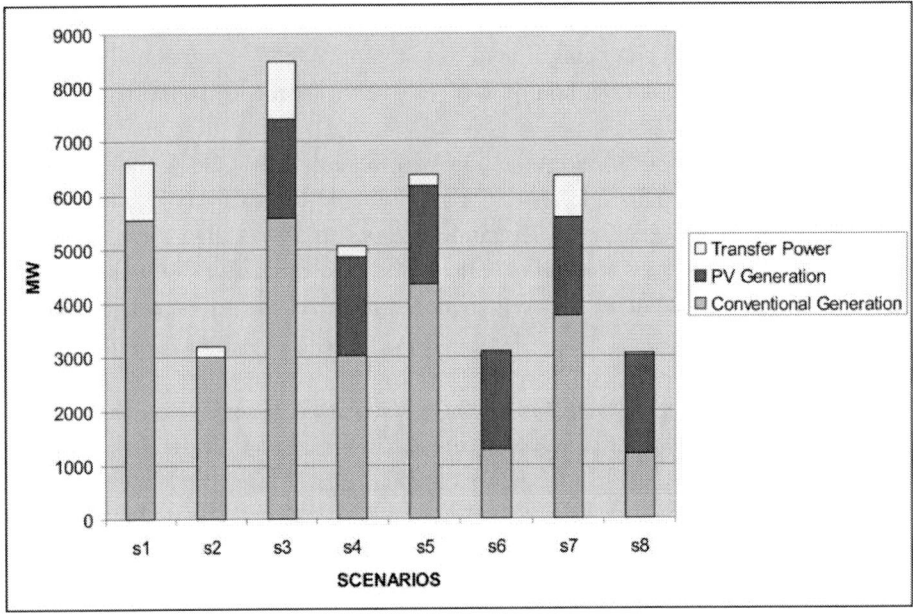

Figure 3.1. Summary of scenarios.

Peak scenarios s5 and s7 have the same load as s1. Conventional generation was hence displaced to accommodate PV generation. Similarly, low-load scenarios s6 and s8 have the same load as s2. In scenarios s5 and s6 units are de-committed with respect to scenarios s1 and s2 (without PV) to accommodate PV generation. However, units are not de-committed to accommodate PV generation in scenarios s7 and s8.

These sets of scenarios present the extreme cases of how the system could be operated depending on the availability of PV production forecast for unit commitment. Specifically, scenarios s5 and s6 assume that accurate forecast of PV generation is considered in the unit commitment, and scenarios s7 and s8 assume that no PV production forecast is considered in the unit commitment.

An additional set of scenarios with power reserve in PV was created. Cases with PV generation (s3 to s8) were modified to accommodate 5% power reserve in the PV systems.

Table 3.2 presents, for each scenario, the power of conventional generation (P_{gen}), PV generation (P_{PV}), power transferred from neighbor areas (P_{trans}), total load (P_{load}), and system losses (P_{loss}). Additionally, Figure 3.1 presents the power production portion of conventional units, PV, and transfer power for the scenarios.

In all scenarios, tap changers of medium-voltage transformers were adjusted to keep voltages within ±0.05 pu. These are observations from load flow analysis of the system with PV penetration:

- Transmission losses are lower with increased PV generation because PV is connected closer to the loads than conventional generation (comparison of scenarios s1 with s5 and s2 with s6). Distribution losses are not represented in detail.
- Accommodating high PV generation during low load is challenging for unit commitment and dispatch. Lower load scenarios s2, s6, and s8 assume that units can be de-committed during lower load periods. If units cannot be de-committed because of operational constraints or cost implications, it would not be possible to accommodate all PV generation and dispatch units above technical minimums. In scenario s8 the same unit commitment as in s2 was considered, but units were dispatched lower to accommodate high PV generation. The technical minimum assumed for all units is 20%. In most systems there are units with technical minimums higher than 20%, accommodating PV generation will then not be possible without de-committing more conventional generation or reducing the PV generation.
- Scenario s6 has only a few conventional units in service. Thus, a significant part of the transmission system is without voltage support.

3.2 Study Cases

Summary tables are presented in next sections to show results of the simulations. The values presented in the summary tables in this section are described in Table 3.3. The description column in the summary tables indicates the following variations:

- No PV. Cases without PV generation.
- IEEE 1547. Conventional PV (Figure 2.8) with disconnection criteria according to IEEE 1547. Undervoltage clearing times set to maximum allowed in this standard.
- Frequency Control. PV with frequency control and MPP tracking operation.
- Frequency Control, 5% reserve. PV with frequency control and with 5% up-reserve.
- No Overfrequency Trips. Conventional PV (Figure 2.8) without overfrequency disconnection.
- Undervoltage Trip. Conventional PV (Figure 2.8) with disconnection criteria according to IEEE 1547. Undervoltage clearing times set below maximum allowed in this standard.

- LVRT. Capability of PV generation. PV is assumed to be able to ride through the applied faults without tripping. Current capability limitations of PV converters are observed during and after faults (Figure 2.9).
- Voltage Control. PV with voltage control as presented in Figure 2.9.
- Anti-Islanding. PV with active anti-islanding (Figure 2.9).
- 30% Motor Load. Load model was modified to account for motor loads.

Key plots are presented in next sections. The complete set of plots for all cases (more than 1000 pages) is not included in this report. An appendix of plots is available on request.

3.2.1 Frequency Performance

3.2.1.1 Generation Trip

Disconnection from a neighboring transmission system was simulated to assess the impact of PV on frequency performance. The disconnection results in a deficit of generating power of 540 MW in peak load cases (s1, s3, s5, and s7) and 200 MW in the lower load cases (s2, s4, s6, and s8). As a reference, the system peak load is 5.4 GW and the lower load is 2.7 GW. For these simulations, the generator G1 (Figure 2.1) connected to bus 39 was split into two generators (G1 and G11) to represent the connection to two neighboring systems. The initial generation of the disconnected generator (G1) was set to obtain the desired power imbalance. Table 3.4 presents a summary of the results and the run cases. The cases include all starting scenarios and different PV behaviors.

The observations from these cases follow:

- The cases where PV displaced conventional generation (5 and 6) had worse performance than the cases without PV (1 and 2). The low-load case (6) with PV resulted in a frequency minimum that was significantly lower than the case without PV (2) because the system inertia is lower. The steady-state frequency value is also negatively affected because fewer units perform frequency control (Figure 3.2). In the high-load case with PV (5), the event is further aggravated because PV generation trips at 59.3 Hz, per IEEE 1547 (Figure 3.3). The system collapses in the simulation. In a real event, significant frequency load shedding would operate.
- If units are committed ignoring PV generation, the frequency performance is similar with (7 and 8) and without PV (1 and 2) because the system inertia is the same and the up reserve is even higher.
- Frequency control of PV improves the steady-state frequency variations. The maximum frequency excursion and the maximum rate of change of frequency are not significantly improved, because of the relatively slow regulating response assumed for PV. Figure 3.4 presents a comparison of a run without PV and with PV and frequency control.

Table 3.3. Summary Table Description

Column	Description
fmax [Hz]	Maximum frequency
fmin [Hz]	Minimum frequency
fss [Hz]	Final steady state frequency
dfmax [Hz/s]	Maximum absolute value of derivative of frequency with respect to time.
tdfmax [sec]	Time of maximum derivative of frequency
V16 (max) [pu]	Maximum per unit voltage at transmission bus 16
v16 (min) [pu]	Minimum per unit voltage at transmission bus 16
v16ss [pu]	Final steady state per unit voltage at transmission bus 16
t16rec [sec]	Time required for voltage at bus 16 to recover (above 0.8 pu) after a fault
Initial PV MW]	Initial PV generation
PV end [MW]	PV generation at the end of the simulation.
Loss sync	Loss of synchronism of units. "0" means no unit lost synchronism, "1" means at least one unit lost synchronism.

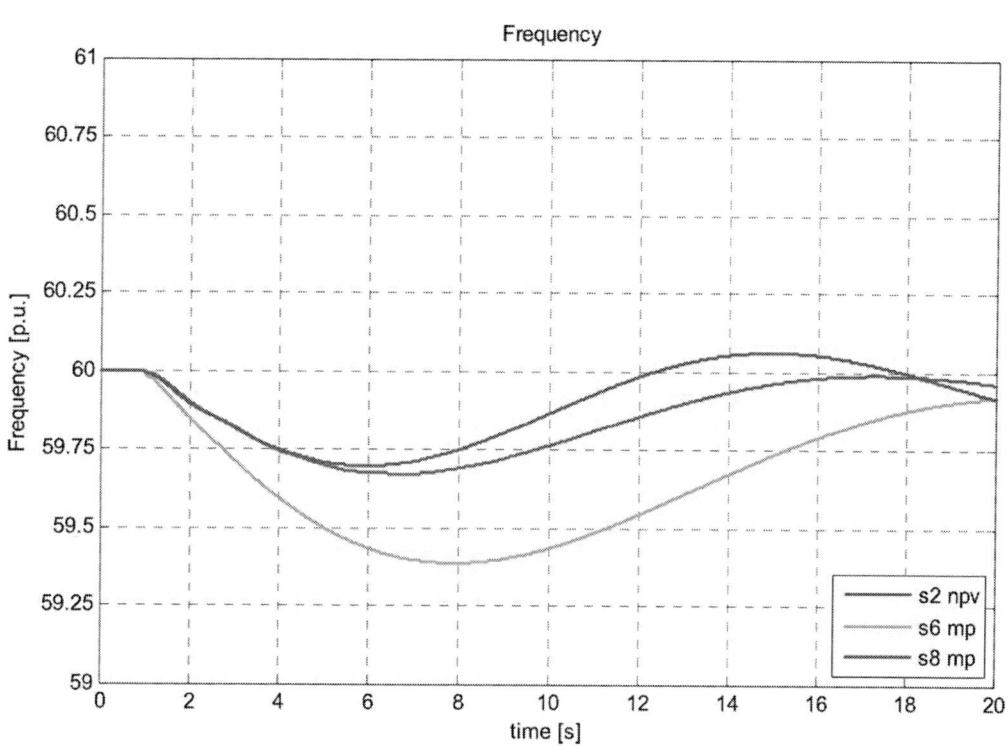

Figure 3.2. Generation trip and lower load cases. Scenarios s2 (red), s6 (green), and s8 (blue).

3.2.1.2 Load Trip

Disconnection of a large load was simulated to assess the impact of PV on frequency performance. The disconnection results in generating power in excess of about 500 MW in peak load cases (s1, s5, and s7) and 250 MW in the lower load cases (s2, s6, and s8). In cases s3 and s4 the imbalance is 30% larger.

Table 3.5 presents a summary of the results and the run cases. The cases include all starting scenarios and different PV behavior.

The observations from these cases follow:

- In cases with PV with IEEE 1547 diconnection criteria (cases 3 to 8), the main impact of PV is associated with trips for frequency higher than 60.5 Hz. In these cases, the maximum frequency excursions are generally better than in cases withoPV (cases 1 and 2) because of the generation trip. The trip of PV causes a considerable frequency drop in cases where major PV generation is disconnected.
- In cases with PV, without overfrequency disconnection and with reduced unit commitment (cases 11 and 12), the reduced number of units on frequency control and the reduced inertia result in higher over-frequencies. Figure 3.5 shows the resulting frequency for the case without PV (red) and for the case with a reduced number of regulating units in the system (blue).
- Frequency regulation has a modest im pact on the final frequency.

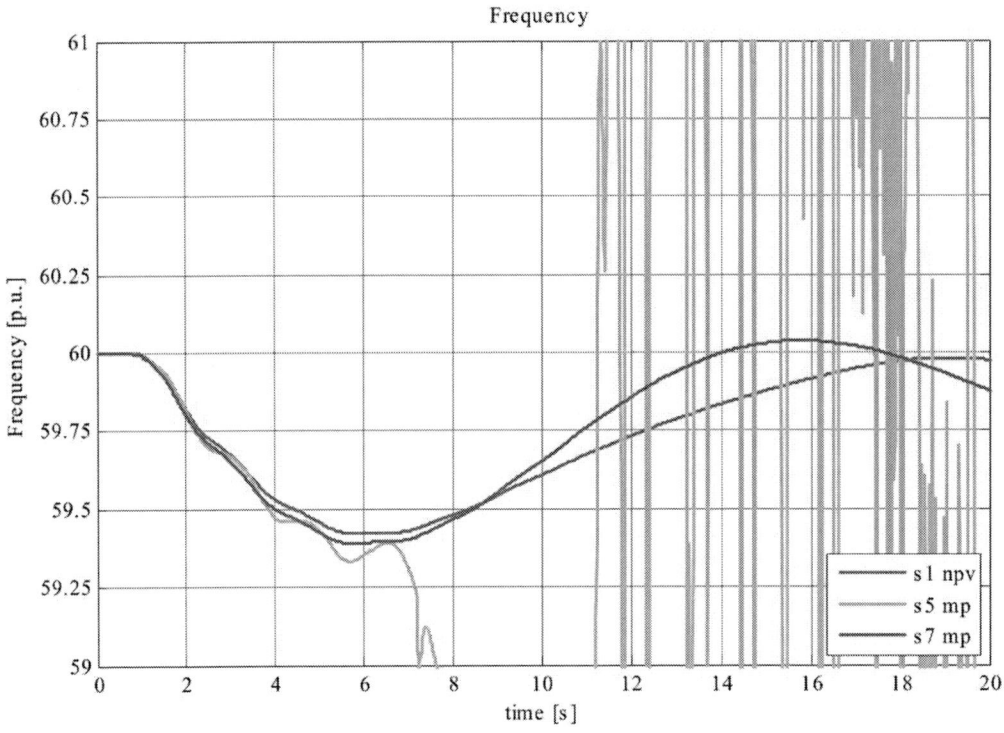

Figure 3.3. Generation trip and high load cases. Scenarios s1 (red), s5 (green), and s7 (blue).

Table 3.4. Frequency Performance-Generation Trip

	CASE	Scen	Description	fmax [Hz]	fmin [Hz]	fss [Hz]	dfmax [Hz/s]	tdfmax [sec]	Comment
1	REG1_s1_g1_npv	s1	No PV	60.00	59.42	59.97	0.26	1.93	0
2	REG1_s2_g1_npv	s2	No PV	60.00	59.67	59.96	0.12	1.83	0
3	REG1_s3_g1_mp	s3	IEEE1547	60.00	59.46	59.94	0.24	1.93	0
4	REG1_s4_g1_mp	s4	IEEE1547	60.00	59.67	59.93	0.12	1.83	0
5	REG1_s5_g1_mp	s5	IEEE1547	-	43.06	-	-	-	collapse
6	REG1_s6_g1_mp	s6	IEEE1547	60.00	59.39	59.92	0.18	1.05	0
7	REG1_s7_g1_mp	s7	IEEE1547	60.04	59.39	59.88	0.26	1.93	0
8	REG1_s8_g1_mp	s8	IEEE1547	60.06	59.70	59.92	0.12	1.73	0
9	REG1_s3r_g1_fc	s3r	Frequency Control. 5% reserve	60.00	59.52	59.90	0.23	1.93	0
10	REG1_s4r_g1_fc	s4r	Frequency Control. 5% reserve	60.00	59.76	59.92	0.10	1.83	0
11	REG1_s5r_g1_fc	s5r	Frequency Control. 5% reserve	60.00	59.40	59.89	0.29	2.03	0
12	REG1_s6r_g1_fc	s6r	Frequency Control. 5% reserve	60.00	59.57	59.94	0.16	1.05	0
13	REG1_s7r_g1_fc	s7r	Frequency Control. 5% reserve	60.03	59.46	59.84	0.24	1.83	0
14	REG1_s8r_g1_fc	s8r	Frequency Control. 5% reserve	60.02	59.77	59.91	0.11	1.63	0

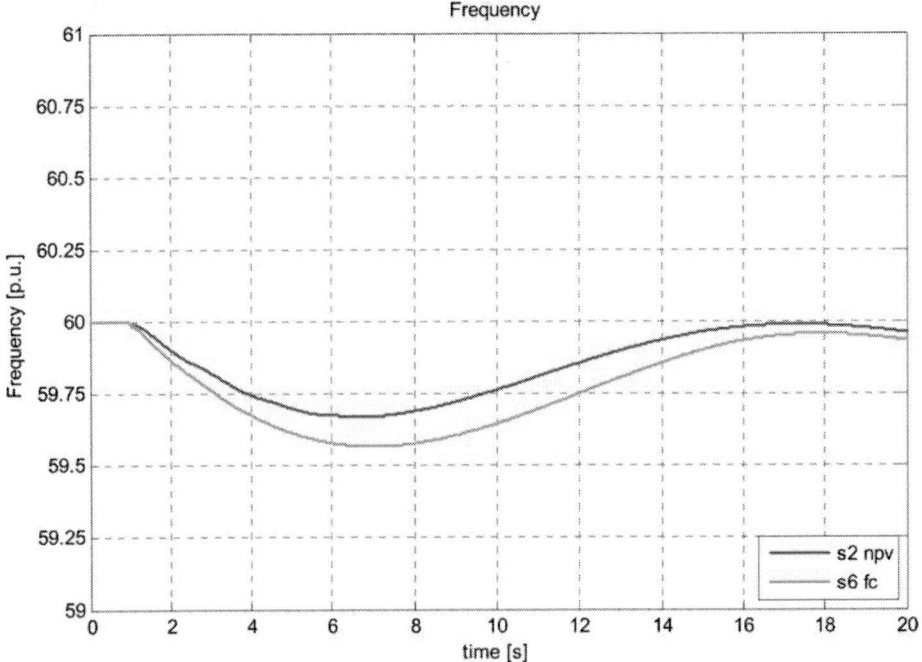

Figure 3.4. Generation trip and low load cases. Scenarios s2 (red) and s6r with frequency control and reserve in PV generation (green).

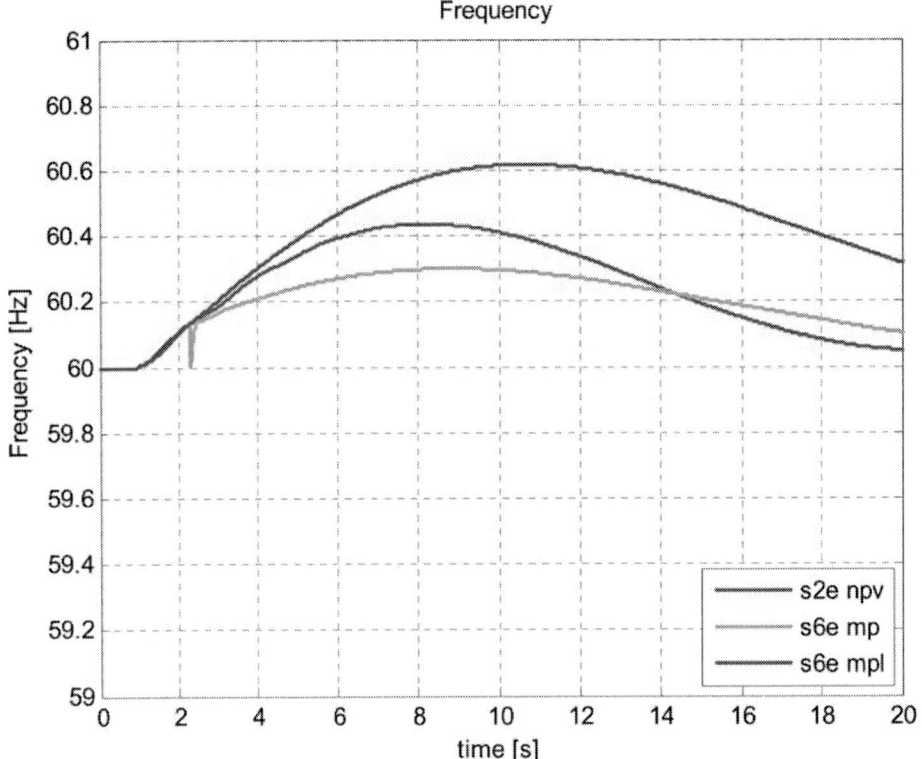

Figure 3.5. Load trip and low load cases. Scenarios s2 (red), s6 with PV acc to IEEE 1547 (green), and s6 without PV disconnection on overfrequency (blue).

Transmission System Performance Analysis for High-Penetration Photovoltaics 23

3.2.2 System Response to Faults

A solid fault at the bus 16 end of the 16-17 line was simulated; 100 ms clearing of 16-17 line was assumed. Table 3.6 presents a summary of the results. The cases include all starting scenarios and different PV behavior. Steady-state and dynamic frequency excursions (f_{min} and f_{ss}) below 59 Hz are highlighted in red in the table. Overvoltages above 1.1 pu (V16$_{max}$) and steady state voltages (V16$_{ss}$) lower than 0.9 pu are also highlighted in red.

The observations from these cases follow:

- The considered fault does not result in significant frequency excursions or prolonged voltage sags for the cases without PV (cases 1 and 2). The red curves in Figure 3.6 and Figure 3.7 indicate system frequency and voltage in bus 16 for cases 1 and 2.
- Sensitive undervoltage disconnection of PV can be detrimental to system reliability. Under the assumption that PV trips for undervoltage, according to IEEE 1547, faster than the maximum clearing time requirement (cases 3 to 8), all PV generation disconnects result in significant frequency excursions, voltage and dynamic stability system collapse. The peak load cases (3, 5, and 7) have oscillatory and voltage stability problems and the lower load cases (4, 6 and 8) have voltage recovery problems. The resulting generation deficit due to PV trips significantly aggravates the contingency. The green curves in Figure 3.6 indicate system frequency and voltage in bus 16 for cases 5 and 6.
- Under the assumption that PV trips for undervoltage according IEEE 1547 at maximum clearing time requirement (cases 9 to 14), the considered fault also results in larger frequency excursions than in the cases without PV. In these cases there are moderate PV trips and considerable frequency excursions, but without system collapse. The blue curves in Figure 3.6 and the green curves in Figure 3.7 indicate system frequency and voltage in bus 16 for these cases.
- In cases that assume LVRT (cases 15 to 20), the frequency drop is significantly less because no generation was lost. The PV active power is reduced during and after the fault becauses of the current limitation of their converters. Figure 3.10 presents the active power output of aggregated PV models and the sum of all PV models for case 15.
- Low load and high PV generation can result in poor voltage recovery if most conventional units are de-committed. In case 12 (associated with scenario s6), the voltage around the high-voltage bus 16 does not recover (green curves in Figure 3.7). Only two generating plants (G2 and G9) are dispatched. The voltage support available in the system is from these generators and the neighboring systems (G1). The area around bus 16 does not have electrically close generators and the system fails to recover. In case 18 (also associated with s6) the voltage at bus 16 does recover because there are no PV trips but only reaches 0.9 pu (blue curves in Figure 3.7).
- Voltage control in PV generation can improve voltage recovery on severe faults. Figure 3.7 shows that case 24 with PV voltage control (black) presents better voltage performance than case 18 (blue).
- Oscillatory stability of the system can be affected with high PV penetration depending on de-committing criteria. The oscillatory stability of the system is

negatively affected in cases with PV and scenario s5 (cases 11 and 17). In the frequency signal of Figure 3.6, the oscillatory component has less damping in cases 11 and 17 (blue and black) than in case 1 without PV (red). Figure 3.8 also shows that case 11 of scenario s5 with PV (green) has worse damping than case 13 of scenario s7 (blue). PV generators do not participate in these oscillations. The reduced number of large generators committed in scenario s5 resulted in lower damping of the electromechanical oscillations observed. That is, the unit commitment had a negative impact on system performance.

- Active anti-islanding can negatively affect oscillatory stability. Comparison of cases with (27 to 32) and without (15 to 20) active anti-islanding does not indicate a noticeable difference in frequency or voltage performance. Further exploration of different (but feasible) anti-islanding settings based on case 29 was performed. Figure 3.9 presents results of case 29 with different anti-islanding settings. The case with five times higher gain than case 29 results in dynamic instability (blue curve in Figure 3.9).
- Motor load in cases 33 to 38 include typical undervoltage tripping settings. Comparing cases 33 to 38 (with motor load) with cases 15 to 20 (without motor load) shows higher maximum voltages can be observed (V16max column in Table 3.6). The undervoltage motor tripping causes these overvoltages.

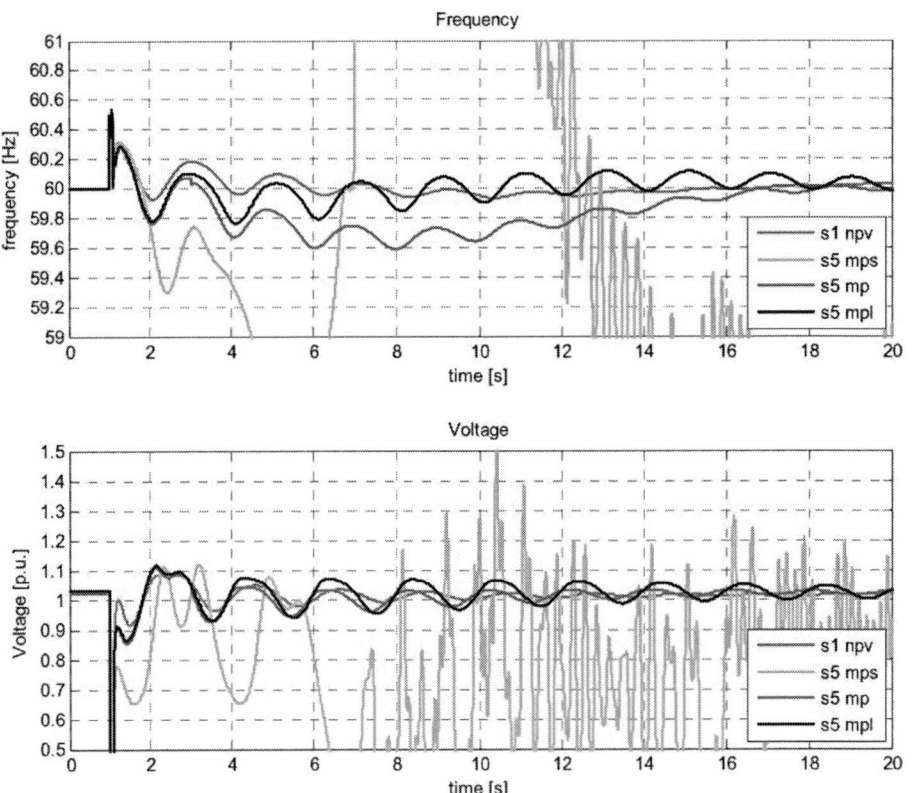

Figure 3.6. Peak load cases with different PV characteristics. Case 1 (red), case 5 (green), case 11 (blue), and case 17 (black).

Table 3.5. Frequency Performance—Load Trip

CASE	Scen	Description	fmax [Hz]	fmin [Hz]	fss [Hz]	dfmax [Hz/s]	tdfmax [sec]	Initial PV [MW]	PV end [MW]
REGI_s1_ll_npv	s1	No PV	60.55	59.99	60.04	0.22	1.83	0	0
REGI_s2_ll_npv	s2	No PV	60.44	60.00	60.05	0.13	1.83	0	0
REGI_s3_ll_mp	s3	IEEE1547	60.37	60.00	60.05	0.27	1.83	1827	1496
REGI_s4_ll_mp	s4	IEEE1547	60.51	59.53	59.72	0.42	5.53	1827	1220
REGI_s5_ll_mp	s5	IEEE1547	60.50	59.75	59.96	0.23	1.83	1827	1334
REGI_s6_ll_mp	s6	IEEE1547	60.30	60.00	60.11	0.58	2.43	1827	1639
REGI_s7_ll_mp	s7	IEEE1547	60.50	59.65	60.04	0.26	5.53	1827	1274
REGI_s8_ll_mp	s8	IEEE1547	60.31	59.96	59.99	0.11	1.83	1827	1827
REGI_s3e_ll_mpl	s3	no overfrequency trips	60.65	60.00	60.06	0.27	1.83	1827	1827
REGI_s4e_ll_mpl	s4	no overfrequency trips	60.59	60.00	60.13	0.19	1.73	1827	1827
REGI_s5e_ll_mpl	s5	no overfrequency trips	60.59	60.00	60.06	0.23	1.83	1827	1827
REGI_s6e_ll_mpl	s6	no overfrequency trips	60.62	60.00	60.32	0.13	1.63	1827	1827
REGI_s7e_ll_mpl	s7	no overfrequency trips	60.53	59.98	60.04	0.21	1.83	1827	1827
REGI_s8e_ll_mpl	s8	no overfrequency trips	60.31	59.96	59.99	0.11	1.83	1827	1827
REGI_s3_ll_fc	s3	Frequency Control	60.52	60.00	60.17	0.26	1.83	1827	1827
REGI_s4_ll_fc	s4	Frequency Control	60.42	60.00	60.17	0.18	1.73	1827	1827
REGI_s5_ll_fc	s5	Frequency Control	60.45	60.00	60.15	0.22	1.83	1827	1827
REGI_s6_ll_fc	s6	Frequency Control	60.36	60.00	60.17	0.12	1.63	1827	1827
REGI_s7_ll_fc	s7	Frequency Control	60.42	60.00	60.14	0.21	1.73	1826.9	1826.9
REGI_s8_ll_fc	s8	Frequency Control	60.25	60.00	60.07	0.11	1.73	1826.9	1826.9

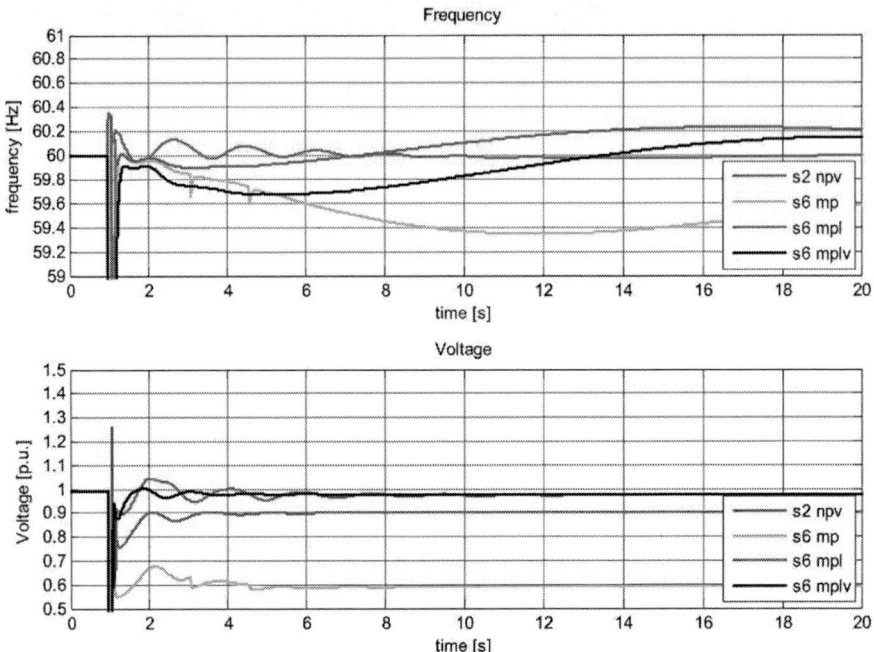

Figure 3.7. Lower load cases with different PV characteristics. Case 2 (red), case 12 (green), case 18 (blue), and case 24 (black).

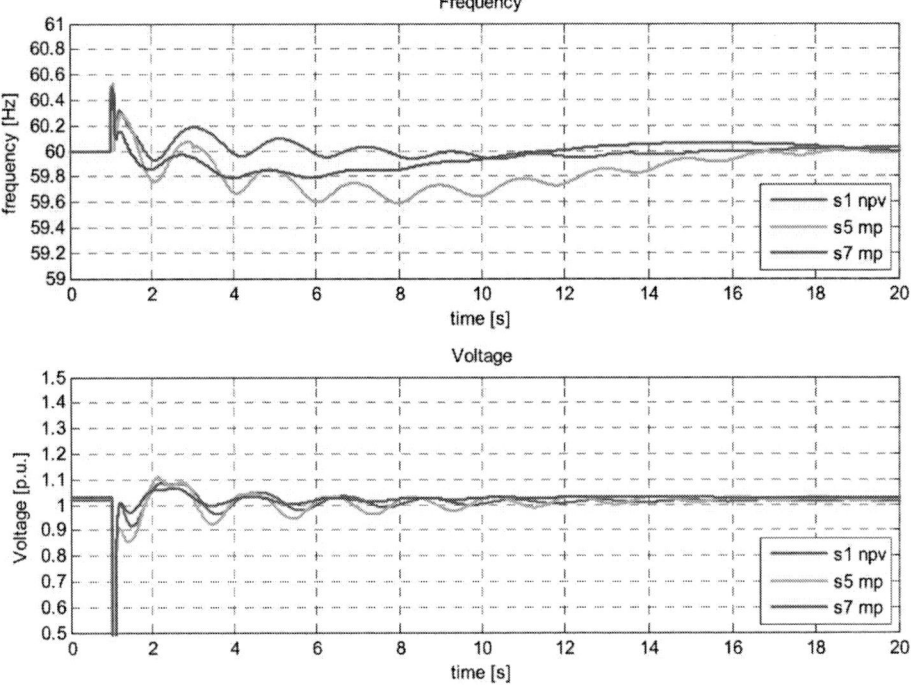

Figure 3.8. Peak load cases with different unit-commitment strategies. Case 1 (red), case 11 (green), and case 13 (blue).

Table 3.6. System Response to Short and Severe Fault

#	CASE	Scen	Cont	Description	Fmax [Hz]	Fmin [Hz]	Fss [Hz]	Dfmax [Hz/s]	tdfmax [sec]	V16 (max) [pu]	V16 (min) [pu]	v16ss [pu]	t16rec [sec]	PV tripped [MW]	Collapse
1	REGI_s1_f3_npv	s1	f3	No PV	60.429	59.928	60.027	3.512	1.016	1.085	0	1.015	0.116	0	
2	REGI_s2_f3_npv	s2	f3	No PV	60.356	59.939	59.998	3.103	1.016	1.045	0	0.975	0.122	0	
3	REGI_s3_f3_mps	s3	f3	undervoltage trip	60.484	57.078	57.452	3.733	1.016	1.035	0	0.902	0.149	1827	osc/volt
4	REGI_s4_f3_mps	s4	f3	undervoltage trip	72.038	38.668	53.837	115.04	16.164	1.271	0	0.417	0.587	1827	volt
5	REGI_s5_f3_mps	s5	f3	undervoltage trip	68.275	57.428	59.078	64.815	7.011	1.534	0	0.925	0.998	1827	volt
6	REGI_s6_f3_mps	s6	f3	undervoltage trip	60	54.756	54.756	147.31	1.016	0.992	0	0.572	19	1827	no volt rec.
7	REGI_s7_f3_mps	s7	f3	undervoltage trip	60.466	58.267	59.65	3.585	1.016	1.129	0	1.039	0.128	1827	osc
8	REGI_s8_f3_mps	s8	f3	undervoltage trip	60.249	57.843	60.106	3.017	1.076	1.047	0	1.027	0.104	1827	slow rec.
9	REGI_s3_f3_mp	s3	f3	IEEE1547	60.499	59.804	60.012	3.733	1.016	1.061	0	0.999	0.116	99	
10	REGI_s4_f3_mp	s4	f3	IEEE1547	60.648	59.716	60.023	3.491	1.016	1.082	0	1.021	0.131	99	
11	REGI_s5_f3_mp	s5	f3	IEEE1547	60.534	59.59	59.989	3.633	1.101	1.108	0	1.026	0.122	287	
12	REGI_s6_f3_mp	s6	f3	IEEE1547	60.326	59.35	59.592	147.31	1.016	1.26	0	0.593	19	782	no volt rec.
13	REGI_s7_f3_mp	s7	f3	IEEE1547	60.508	59.786	59.996	3.585	1.016	1.066	0	1.023	0.116	99	
14	REGI_s8_f3_mp	s8	f3	IEEE1547	60.287	59.751	59.999	2.225	1.131	1.053	0	1.012	0.101	99	
15	REGI_s3_f3_mpl	s3	f3	Lvrt	60.499	59.851	60.02	3.733	1.016	1.067	0	1.004	0.116	0	
16	REGI_s4_f3_mpl	s4	f3	Lvrt	60.648	59.778	60.055	3.491	1.016	1.091	0	1.026	0.131	0	
17	REGI_s5_f3_mpl	s5	f3	Lvrt	60.534	59.764	59.981	3.633	1.101	1.115	0	1.029	0.122	0	
18	REGI_s6_f3_mpl	s6	f3	Lvrt	60.326	59.91	60.207	147.31	1.016	1.26	0	0.9	0.101	0	
19	REGI_s7_f3_mpl	s7	f3	Lvrt	60.508	59.837	60.005	3.585	1.016	1.071	0	1.026	0.116	0	
20	REGI_s8_f3_mpl	s8	f3	Lvrt	60.287	59.822	59.996	2.225	1.131	1.054	0	1.013	0.101	0	
21	REGI_s3_f3_mplv	s3	f3	Lvrt. Voltage control	60.516	59.78	60.013	3.736	1.016	1.057	0	1.005	0.113	0	
22	REGI_s4_f3_mplv	s4	f3	Lvrt. Voltage control	60.575	59.734	60.087	4.99	1.101	1.088	0	1.031	0.125	0	
23	REGI_s5_f3_mplv	s5	f3	Lvrt. Voltage control	60.538	59.69	60.049	3.725	1.016	1.102	0	1.019	0.119	0	
24	REGI_s6_f3_mplv	s6	f3	Lvrt. Voltage control	60.147	59.89	60.147	497.57	1.016	1.002	0	0.977	0.1	0	
25	REGI_s7_f3_mplv	s7	f3	Lvrt. Voltage control	60.507	59.811	59.991	3.601	1.016	1.064	0	1.029	0.113	0	

Table 3.6 (Continued)

#	CASE	Scen	Cont	Description	Fmax [Hz]	Fmin [Hz]	Fss [Hz]	Dfmax [Hz/s]	tdfmax [sec]	V16 (max) [pu]	V16 (min) [pu]	v16ss [pu]	t16rec [sec]	PV tripped [MW]	Collapse
26	REGI_s8_f3_mplv	s8	f3	Lvrt. Voltage control	60.258	59.812	59.987	2.296	1.016	1.059	0	1.018	0.1	0	
27	REGI_s3_f3_mpla	s3	f3	Lvrt. Anti-Islanding	60.525	59.838	60.018	3.632	1.016	1.065	0	1.004	0.116	0	
28	REGI_s4_f3_mpla	s4	f3	Lvrt. Anti-Islanding	60.616	59.773	60.052	3.442	1.016	1.086	0	1.026	0.122	0	
29	REGI_s5_f3_mpla	s5	f3	Lvrt. Anti-Islanding	60.556	59.747	59.994	3.794	1.016	1.122	0	1.044	0.119	0	
30	REGI_s6_f3_mpla	s6	f3	Lvrt. Anti-Islanding	60.24	59.901	60.205	303.02	1.016	1.08	0	0.901	0.104	0	
31	REGI_s7_f3_mpla	s7	f3	Lvrt. Anti-Islanding	60.508	59.826	60.002	3.333	1.101	1.069	0	1.026	0.116	0	
32	REGI_s8_f3_mpla	s8	f3	Lvrt. Anti-Islanding	60.314	59.819	59.993	2.325	1.131	1.06	0	1.013	0.101	0	
33	REGI_s3_f3_mplm	s3	f3	Lvrt. 30% motor load	61.135	59.997	60.179	1.787	1.101	1.176	0	1.066	0.185	0	
34	REGI_s4_f3_mplm	s4	f3	Lvrt. 30% motor load	61.093	59.924	60.334	3.242	1.101	1.234	0	1.081	0.185	0	
35	REGI_s5_f3_mplm	s5	f3	Lvrt. 30% motor load	61.239	59.997	60.203	2.016	1.101	1.236	0	1.071	0.188	0	
36	REGI_s6_f3_mplm	s6	f3	Lvrt. 30% motor load	61.098	59.734	60.73	8.224	1.101	1.127	0	1.085	0.185	0	
37	REGI_s7_f3_mplm	s7	f3	Lvrt. 30% motor load	60.788	59.997	60.069	1.747	1.191	1.142	0	1.059	0.182	0	
38	REGI_s8_f3_mplm	s8	f3	Lvrt. 30% motor load	60.31	59.903	59.977	1.331	1.191	1.082	0	1.022	0.131	0	

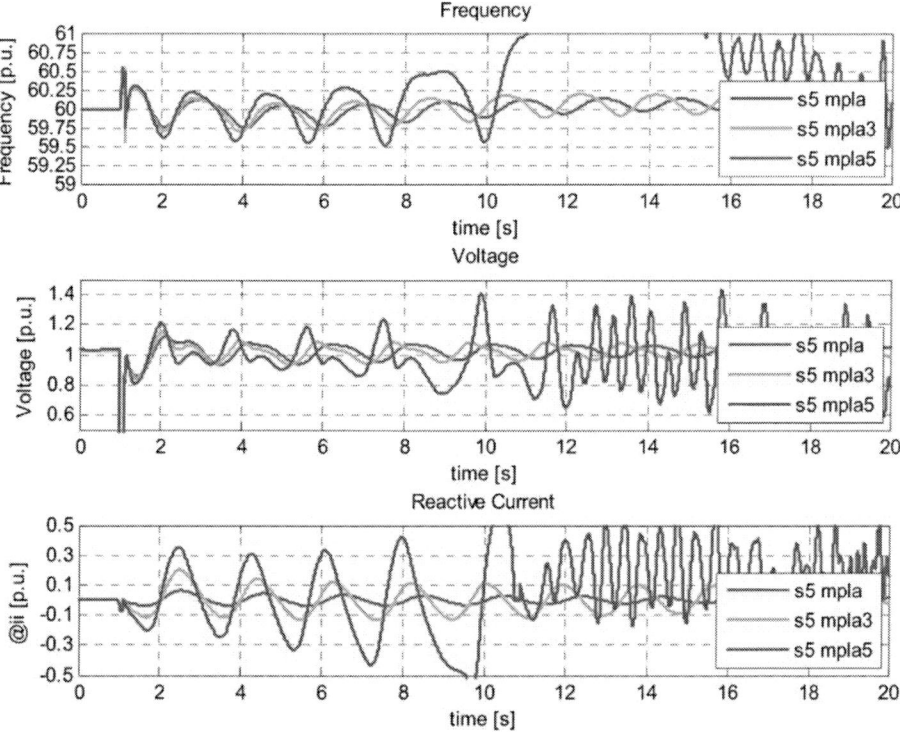

Figure 3.9. Peak load cases with different anti-islanding settings. Case 29 (red), case 29 with three times higher anti-islanding gain (green), and case 29 with five times higher gain (blue)

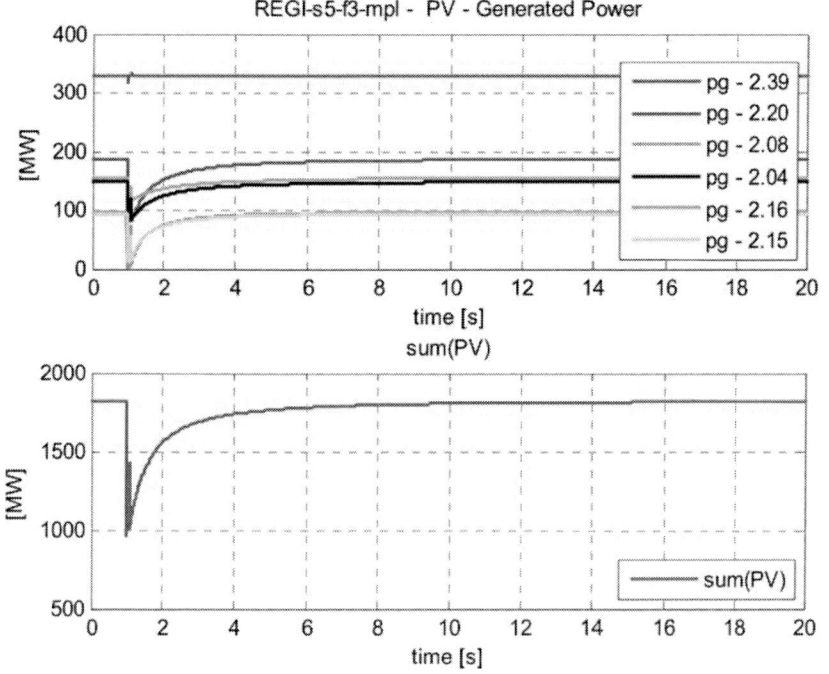

Figure 3.10. Power reduction of PV generation during and after faults (PV LVRT)

4.0 GAP ANALYSIS

The design and operation of systems with high PV penetration will require improved understanding of PV behavior during and after system faults. Present standards (IEEE 1547) assume relatively fast disconnection of PV. System planning should also consider systems that do not trip during faults and the performance that can be expected from PV technologies.

Unit commitment strategy has a significant impact on system performance at high PV penetration levels. A good understanding of the constraints on unit commitment caused by performance and generation flexibility is crucial to enable the operation of systems with high penetration of PV. Other critical gap is associated with understanding of potential additional regulation and load-following requirements because of PV.

5.0 RECOMMENDATIONS FOR FUTURE RESEARCH

Recommended future research in this area includes:

- Develop models that are accurate enough to estimate aggregated behavior of PV systems for system planning. The behavior of aggregated PV during and after faults is most relevant. Converter technology and control can result in considerable differences between systems.
- Develop guidelines for enhancing transmission planning databases to accommodate such models, including aggregated representation of medium- and low-voltage networks.
- Improve understanding of and provide guidelines to quantify the performance and economic impact of PV penetration on regulation and load-following requirements.
- Develop methodologies for estimating the required flexibility of the generation assets to meet regulation and load-following requirements; in particular, the requirements for generating units to ramp production up or down and to stop and start.
- Develop a unit commitment and dispatch strategy for conventional units in systems with high-penetration PV. The proposed strategy should include reliability requirements, operational costs, regulation, and load-following costs. The use of PV forecasts in unit commitment is also instrumental to increase the value of PV generation for high penetration. The approach may require that part of the unit scheduling be done a few hours in advance (instead of 24 to 48 hours currently required) to improve PV production forecast accuracy. The extension of such research to systems with PV and wind generation is also recommended.
- Provide guidance to quantify the value in terms of performance and the economic benefit of potentially mitigating measures of PV power variability (modifications of control zone constraints, flexible conventional generation, centralized and local energy storage, forecast, etc.). These guidelines could be applied to different systems and generation resources. The extension of such research to systems with PV and wind generation is also recommended.
- Develop methods to reliably forecast PV generation at regional levels.
- Develop methods to estimate the actual PV generation to help with system operation.

- Develop active anti-islanding schemes that do not affect regional system performance.
- Develop a strategy and specification for PV voltage control.
- Develop potential remuneration mechanisms for ancillary services associated with voltage support.

CONCLUSIONS AND RECOMMENDATIONS

These are the observations from this effort:

6.1 Observations of System Aspects

- Unit commitment strategy has significant a impact on system performance at high PV penetration levels:
    - System inertia and frequency regulation are reduced as conventional generation is de-committed. This results in power imbalances creating larger frequency excursions.
    - Thermal units operate at less efficient load levels if conventional generation is not de-committed.
    - Reactive power support in the transmission system is reduced as conventional generation is de-committed. This is of particular relevance for high PV generation and moderate load conditions. This results in longer voltage recovery times after faults and a higher risk of voltage collapse.
    - Dynamic stability of the system can be affected.
- Transmission losses are lower with increased PV generation because PV is connected closer to the loads than is conventional generation.
- High-penetration PV will require flexible generation. The ability of a system to accommodate PV generation can be limited under moderate load conditions and conventional units with high technical minimum or stop/start limitations.
- With substantial PV penetration compliant with IEEE 1547, there is considerable reduction in system reliability. Additional risks are associated with the extensive loss of PV generation during transmission faults. It is recommended to:
    - Require minimum undervoltage clearing times (instead of maximum clearing times) to avoid significant PV trips during, at least, transmission faults with primary clearing.
    - Set underfrequency tripping below frequency load shedding stages.
- If PV performs frequency control, the steady-state frequency performance is similar to the system without PV. Depending on the bandwidth of the PV frequency control, minimum frequency excursions could be reduced. Frequency control in PV generation does not compensate for the reduced system inertia with PV generation.
- PV generation could provide primary frequency control for frequency excursions above nominal without significantly reducing energy production.

- PV generation could provide primary frequency control for frequency decrements as well, but it requires operation with a primary reserve. This could considerably reduce energy production if operation with a reserve is required frequently. There is also a communication challenge of accessing each PV system to require a specific operating point.
- Anti-islanding schemes can affect the oscillatory stability of the bulk power system.

6.2 Observations of Photovoltaics Potential Performance

- The LVRT capability of PV would reduce the negative impact on the reliability of high-penetration PV.
- Even if the PV stays connected during and after a system fault, voltage sags may reduce prolonged PV power output. The power reduction depends on the specific converter/panel control and the voltage recovery in the system.
- For high PV penetration and if significant conventional generation is de-committed, PV voltage control can partially compensate for reduced voltage support in the transmission system.
- Voltage regulation tends to counteract the operation of anti-islanding schemes (of the type studied here). There is potential for voltage regulation to be tuned less responsively to ensure effective disconnection during islanding and steady-state voltage control.

6.3 Relevant Aspects That Were not Analyzed

- The dispatch of fewer regulating resources to accommodate variable PV generation can increase the requirement of load-following reserve in the system and for individual units. Additionally, the variability of PV generation may also increase load-following requirements.
- PV systems perform periodic sweeps to characterize the nonlinear relationship between the voltage and the current of the PV modules and the actual solar irradiation. The sweeps result in power output reductions between zero and maximum available power during many seconds. In some cases (depending on the PV system control), a sweep is also performed after significant voltage variations. Many PV systems simultaneously performing a sweep after a transient voltage sag can reduce transient generation considerably.
- The implementation of voltage control on individual PV systems is challenging. There is potential for undesirable interactions between PV systems connected to same feeder and phase and between PV systems and other voltage regulating devices.
- High penetration of PV in distribution feeders can complicate frequency load shedding. Present frequency load shedding schemes are based on frequency relays that disconnect complete feeders. The operation of such relays at times of the day when significant PV generation is produced will result in less effective megawatts of load disconnected.

REFERENCES

[1] IEEE 39 Bus System Benchmark Model. Available at http://psdyn.ece.wisc.edu/IEEE_benchmarks/index.htm
[2] GE PSLF Manual
[3] Ye, Z., Walling, R., Zhou, R., Li, L. & Wang, T.: *Study and Development of Anti-Islanding Control for Grid-Connected Inverter*. General Electric Global Research Center, NREL/SR-560-36243, 2004. Available at http://www.nrel.gov/docs/fy04osti/36243.pdf

APPENDIX A: PSLF LOAD FLOW RESULTS

Table A.1. Load Flow Results

BUS-NO	PSLF V-PU	PSLF DEG	BENCHMARK V (PU)	BENCHMARK Angle (deg)	BUS-NO	PSLF V-PU	PSLF DEG	BENCHMARK V (PU)	BENCHMARK Angle (deg)
1	1.0474	-10.03	1.047	-8.44	21	1.0317	-5.37	1.032	-3.78
2	1.0487	-7.34	1.049	-5.75	22	1.0498	-0.92	1.05	0.67
3	1.0302	-10.19	1.03	-8.6	23	1.0448	-1.12	1.045	0.47
4	1.0039	-11.19	1.004	-9.61	24	1.0373	-7.66	1.037	-6.07
5	1.0053	-10.2	1.005	-8.61	25	1.0576	-5.95	1.058	-4.36
6	1.0077	-9.54	1.008	-7.95	26	1.0521	-7.11	1.052	-5.53
7	0.997	-11.71	0.997	-10.12	27	1.0377	-9.08	1.038	-7.5
8	0.996	-12.2	0.996	-10.62	28	1.0501	-3.6	1.05	-2.01
9	1.0282	-11.91	1.028	-10.32	29	1.0499	-0.84	1.05	0.74
10	1.0171	-7.01	1.017	-5.43	30	1.0475	-4.92	1.048	-3.33
11	1.0127	-7.87	1.013	-6.28	31	0.982	-1.59	0.982	0
12	1.0001	-7.83	1	-6.24	32	0.9831	0.98	0.983	2.57
13	1.0143	-7.69	1.014	-6.1	33	0.9972	2.61	0.997	4.19
14	1.0117	-9.24	1.012	-7.66	34	1.0123	1.59	1.012	3.17
15	1.0154	-9.32	1.015	-7.74	35	1.0493	4.04	1.049	5.63
16	1.0318	-7.77	1.032	-6.19	36	1.0635	6.74	1.064	8.32
17	1.0335	-8.89	1.034	-7.3	37	1.0278	0.83	1.028	2.42
18	1.0309	-9.81	1.031	-8.22	38	1.0265	6.22	1.027	7.81
19	1.0499	-2.61	1.05	-1.02	39	1.03	-11.64	1.03	-10.05
20	0.9912	-3.6	0.991	-2.01					

APPENDIX B: DYNAMIC MODELS

Table B-1. Generator Dynamic Data

no.	bus	kv	Mbase	Ld	Lpd	Lppd	Lq	Lpq	Lppq	Ll	Ra	Tpdo	Tppdo	Tpqo	Tppqo	S1	S12	H
10	30	22	1290	1.29	0.4	0.082	0.89	0.103	0.082	0.013	0	10.2	0.03	0.2	0.04	0.05	0.3	3.255
2	31	22	574	1.693	0.4	0.2	1.618	0.976	0.2	0.035	0	6.56	0.03	1.5	0.04	0.05	0.3	5.28
3	32	22	753	1.879	0.4	0.2	1.785	0.66	0.2	0.03	0	5.7	0.03	1.5	0.04	0.05	0.3	4.75
4	33	22	917	2.403	0.4	0.2	2.367	1.523	0.2	0.03	0	5.69	0.03	1.5	0.04	0.05	0.3	3.117
5	34	22	303	2.03	0.4	0.2	1.879	0.503	0.2	0.054	0	5.4	0.03	0.44	0.04	0.05	0.3	8.58
6	35	22	800	2.032	0.4	0.2	1.928	0.651	0.2	0.022	0	7.3	0.03	0.4	0.04	0.05	0.3	4.35
7	36	22	816	2.408	0.4	0.2	2.384	1.518	0.2	0.032	0	5.66	0.03	1.5	0.04	0.05	0.3	3.23
8	37	22	702	2.035	0.4	0.2	1.965	0.639	0.2	0.028	0	6.7	0.03	0.41	0.04	0.05	0.3	3.46
9	38	22	702	1.478	0.4	0.2	1.439	0.411	0.2	0.03	0	4.79	0.03	1.96	0.04	0.05	0.3	4.92
Representation of connection to large grid																		
1	39	1	6667	1.333	0.4	0.2	1.267	0.533	0.2	0.003	0	7	0.03	0.7	0.04	0.05	0.3	7.5

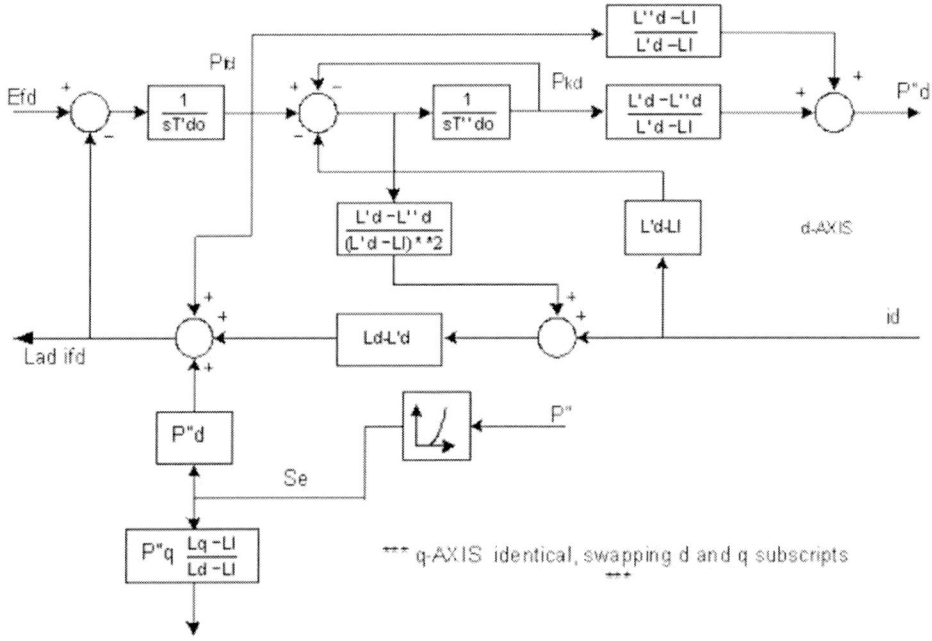

Figure B.1. D-component of PSLF – "genrou" model.

Table B.2. AVR Parameters

Unit No.	Ta/Tb	Tb	K	Te	Emin	Ema	Kc	Tc	Edf min	Edf max	Tr	Ta	Vref
1	0.1	10	200	0.015	-5	5	0.08	0	-5	5	0.01	1	1.03
2	0.1	10	200	0.015	-5	5	0.08	0	-5	5	0.01	1	0.982
3	0.1	10	200	0.015	-5	5	0.08	0	-5	5	0.01	1	0.9831
4	0.1	10	200	0.015	-5	5	0.08	0	-5	5	0.01	1	0.9972
5	0.1	10	200	0.015	-5	5	0.08	0	-5	5	0.01	1	1.0123
6	0.1	10	200	0.015	-5	5	0.08	0	-5	5	0.01	1	1.0493
7	0.1	10	200	0.015	-5	5	0.08	0	-5	5	0.01	1	1.0635
8	0.1	10	200	0.015	-5	5	0.08	0	-5	5	0.01	1	1.0278
9	0.1	10	200	0.015	-5	5	0.08	0	-5	5	0.01	1	1.0265
10	0.1	10	200	0.015	-5	5	0.08	0	-5	5	0.01	1	1.0475

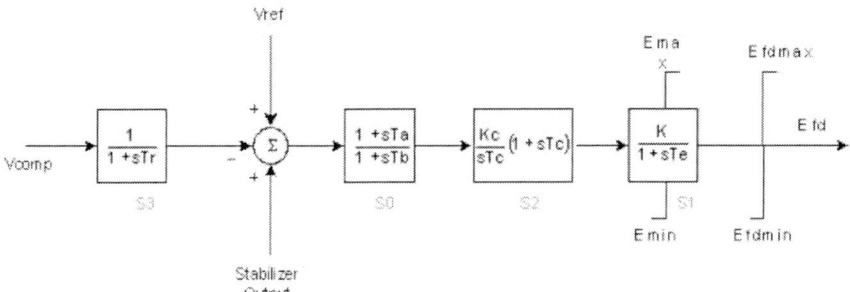

Figure B.2. PSLF – AVR model.

Table B.3. Stabilizer Parameters

Unit No.	J1	K1	J2	K2	Tw1	Tw2	Tw3	Tw4	T6	T7	Ks2	Ks4	T8	T9	n	m	Ks1	T1	T2	T3	T4	Vst max	Vst min	a	Ta	Tb
1	1	0	3	0	10	0	2	0	0	0	0	0	0.5	0.1	0	0	1	5	0.6	3	0.5	0.2	-0.2	1	0	0
2	1	0	3	0	10	0	2	0	0	0	0	0	0.5	0.1	0	0	0.5	5	0.4	1	0.1	0.2	-0.2	1	0	0
3	1	0	3	0	10	0	2	0	0	0	0	0	0.5	0.1	0	0	0.5	3	0.2	2	0.2	0.2	-0.2	1	0	0
4	1	0	3	0	10	0	2	0	0	0	0	0	0.5	0.1	0	0	2	1	0.1	1	0.3	0.2	-0.2	1	0	0
5	1	0	3	0	10	0	2	0	0	0	0	0	0.5	0.1	0	0	1	1.5	0.2	1	0.1	0.2	-0.2	1	0	0
6	1	0	3	0	10	0	2	0	0	0	0	0	0.5	0.1	0	0	4	0.5	0.1	0.5	0.05	0.2	-0.2	1	0	0
7	1	0	3	0	10	0	2	0	0	0	0	0	0.5	0.1	0	0	7.5	0.2	0.02	0.5	0.1	0.2	-0.2	1	0	0
8	1	0	3	0	10	0	2	0	0	0	0	0	0.5	0.1	0	0	2	1	0.2	2	0.1	0.2	-0.2	1	0	0
9	1	0	3	0	10	0	2	0	0	0	0	0	0.5	0.1	0	0	2	1	0.5	2	0.1	0.2	-0.2	1	0	0
10	1	0	3	0	10	0	2	0	0	0	0	0	0.5	0.1	0	0	1	1	0.05	3	0.5	0.2	-0.2	1	0	0

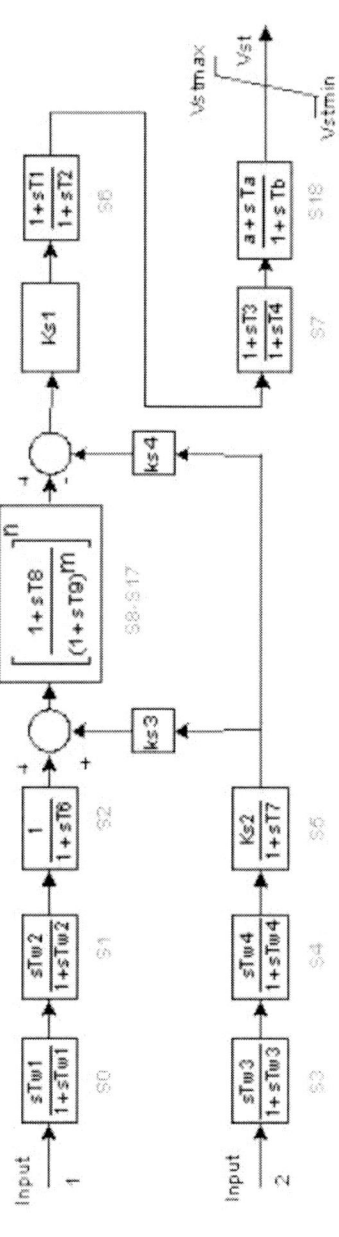

Figure B.3. PSLF – "pss2a" model.

$$P = P_o (p1v^2 + p2v + p3 + p4 (1 + lpd\, f))$$
$$Q = Q_o (q1v^2 + q2v + q3 + q4 (1 + lqd\, f))$$

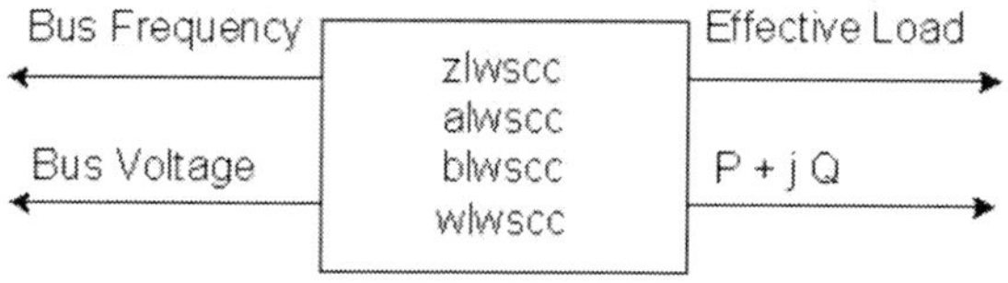

```
Bus Frequency  →          zlwscc         Effective Load →
                          alwscc
Bus Voltage   →           blwscc         P + jQ         →
                          wlwscc
```

Figure B.4. PSLF – "wlwscc" load model polynomial representation.

APPENDIX C: MODIFICATIONS TO IEEE 39 BUS SYSTEM

Load Flow Results of Extended Model

Table C.1. Load Flow Result of Extended Model

BUS-NO	KV	PSLF V-PU	PSLF DEG	IEEE V (PU)	IEEE Angle (deg)	BUS-NO	KV	PSLF V-PU	PSLF DEG
1	345	1.0479	-20.78	1.0474	-8.44	103	22	1.0489	-24.34
2	345	1.0501	-18.16	1.0487	-5.75	203	0.44	1.0603	-29.12
3	345	1.0312	-20.65	1.0302	-8.6	104	22	1.0371	-23.48
4	345	1.0007	-20.1	1.0039	-9.61	204	0.44	1.0175	-27.06
5	345	0.9993	-17.73	1.0053	-8.61	107	22	1.0412	-22.76
6	345	1.0013	-16.71	1.0077	-7.95	207	0.44	1.023	-26.34
7	345	0.9911	-19.39	0.997	-10.12	108	22	1.0416	-23.48
8	345	0.9904	-20.08	0.996	-10.62	208	0.44	1.0249	-27.15
9	345	1.0254	-21.5	1.0282	-10.32	115	22	1.0036	-23.47
10	345	1.0134	-15.02	1.0172	-5.43	215	0.44	0.9709	-26.83
11	345	1.0082	-15.61	1.0127	-6.28	116	22	1.0458	-22.72
12	100	0.9961	-15.84	1.0002	-6.24	216	0.44	1.0484	-27.24
13	345	1.0106	-15.97	1.0143	-6.1	118	22	1.0385	-24.36
14	345	1.0087	-18.25	1.0117	-7.66	218	0.44	1.032	-28.63
15	345	1.0157	-20.03	1.0154	-7.74	120	22	1.0314	-19.42
16	345	1.034	-19.06	1.0318	-6.19	220	0.44	1.0258	-23.85
17	345	1.0355	-20.07	1.0336	-7.3	121	22	1.0264	-20.26
18	345	1.0327	-20.75	1.0309	-8.22	221	0.44	1.0017	-23.7
19	345	1.051	-14.37	1.0499	-1.02	123	22	1.0445	-16.11
20	100	0.9922	-15.72	0.9912	-2.01	223	0.44	1.028	-19.74
21	345	1.0338	-16.88	1.0318	-3.78	124	22	1.031	-22.31
22	345	1.0512	-12.5	1.0498	0.67	224	0.44	1.004	-25.48
23	345	1.0465	-12.74	1.0448	0.47	125	22	1.0207	-20.74
24	345	1.0406	-19.06	1.0373	-6.07	225	0.44	1.0093	-25.1
25	345	1.0569	-17.03	1.0576	-4.36	126	22	1.0205	-22.54
26	345	1.0532	-18.72	1.0521	-5.53	226	0.44	1.0166	-27.23
27	345	1.0396	-20.58	1.0377	-7.5	127	22	1.0412	-24.08
28	345	1.0509	-15.69	1.0501	-2.01	227	0.44	1.0292	-28.02
29	345	1.0506	-12.97	1.0499	0.74	128	22	1.0184	-19.52

Table C.1. Load Flow Result of Extended Model

BUS-NO	PSLF			IEEE		BUS-NO	PSLF		
	KV	V-PU	DEG	V (PU)	Angle (deg)		KV	V-PU	DEG
30	22	1.0475	-15.74	1.0475	-3.33	228	0.44	1.0131	-24.18
31	22	0.982	-1.59	0.982	0	129	22	1.0195	-16.82
32	22	0.9831	-7	0.9831	2.57	229	0.44	1.018	-21.62
33	22	0.9972	-9.16	0.9972	4.19	131	22	1.036	-4.78
34	22	1.0123	-10.53	1.0123	3.17	231	0.44	1.009	-7.82
35	22	1.0493	-7.54	1.0493	5.63	139	22	1.0333	-25.96
36	22	1.0635	-4.9	1.0635	8.32	239	0.44	1.023	-30.14
37	22	1.0278	-10.24	1.0278	2.42				
38	22	1.0265	-5.91	1.0265	7.81				
39	345	1.03	-22.35	1.03	-10.05				

Governors

Table C.2. Modified Governor Mbase Values

BUS-NO	ID	PGEN	IREG	MBASE	MBGOV
30	10	250	2	1290	349
31	2	520.8	6	573.9	665
32	3	650	10	753.3	811
33	4	632	19	917.4	797
34	5	508	20	303	612
35	6	650	22	800	816
36	7	560	23	816.3	676
37	8	540	25	701.8	642
38	9	830	29	701.8	987
39	1	1000	39	6666.7	1194

APPENDIX D: SINGLE LINE DIAGRAMS OF STARTING SCENARIOS

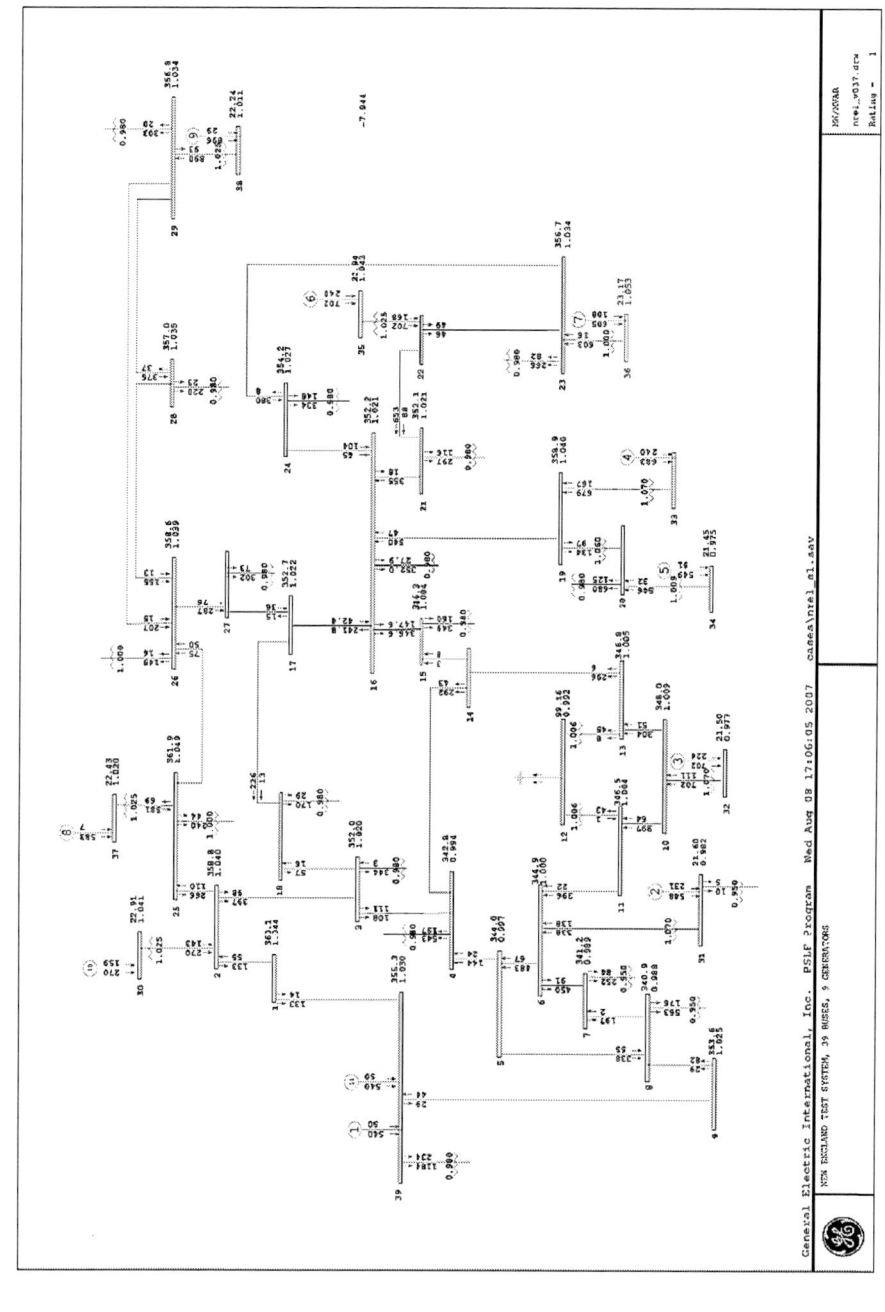

Figure D.1. Scenario s1 peak load (IEEE benchmark load). No PV generation.

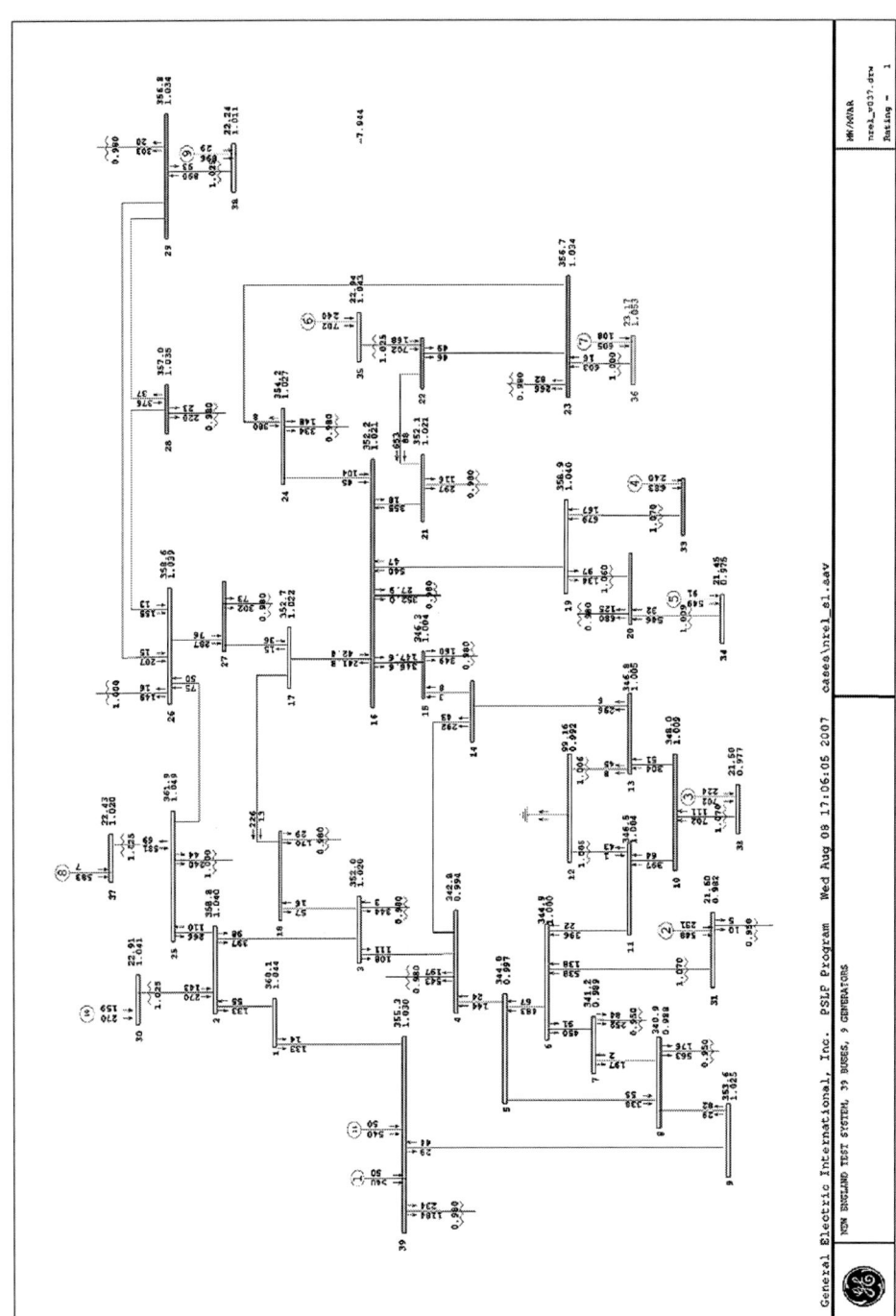

Figure D.2. Scenario s2 low load (50% of peak). No PV generation.

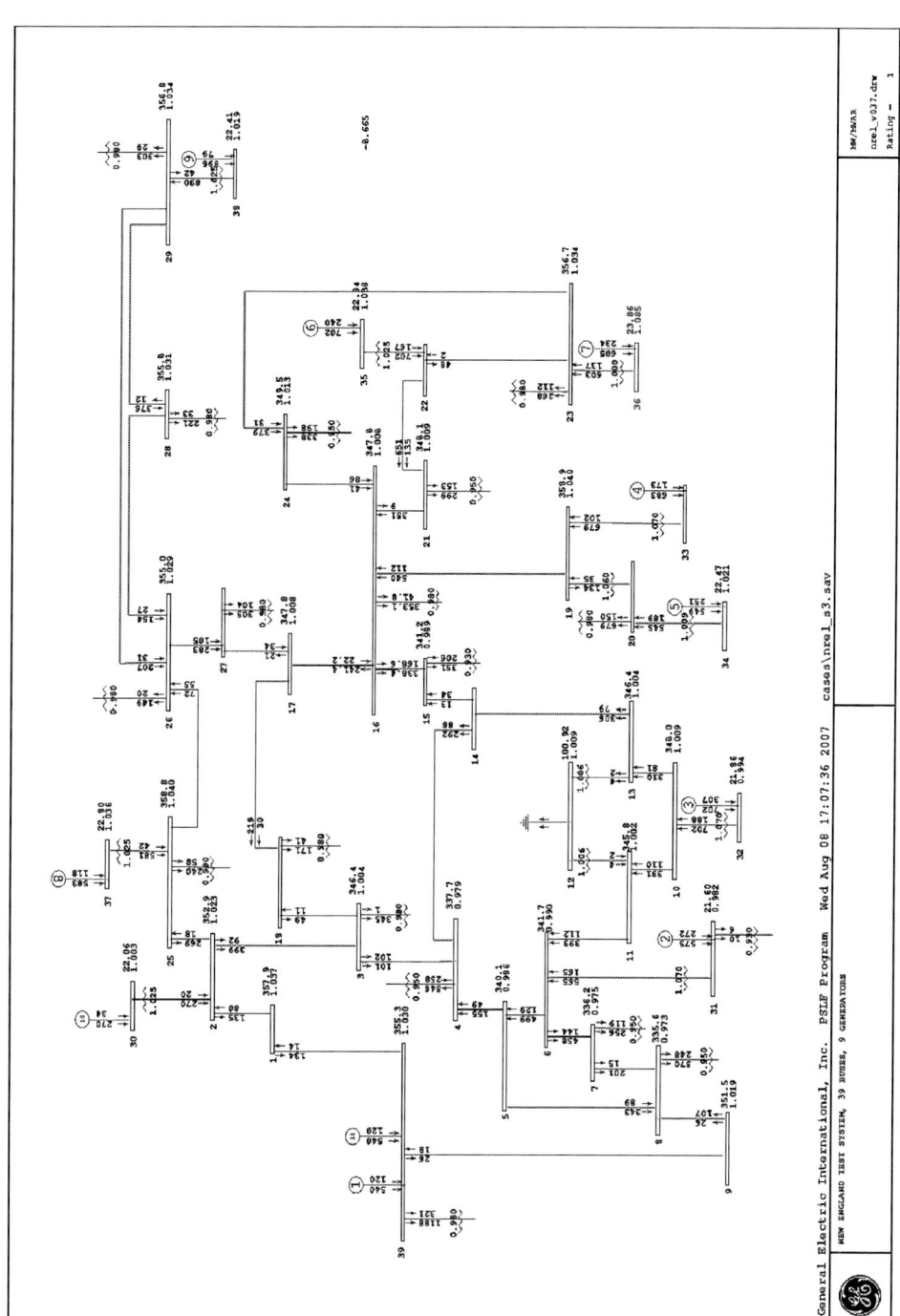

Figure D.3. Scenario s3 peak (plus 30%). 30% PV generation.

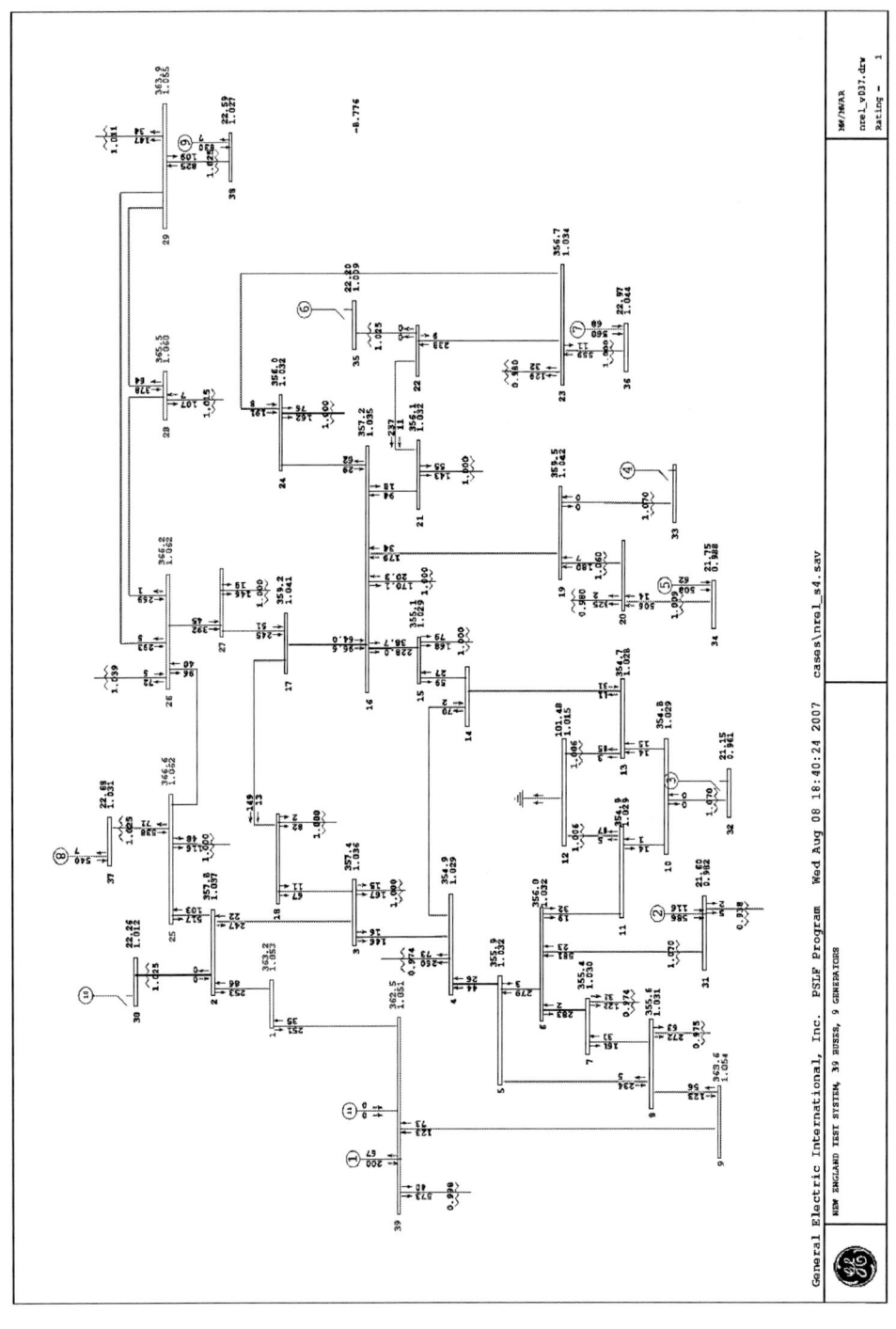

Figure D.4. Scenario s4 low load plus 30%, 30% PV generation.

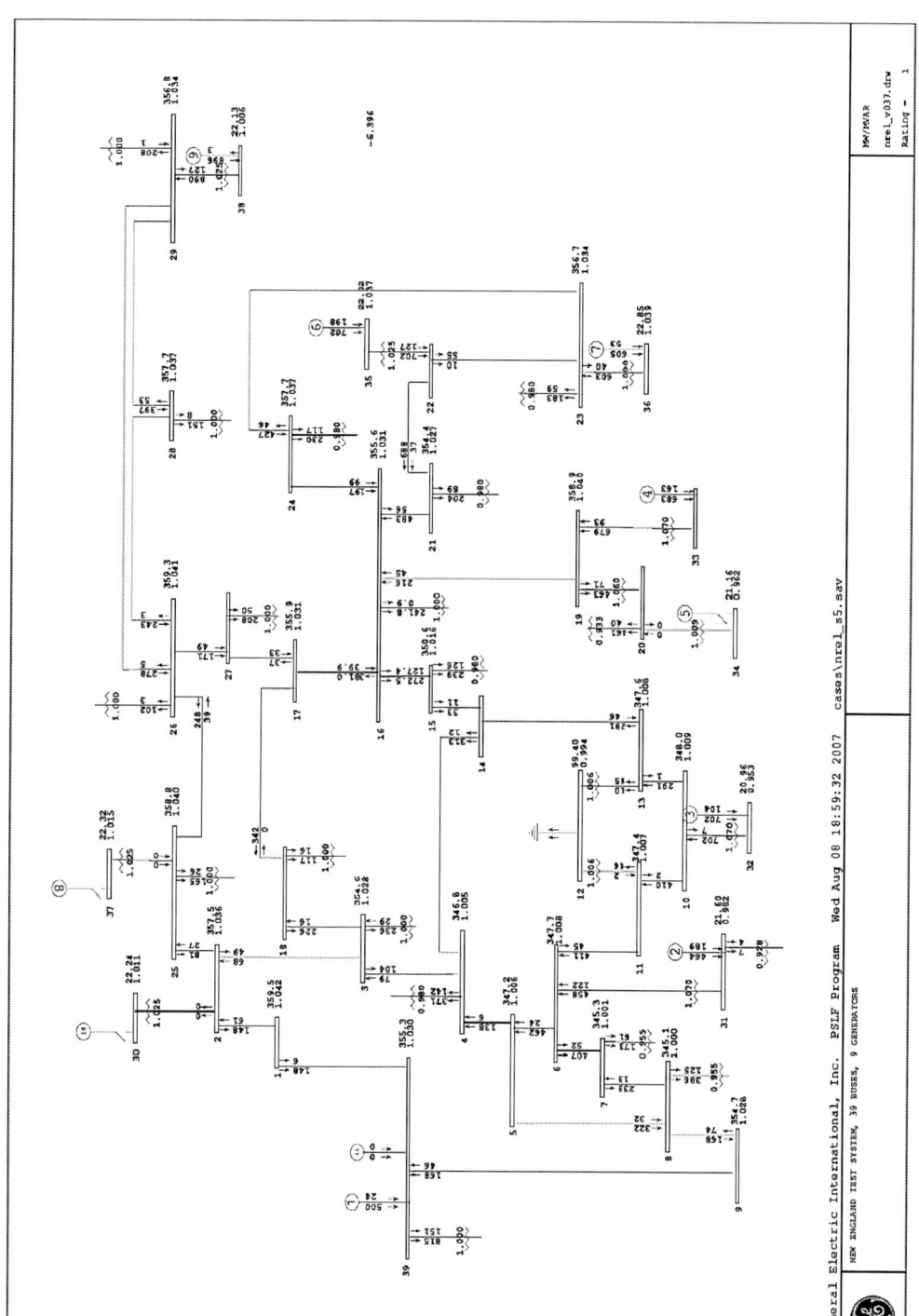

Figure D.5. Scenario s5 peak load. 30% PV generation. Conventional generation de-committed.

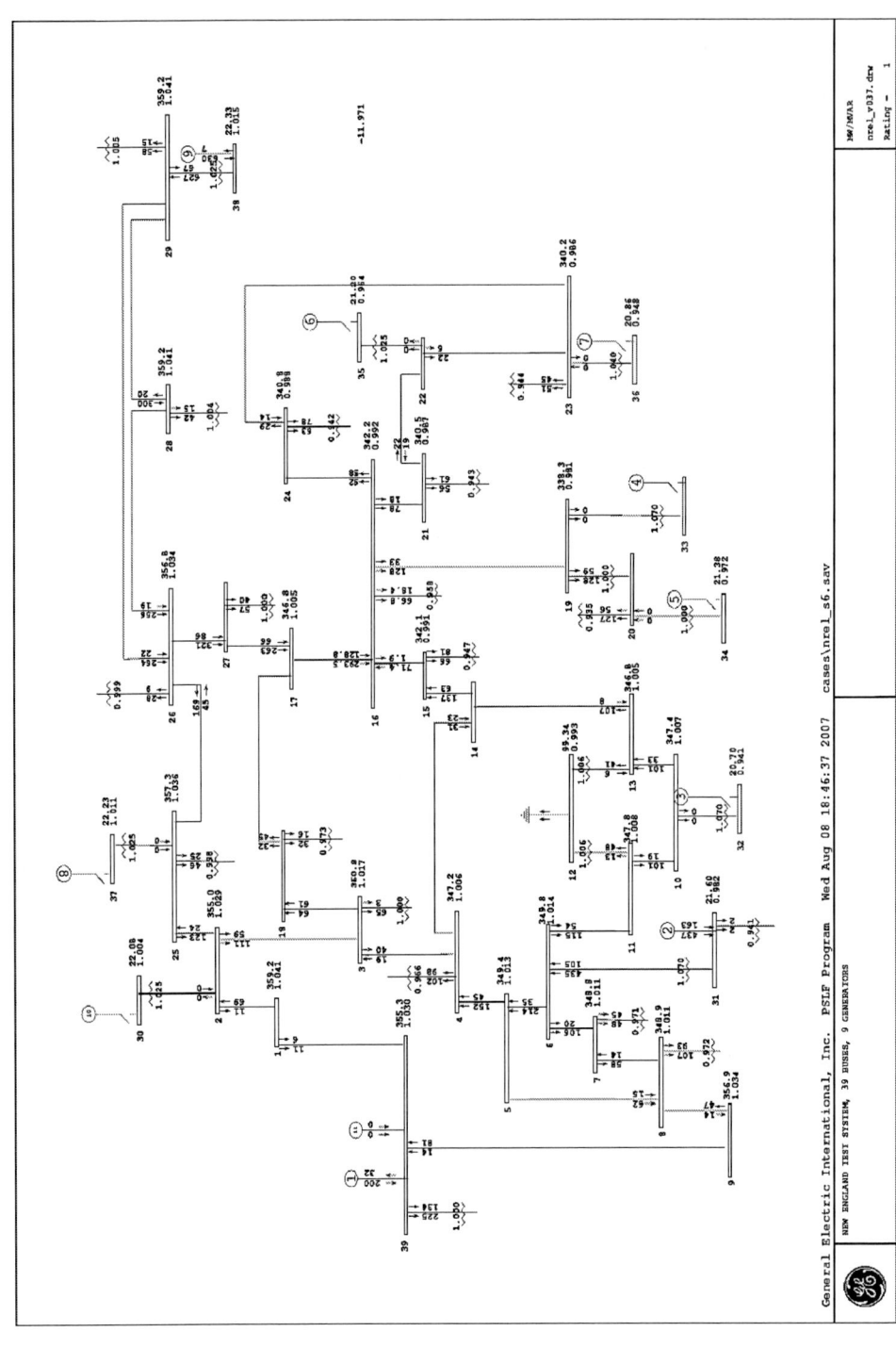

Figure D.6. Scenario s6 low load. 30% PV generation. Conventional generation de-committed.

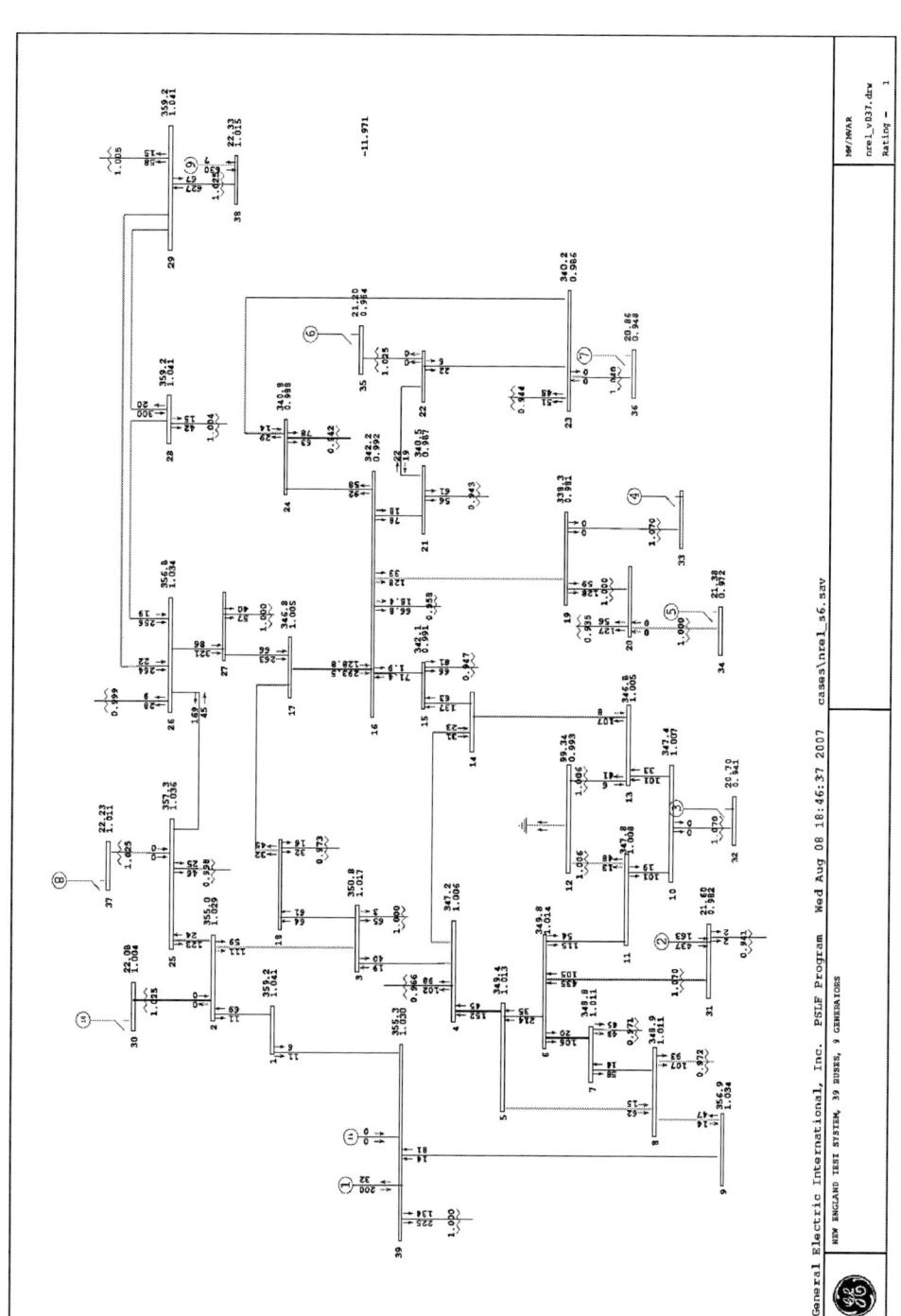

Figure D.7. Scenario s7 peak load. 30% PV generation. Conventional generation not de-committed.

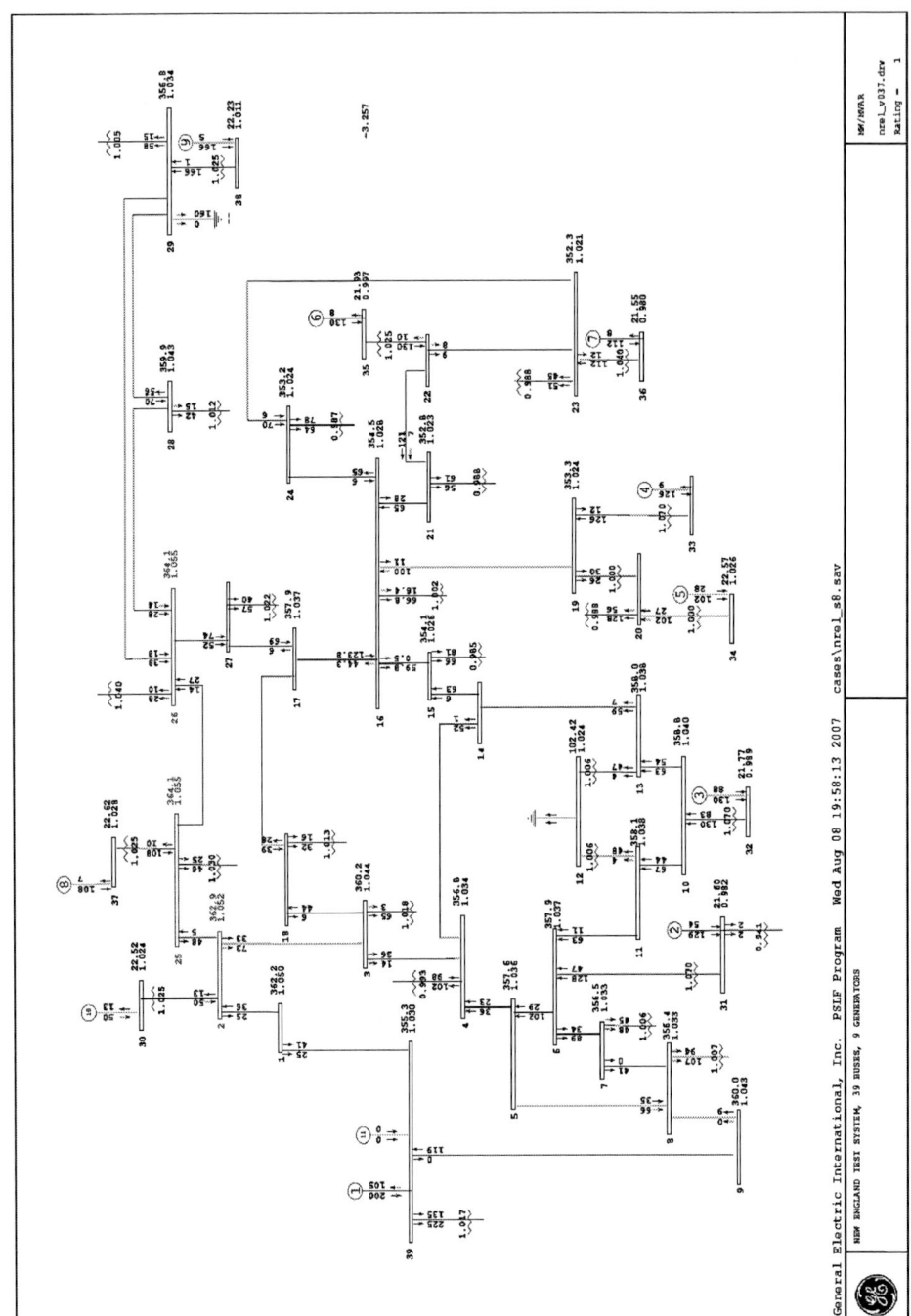

Figure D.8. Scenario s8 low load. 30% PV generation. Conventional generation not de-committed.

APPENDIX E: PV MODEL VERIFICATION

Figure E.1 shows the simple simulation system used to test the model and controls of the PV model. The source is a generator significantly larger than the PV. Loads were also included in the setup as needed to create frequency excursions needed in the analysis.

Maximum Power Point Tracking

Step changes of solar irradiation were simulated to calibrate the responsiveness of the MPP tracking loop. The results of a step change from 0.6 to 1 pu are presented in Figure E.2. It takes about three seconds to reach 0.96 pu power output (90% of requested change). Significantly more time is required to achieve the final value (1 pu) due to the non-linear relationship between power and the voltage of the panel (section 0). The signals presented are system frequency (F_{PV}), PV voltage (V_{PV}), PV power (P_{PV}), active current component (I_r) and reactive current component (I_i)

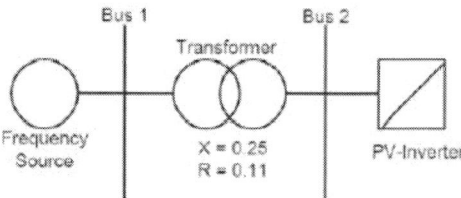

Figure E.1. PV-model test.

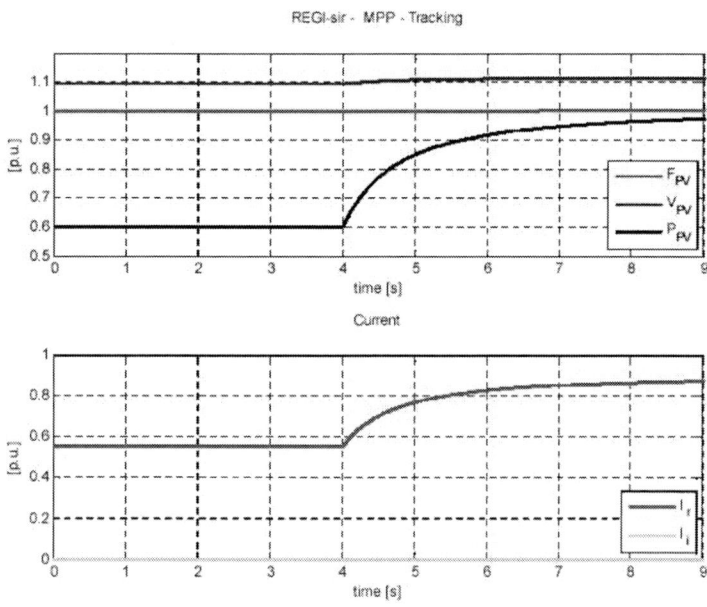

Figure E.2. MPP-tracking response to a step change in solar irradiation.

Frequency Control

Frequency ramps were applied to the PV system to test the frequency control response. Figure E.3 presents the response of the system to a frequency increase after 2 seconds and a later frequency reduction (starting at 5 seconds). The signals presented are system frequency (F_{PV}), PV voltage (V_{PV}), PV power (P_{PV}), active current component (I_r) and reactive current component (I_i), control output to achieve maximum available power (k_{PAC}), MPP control output, frequency control output (k_{FC}), active current limit output (k_{ACL}), and control mode (CM). The control mode signal is 1 for frequency control, -1 for MPP control, and 0 for current limiter control. The control signals refer to Figure 2.9. The PV frequency control was set to reach steady state in 2 to 3 seconds.

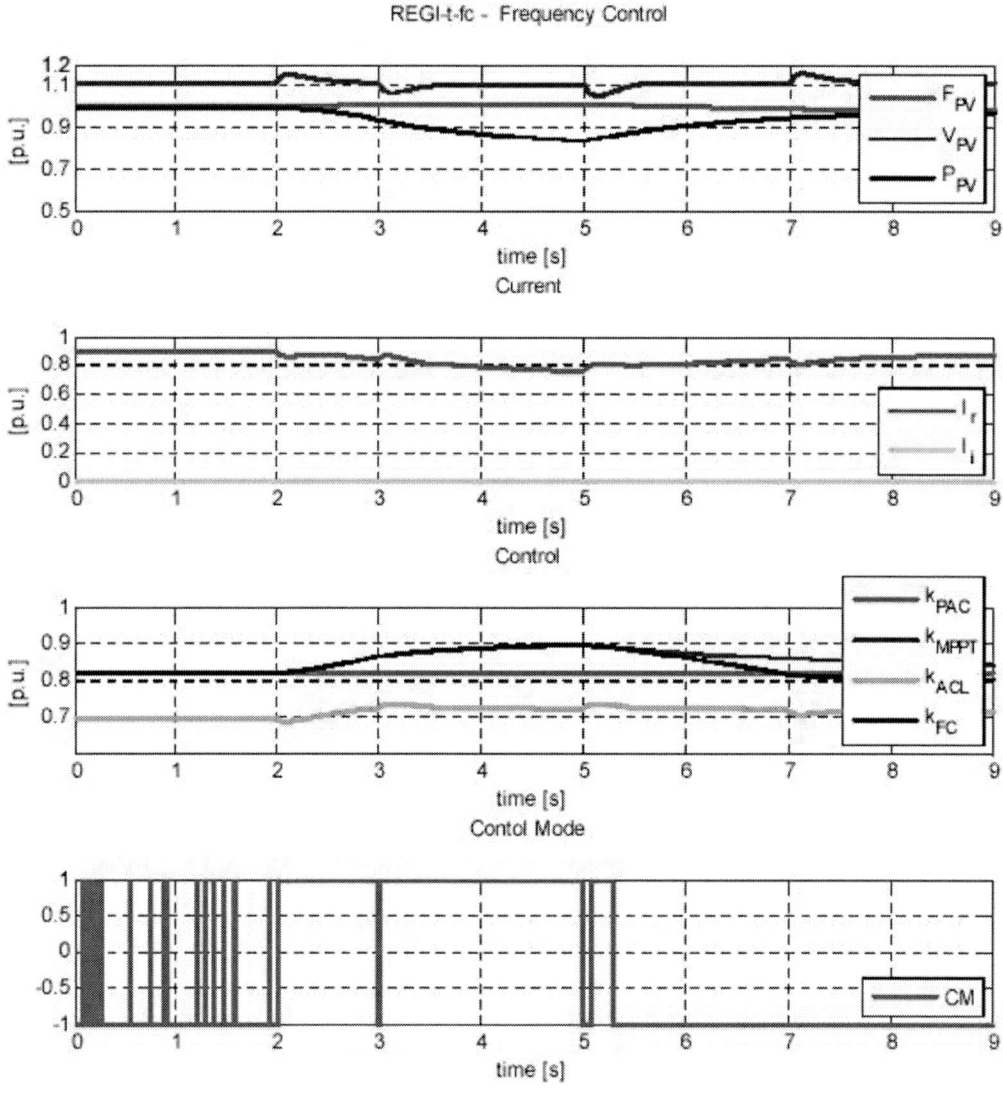

Figure E.3. PV response to frequency changes.

A similar run with initial reserve is presented in Figure E.4. The initial power is 20% below available power.

Voltage Control

Step changes in voltage reference were performed to test the voltage control. Figure E.5 and Figure E.6 present the responses for two different settings. PV voltage (V_{PV}) and reactive current component (I_i) are presented. The different settings are used later in the anti-islanding section.

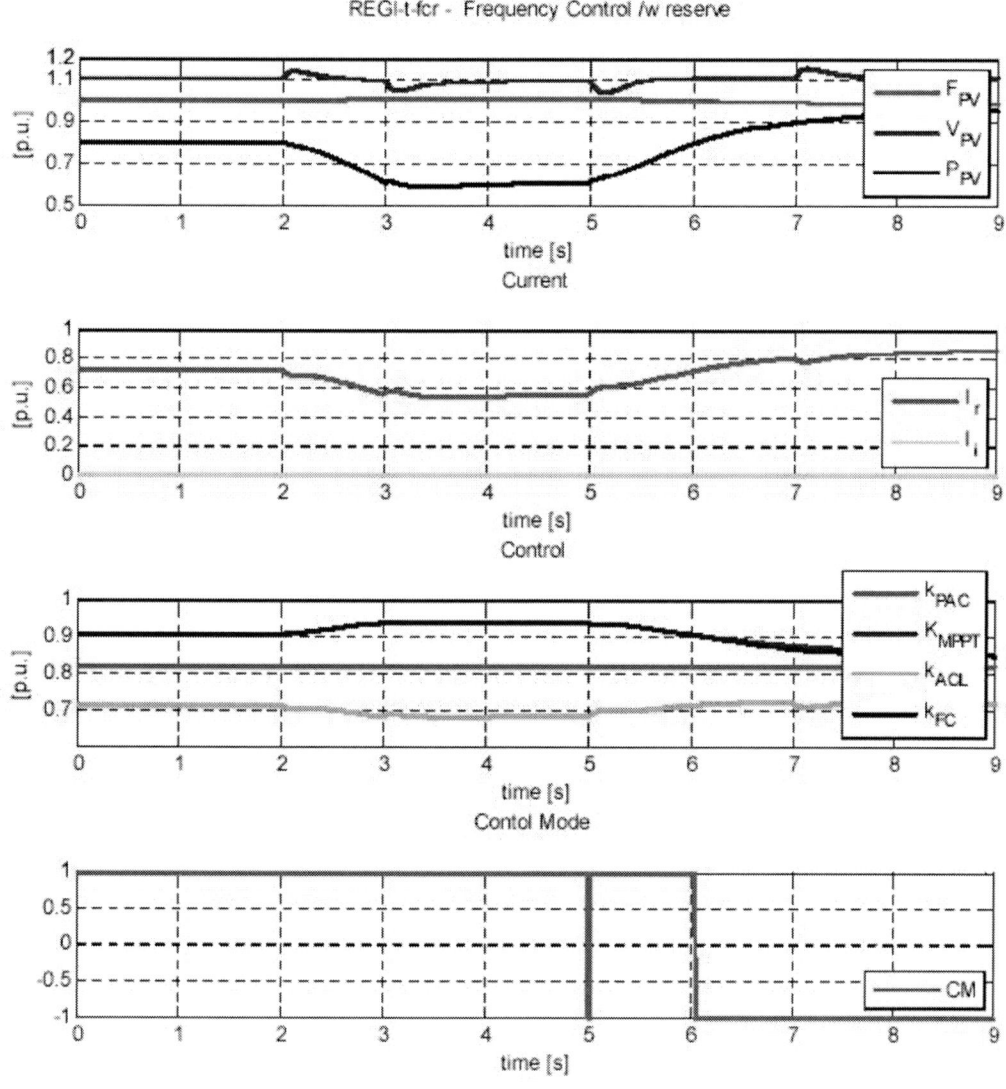

Figure E.4. PV response to frequency changes with initial reserve.

Fault Response

Different faults were applied to the test system to verify the PV response. All these cases assumed that the PV does not trip for voltage sags. Figure E.7 presents the response to a bus fault applied after two seconds and cleared one second later. MPP tracking and an active current limiter are active as presented in Figure 2.8. PV active power is reduced during the fault to limit the current (ACL control in Figure 2.8). The MPP tracking control brings the active power back to its initial value after the fault.

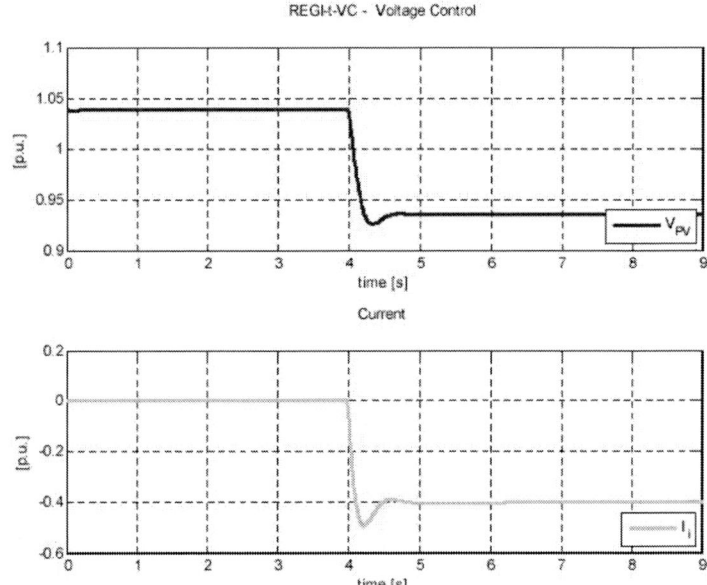

Figure E.5. Response to voltage reference step change.

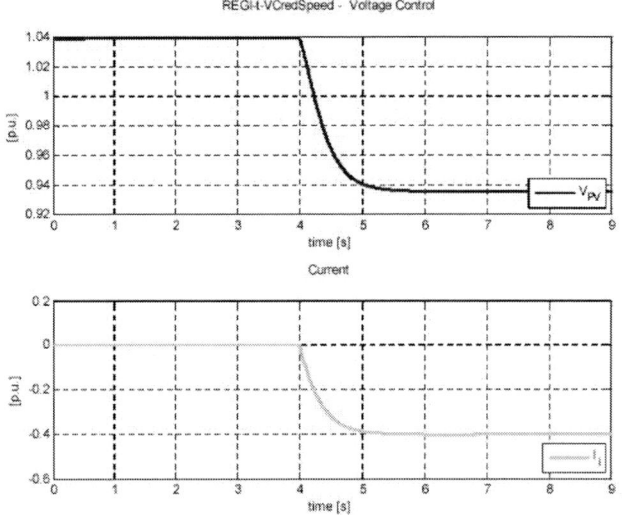

Figure E.6. Response to voltage reference step change, slower settings.

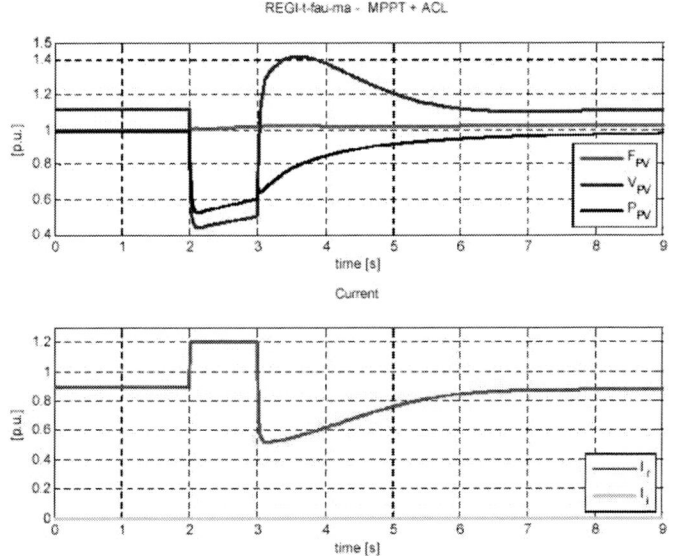

Figure E.7. PV response to voltage sag.

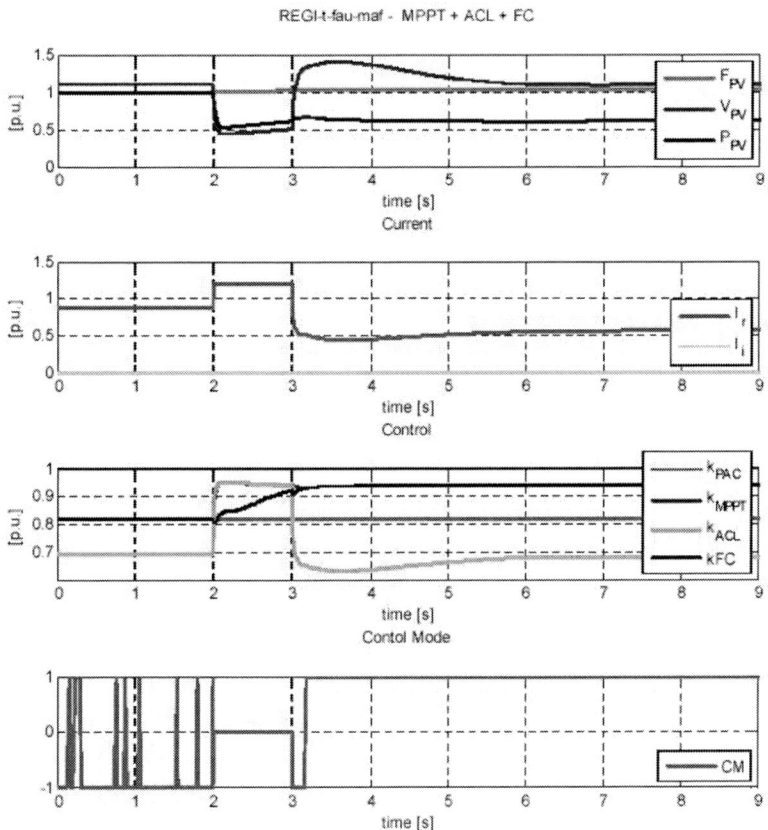

Figure E.8. PV response to voltage sag.

Figure E.8 presents the response to the same fault with the frequency control active (*FC* in Figure 2-9). The signals presented are system frequency (*FPV*), PV voltage (*V$_{PV}$*), PV power (*P$_{PV}$*), active current component (*I$_r$*) and reactive current component (*I$_i$*), control output to achieve maximum available power (*k$_{PAC}$*), MPP control output, frequency control output (*k$_{FC}$*), active current limit output (*k$_{ACL}$*), and control mode (*CM*). The control mode signal is 1 for frequency control, -1 for MPP control, and 0 for current limiter control. The control signals refer to Figure 2.9. The frequency increase at the end of the simulation results in the frequency control governing the module. All control mode transitions are smooth.

As mentioned in section 2.2.8, a system without fast DC link control was considered. Figure E.9 presents the response to the same fault with a PV converter more sensitive to voltage changes. After the fault, the power output drops to zero because the DC voltage increases proportionally with the terminal voltage, bringing the module to the open circuit voltage. The PV power-out recovers relatively fast when the terminal voltage returns to its initial value.

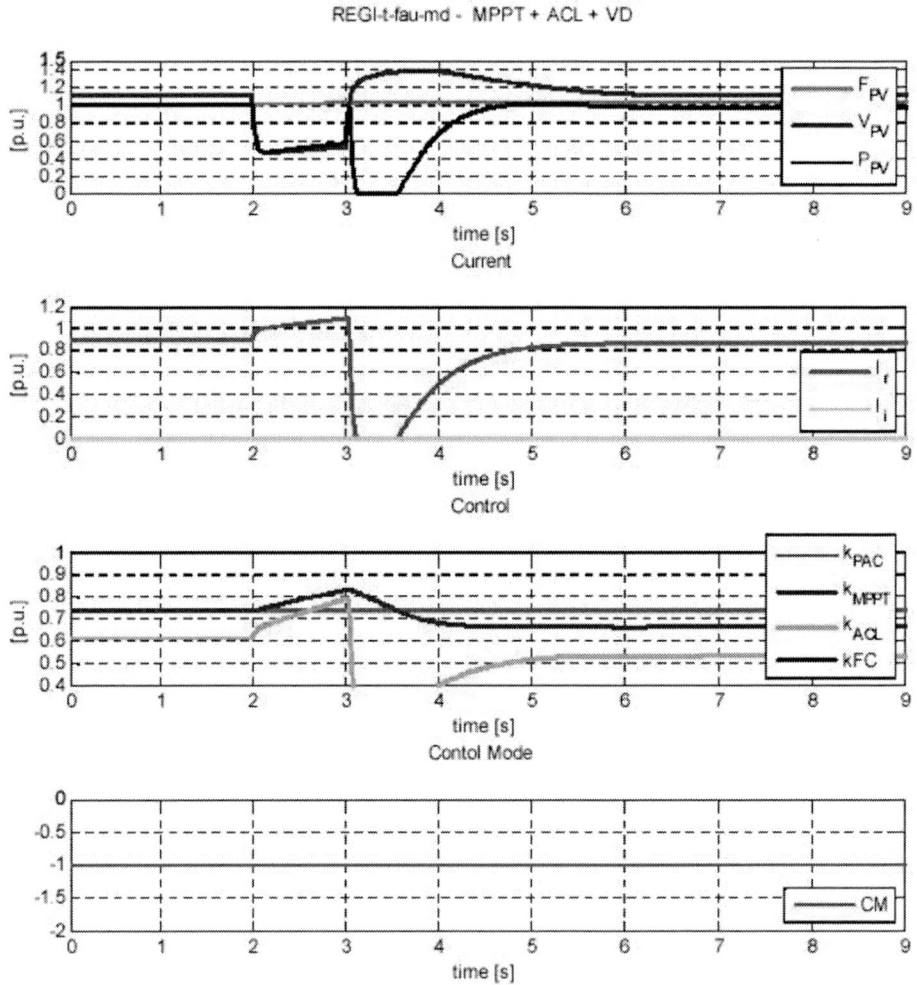

Figure E.9. PV response to voltage sag without fast DC link voltage control.

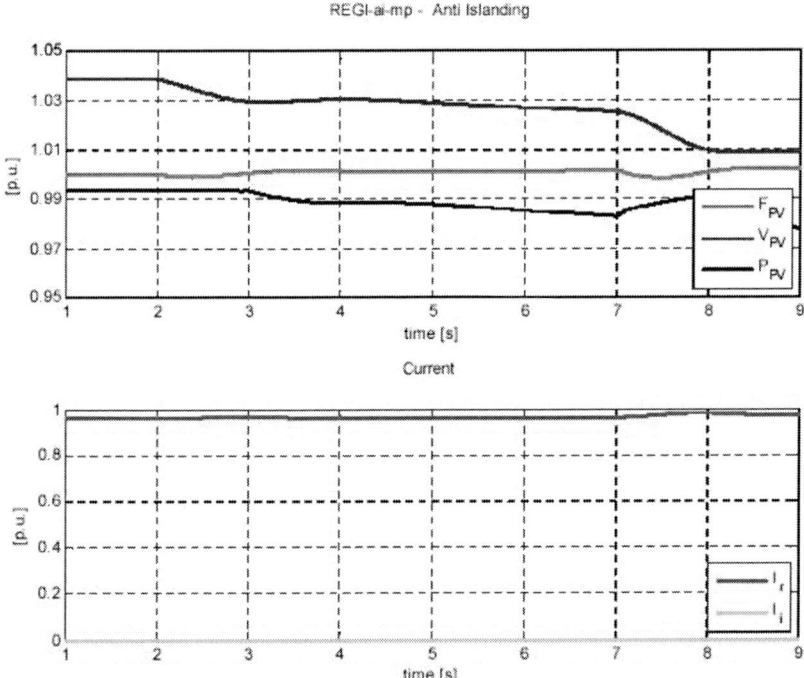

Figure E.10. PV islanding without active anti-islanding.

Active Anti–Islanding

The PV was islanded with a motor load equal to the PV output to evaluate tune the anti-islanding loop in Figure 2.9. Figure E.10 presents the PV response without anti-islanding. The PV remains in stable operation after disconnection from bulk system (after 2 seconds).

Figure E.11 presents the response with the anti-islanding loop enabled. The PV trips about 0.6 seconds after disconnection from bulk system. The anti-islanding action effectively increases the voltage, destabilizing the frequency.

Figure E.12 presents the response with active anti-islanding and voltage control enabled. The PV does not trip. The voltage control compensates the anti-islanding reactive current request.

Figure E.13 presents the response with active anti-islanding and voltage control enabled. The voltage control is set slower (as in Figure E.6). The PV trips in about 0.6 seconds. The voltage control is slower and cannot compensate for the anti-islanding destabilizing request. Figure E.14 presents a similar case with five times higher gain than the anti-islanding loop. The PV trips in about 0.3 seconds.

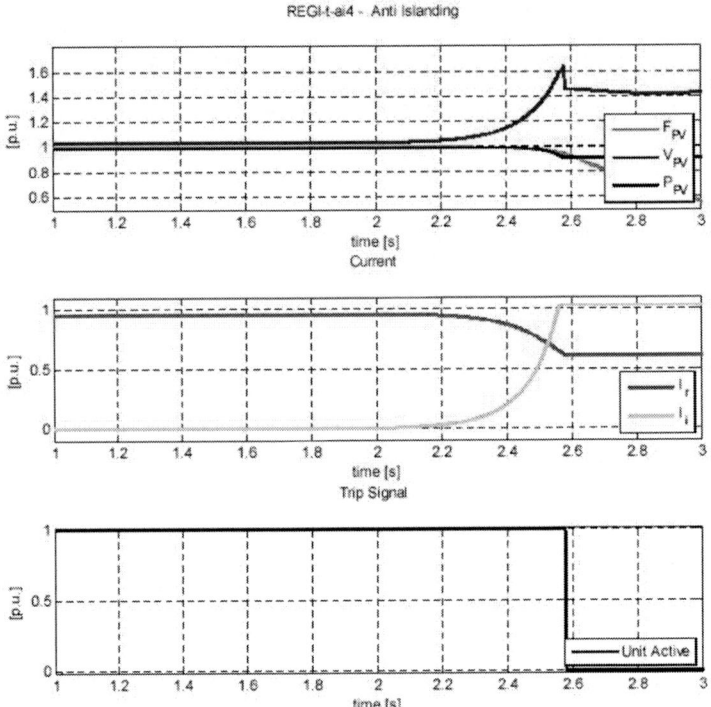

Figure E.11. PV islanding with active anti-islanding.

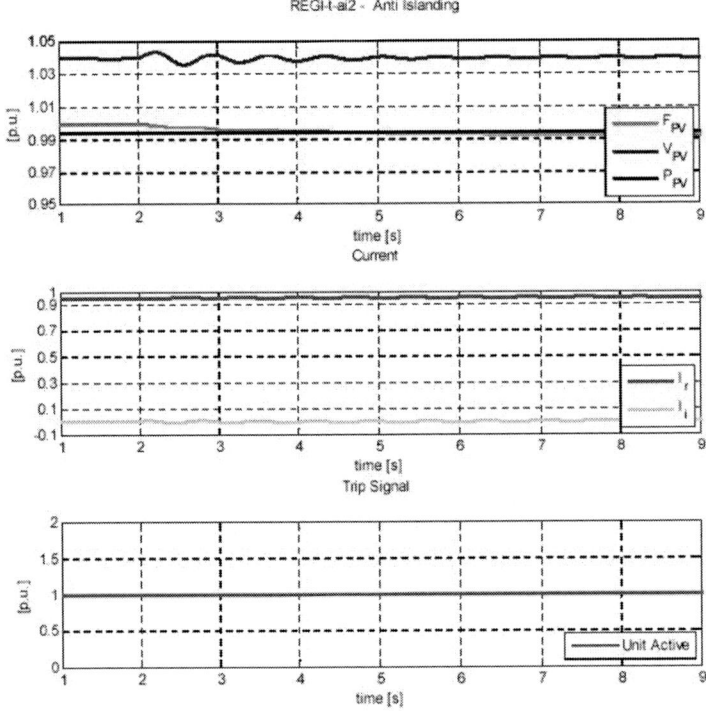

Figure E.12. PV islanding with active anti-islanding and voltage control.

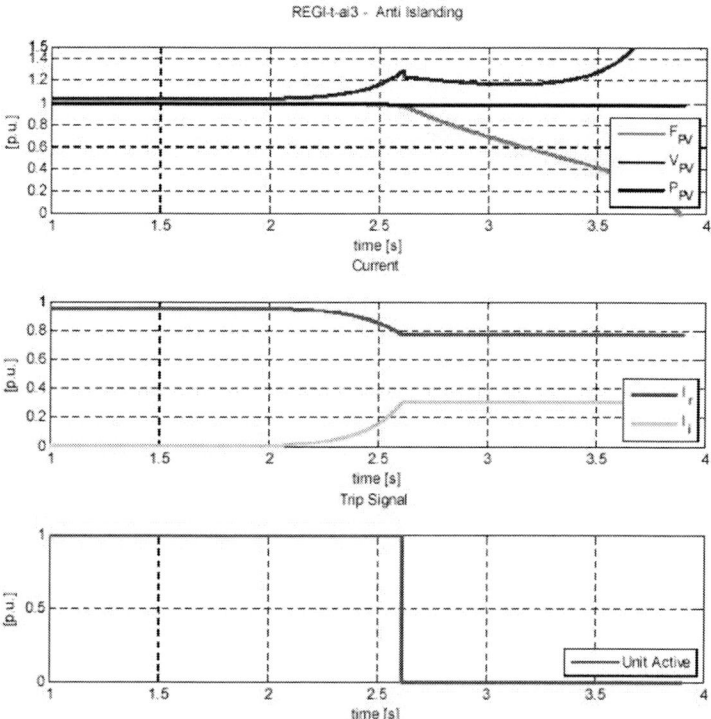

Figure E.13. PV islanding with active anti-islanding and slower voltage control.

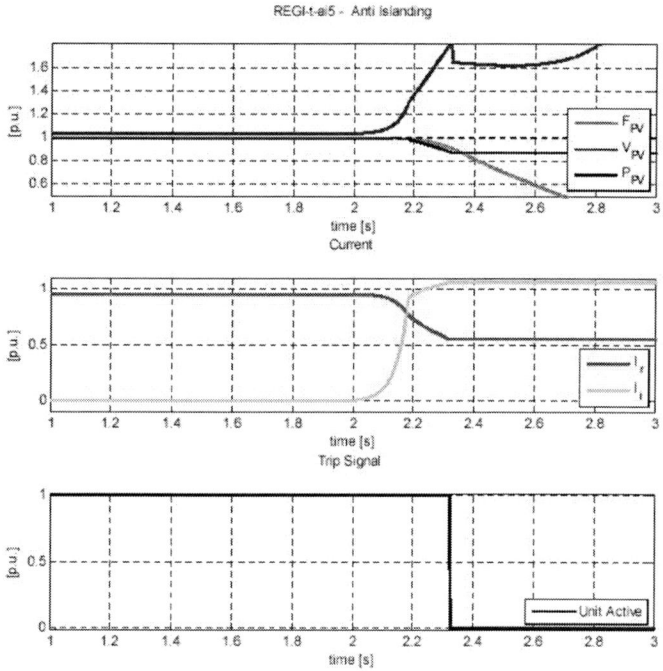

Figure E.14. PV islanding with active anti-islanding (higher gain) and slower voltage control.

REPORT DOCUMENTATION PAGE		Form Approved OMB No. 0704-0188
	The public reporting burden for this collection of information is estimated to average 1 hour per response, including the time for reviewing instructions, searching existing data sources, gathering and maintaining the data needed, and completing and reviewing the collection of information. Send comments regarding this burden estimate or any other aspect of this collection of information, including suggestions for reducing the burden, to Department of Defense, Executive Services and Communications Directorate (0704-0188). Respondents should be aware that notwithstanding any other provision of law, no person shall be subject to any penalty for failing to comply with a collection of information if it does not display a currently valid OMB control number. **PLEASE DO NOT RETURN YOUR FORM TO THE ABOVE ORGANIZATION.**	
1. REPORT DATE *(DD-MM-YYYY)* February 2008	2. REPORT TYPE Subcontract report	3. DATES COVERED *(From - To)*
4. TITLE AND SUBTITLE Transmission System Performance Analysis for High-Penetration Photovoltaics		5a. CONTRACT NUMBER DE-AC36-99-GO10337
		5b. GRANT NUMBER
		5c. PROGRAM ELEMENT NUMBER
6. AUTHOR(S) S. Achilles, S. Schramm, and J. Bebic		5d. PROJECT NUMBER NREL/SR-581-42300
		5e. TASK NUMBER PVB7.6401
		5f. WORK UNIT NUMBER
7. PERFORMING ORGANIZATION NAME(S) AND ADDRESS(ES) GE Global Research 1 Research Circle Niskayuna, NY 12309		8. PERFORMING ORGANIZATION REPORT NUMBER ADC-7-77032-01
9. SPONSORING/MONITORING AGENCY NAME(S) AND ADDRESS(ES) National Renewable Energy Laboratory 1617 Cole Blvd. Golden, CO 80401-3393		10. SPONSOR/MONITOR'S ACRONYM(S) NREL
		11. SPONSORING/MONITORING AGENCY REPORT NUMBER NREL/SR-581-42300
12. DISTRIBUTION AVAILABILITY STATEMENT National Technical Information Service U.S. Department of Commerce 5285 Port Royal Road Springfield, VA 22161		
13. SUPPLEMENTARY NOTES NREL Technical Monitor: Ben Kroposki		
14. ABSTRACT *(Maximum 200 Words)* This study is an assessment of the potential impact of high levels of penetration of photovoltaic (PV) generation on transmission systems. The effort used stability simulations of a transmission system with different levels of PV generation and load.		
15. SUBJECT TERMS photovoltaics; PV; transmission; renewable systems interconnection; GE Global Research; National Renewable Energy Laboratory; NREL		
16. SECURITY CLASSIFICATION OF: a. REPORT Unclassified b. ABSTRACT Unclassified c. THIS PAGE Unclassified	17. LIMITATION OF ABSTRACT UL	18. NUMBER OF PAGES
		19a. NAME OF RESPONSIBLE PERSON
		19b. TELEPHONE NUMBER *(Include area code)*

End Note

[1] IEEE 1547 Standard for Interconnecting Distributed Resources with Electric Power Systems, 2003.

In: Renewable Energy Grid Integration
Editor: Mitchell B. Ferguson

ISBN: 978-1-60741-325-7
© 2011 Nova Science Publishers, Inc.

Chapter 2

DISTRIBUTED PHOTOVOLTAIC SYSTEMS DESIGN AND TECHNOLOGY REQUIREMENTS

Chuck Whitaker, Jeff Newmiller, Michael Ropp and Benn Norris

ABSTRACT

To facilitate more extensive adoption of renewable distributed electric generation, the U.S. Department of Energy launched the Renewable Systems Interconnection (RSI) study during the spring of 2007. The study addressed the technical and analytical challenges that must be addressed to enable high penetration levels of distributed renewable energy technologies. Interest in PV systems is increasing and the installation of large PV systems or large groups of PV systems that are interactive with the utility grid is accelerating, so the compatibility of higher levels of distributed generation needs to be ensured and the grid infrastructure protected. The variability and nondispatchability of today's PV systems affect the stability of the utility grid and the economics of the PV and energy distribution systems. Integration issues need to be addressed from the distributed PV system side and from the utility side. Advanced inverter, controller, and interconnection technology development must produce hardware that allows PV to operate safely with the utility and act as a grid resource that provides benefits to both the grid and the owner.

ACKNOWLEDGMENTS

The authors wish to acknowledge the extensive contributions of the following people to this report:

- Jovan Bebic, General Electric Global Research Division
- Mike Behnke, BEW Engineering
- Ward Bower, Sandia National Laboratories
- John Bzura, National Grid
- Tom Key, Electric Power Research Institute.

ACRONYMS

AC	alternating current
ADSL	asymmetric digital subscriber line
BPL	broadband over power line
DG	distributed generation, distributed generator
EMS	energy management system
GE	General Electric
IEC	International Electro-technical Committee
IEEE	Institute of Electrical and Electronics Engineers
LAN	local area network
LTC	load tap changing
LV	low voltage
MPP	maximum power point
MTBF	mean time before failure
MV	medium voltage
NDZ	nondetection zone
NREL	National Renewable Energy Laboratory
OF	over frequency
OV	over voltage
PLCC	power line carrier communications
PV	photovoltaic
RSI	Renewable Systems Integration
SEGIS	solar energy grid integration system
SFS	Sandia Frequency Shift
SVC	static VAr compensator
SVR	step voltage regulator
SVS	Sandia Voltage Shift
UF	under frequency
UPS	uninterruptible power supply
UV	under voltage
VAr	volt-ampere reactive
VPCC	point of common coupling voltage
WECC	Western Electricity Coordinating Council

EXECUTIVE SUMMARY

Distributed photovoltaic (PV) systems currently make an insignificant contribution to the power balance on all but a few utility distribution systems. Interest in PV systems is increasing and the installation of large PV systems or large groups of PV systems that are interactive with the utility grid is accelerating, so the compatibility of higher levels of distributed generation needs to be ensured and the grid infrastructure protected. The variability and nondispatchability of today's PV systems affect the stability of the utility grid and the economics of the PV and energy distribution systems.

Integration issues need to be addressed from the distributed PV system side and from the utility side. Advanced inverter, controller, and interconnection technology development must produce hardware that allows PV to operate safely with the utility and act as a grid resource that provides benefits to both the grid and the owner. Advanced PV system technologies include inverters, controllers, related balance-of-system, and energy management hardware that are necessary to ensure safe and optimized integrations, beginning with today's unidirectional grid and progressing to the smart grid of the future.

Recommendations

- Develop solar energy grid integration systems (see Figure below) that incorporate advanced integrated inverter/controllers, storage, and energy management systems that can support communication protocols used by energy management and utility distribution level systems.
- Develop advanced integrated inverter/controller hardware that is more reliable with longer lifetimes, e.g., 15 years mean time before failure and a 50% cost reduction. The ultimate goal is to develop inverter hardware with lifetimes equivalent to PV modules.

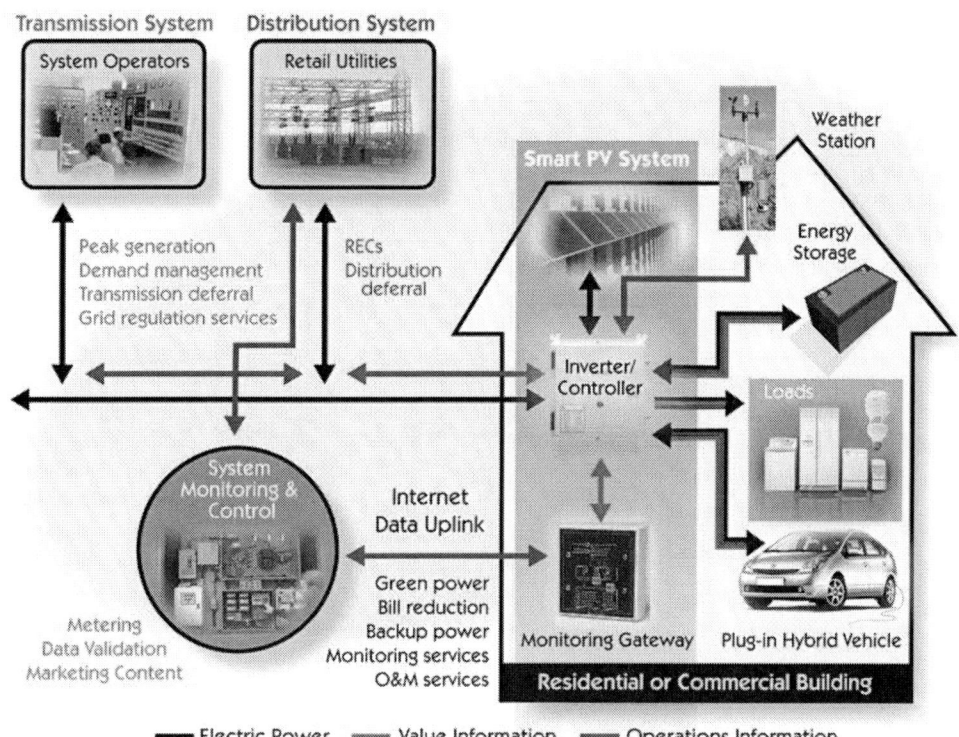

The solar energy grid integration system integrated with advanced distribution systems.

- Research and develop regulation concepts to be embedded in inverters, controllers, and dedicated voltage conditioner technologies that integrate with power system voltage regulation, providing fast voltage regulation to mitigate flicker and faster voltage fluctuations caused by local PV fluctuations.
- Investigate DC power distribution architectures as an into-the-future method to improve overall reliability (especially with microgrids), power quality, local system cost, and very high-penetration PV distributed generation.
- Develop advanced communications and control concepts that are integrated with solar energy grid integration systems. These are key to providing sophisticated microgrid operation that maximizes efficiency, power quality, and reliability.
- Identify inverter-tied storage systems that will integrate with distributed PV generation to allow intentional islanding (microgrids) and system optimization functions (ancillary services) to increase the economic competitiveness of distributed generation.

1. INTRODUCTION

The installed capacity of grid-connected photovoltaic (PV) power system installations has grown dramatically over the last five years (see Figure 1.1). The capacity is still less than 1% of the peak electricity load on the utility grid, but at this growth rate, a 5% or 10% level may be less than a decade away. Such penetration levels are significantly higher than the currently- assumed limits under which net energy metering is allowed, and reaching those levels is likely to require significant changes to current inverter technology and regulations to maintain reliable and economical grid operation.

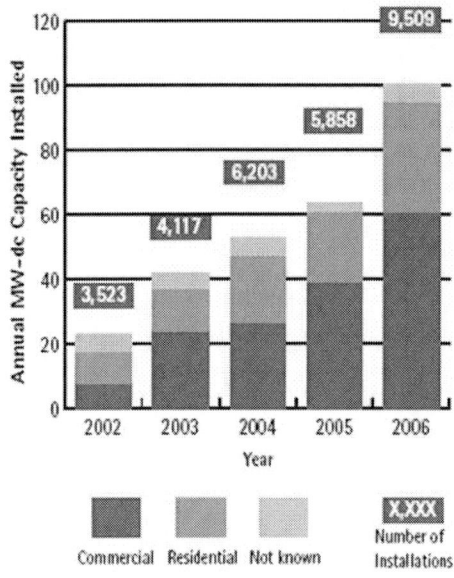

Figure 1.1. U.S. grid-tied PV installations.

At the scale of the entire interconnected electric power grid, generated electric power must be consumed within milliseconds of being generated. Excess power can be accumulated with energy storage systems such as pumped hydro, but conventional energy storage systems respond much more slowly than the load changes, so peaking generation is throttled back to stabilize the power flow into and out of the grid. In addition, when the load on the utility grid reaches new peak levels, the system operators must start activating every available generating source, and even minor throttling back of generation may cause the grid voltage to collapse.

Grid-connected PV power system designs focus on converting as much irradiant power as possible into real power (current flowing into the grid in phase with the utility-defined voltage). This design goal is appropriate for a technology that has insufficient installed capacity to approach the typical loads supplied by the electric power utility infrastructure. However, as the installed capacity of this technology grows, this assumption will at some point no longer hold true; in some small areas of the electric power distribution system (for example, some rural feeders), solar electric power generation has already approached or exceeded the local daytime load and the electric utilities have begun to modify their physical infrastructure (e.g., wire size and voltage control settings) to adapt to this new power flow pattern. If this trend continues, PV power systems will be required to provide increasing levels of grid support services and to participate to a greater extent in utility dispatch and operations processes.

Stand-alone PV power systems must already deal with issues of this type, albeit on a much smaller scale. Inverters in stand-alone systems must regulate their output alternating current (AC) bus voltages by supplying current as needed to maintain voltage. Battery energy storage is usually included to address power demand surges, store generated power during low demand, and continue to supply power to the load during cloudy or nighttime conditions. The technology is available to incorporate similar features into grid-tied PV inverters, but doing so would drive up the cost of PV electric power compared to real-power-optimized grid- connected PV power systems.

The parallels are striking between the surge and demand variability characteristics of off-grid PV power systems and the peaking load and demand variation concerns faced by an electric power grid's system operator. However, there are some significant differences between these systems. One difference is that the off-grid PV power system has a relatively small number of loads, many of which are significant by comparison with the generating capacity, so variations in load tend to be relatively large and abrupt [1]. Conversely, the electric power grid has billions of loads that are tiny by comparison with the generating capacity, so variations tend to be smooth. The other difference is that the conventional electric power grid can store energy by reducing consumption of generating plant fuel. Solar power cannot be conserved this way for later use, so the off-grid PV power system usually includes an energy storage subsystem to keep some of that unused power for later low-light conditions. When the storage is full the PV power conversion is throttled back and available energy is discarded. Grid-connected PV power systems avoid the capital costs and roundtrip inefficiency of electric power storage in favor of dependence on conventional power sources as the backup power supply, because there are no incentives or regulations directing them to do otherwise.

If grid-connected PV power systems were negligibly cheap, the system operator would prefer to curtail power production (and waste available irradiant power) when demand drops. However, all generating plants have some capital cost and their owners would prefer to operate them at full capacity to maximize revenue per year (or month or day). The system operator must throttle back generation or increase the rate of storage (e.g., water pumping) when demand drops, or the frequency will climb too high. This condition is managed by the system operator, who uses a combination of carrot (pricing) and stick (regulation) actions. As these types of actions that curtail power production are applied to grid-connected PV power systems, designers will have to choose between discarding available power and adding storage.

Localized voltage regulation issues will be one of the first impacts on grid operation as penetration grows. Power flowing toward the substation can result in increased voltage levels. In particular, capacitor banks and voltage regulators that normally boost voltage slightly may now push voltages above standard voltage limits. The power level at which such effects become detrimental may vary greatly from one feeder to the next, depending on the size and location of capacitor banks and voltage regulators, as well as on the resistance and impedance of the distribution system wires (a long or undersized power line is more likely to be sensitive to power variations). Possible approaches to resolving this issue are to curtail real power generation during peak times (with diversion to storage or power dissipation, wasted or otherwise used) and to control reactive power (voltage regulation, which is currently prohibited for distributed generation [DG]).

Another issue that could become significant as penetration of PV power production increases is voltage flicker. This effect occurs when one generating source reactive power output increases or (more commonly) decreases faster than the remaining generators can compensate. Rapid changes in irradiance (up to 15% per second) as clouds pass over will lead to PV power transients that are expected to tax the ability of rotating machine generators to react and restore system voltage.

One architectural change that is anticipated in the grid of the future is microgrid operation, in which campus- to neighborhood-sized areas served by common distribution system equipment are designed to disconnect and run independently from the rest of the grid if a supply disruption occurs. Technical solutions to certain grid-connected problems such as voltage regulation and power throttling are also useful for operating a microgrid.

Finally, with so many additional functions allocated to the inverter, the inverter becomes ever more critical to the system function, and the reliability of current technology inverters becomes a significant issue. Combining discrete components into prepackaged integrated components, reducing operating temperatures of components, and increasing electrical rating margins are expected to be key steps for increasing the mean time before failure (MTBF) from fewer than 10 years to 15 years or more.

2. STATUS OF PHOTOVOLTAIC SYSTEM DESIGNS

Major categories of PV system designs include grid-connected without storage, grid-connected with storage, and off-grid.

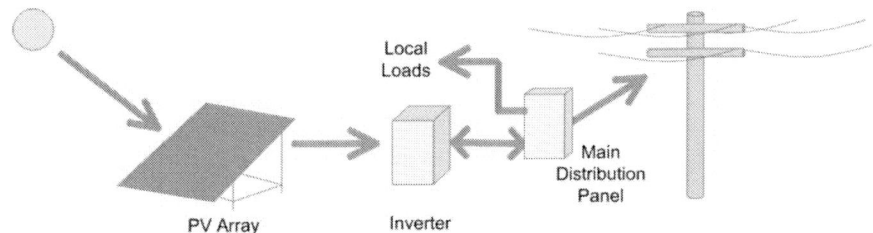

Figure 2.1. Grid-connected PV power system with no storage.

2.1. Grid-Connected with No Storage

The major elements of a grid-connected PV system that does not include storage are shown in Figure 2.1. The inverter may simply fix the voltage at which the array operates, or (more commonly) use a maximum power point (MPP) tracking function to identify the best operating voltage for the array. The inverter operates in phase with the grid (unity power factor), and generally delivers as much power as it can to the electric power grid given the sunlight and temperature. The inverter acts as a current source; it produces a sinusoidal output current but does not act to regulate its terminal voltage in any way.

The utility connection can be made by connecting to a circuit breaker on a distribution panel or by a service tap between the distribution panel and the utility meter. Either way, the PV generation reduces the power taken from the utility power grid, and may provide a net power flow into the utility power grid if the interconnection rules permit.

A simplified equivalent circuit of the same basic grid-connected system is shown in Figure 2.2. The PV system typically appears to the grid as a controlled current source, local loads may consist of resistive, inductive, and capacitive elements, and the utility source is represented by its Thevenin-equivalent model (voltage source Utility_V with series impedance Utility _Z). The local loads within a single residence rarely include much capacitance, but if a whole neighborhood is modeled at once, voltage support capacitors maintained by the utility may contribute significantly to the local load mix. This leads to conditions that could fool the inverter into running, even if the utility becomes disconnected (unintentional islanding). The utility source impedance models such things as the impedances of transformers and cables. The inverter handles all grid interface functions (synchronization, over/undervoltage [OV/UV] and over/underfrequency [OF/UF] disconnects, anti-islanding) and PV array control functions (MPP tracking).

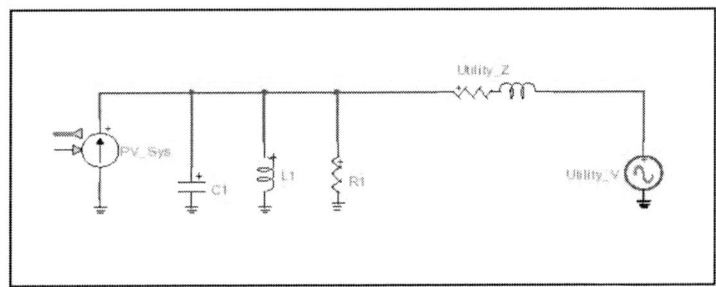

Figure 2.2. Schematic drawing of a modern grid-connected PV system with no storage.

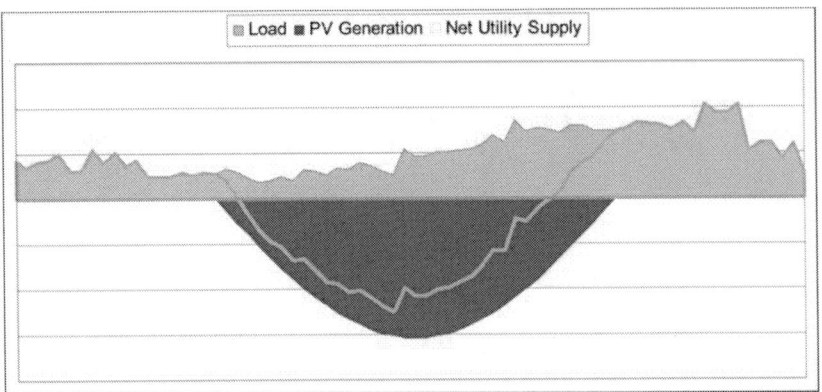

Figure 2.3. Power flows required to match PV energy generation with load energy consumption.

The ratio of PV system size to local load demand may be small enough that reverse power flow from the PV to the utility never occurs, but at high penetration the magnitude of the reverse power flow at midday is likely to exceed the magnitude of the nighttime load power. As shown in Figure 2.3, if we try to make the generation energy (area of red hump) equal to the load energy (blue area), the daytime power production (peak of red generation hump at solar noon) is likely to exceed the peak load power flow because most loads draw power all night when the PV system cannot supply power. For this residential load example, the peak load power flow is a double peak in late evening, which highlights the time misalignment that can occur between residential load and PV generation. Fortunately, commercial loads peak in the early afternoon, so the total PV generation in a utility system can reduce the peak system load, even though it may have no impact on the peak load at the residence where the PV is installed.

As part of this work, an extensive literature search was conducted to assess the current body of knowledge of expected problems associated with high penetration levels of grid-tied PV. The results of that literature survey are presented here.

Several studies have been conducted to examine the possible impacts of high levels of utility penetration of this type of PV system. One of the first issues studied was the impact on power system operation of PV system output fluctuations caused by cloud transients. A 1985 study [2] in Arizona examined cloud transient effects if the PV were deployed as a central-station plant and found that the maximum tolerable system-level penetration level of PV was approximately 5%. The limit was imposed by the transient following capabilities (ramp rates) of the conventional generators. Another paper published in that same year [3] about the operating experience of the Southern California Edison central station PV plant at Hesperia, California, reported no such problems, but suggests that this plant had a very stiff connection to the grid and represented a very low PV penetration level at its point of interconnection.

In 1988, another study dealt with voltage regulation issues with the Public Service Company of Oklahoma system when clouds passed over an area with high PV penetration levels, if the PV were distributed over a wide area (south Tulsa, Oklahoma) [4]. At penetration levels of 15%, cloud transients caused significant but solvable power swing issues at the system level, and thus 15% was deemed to be the maximum system-level penetration.

In 1989, a paper describing a study on harmonics at the Gardner, Massachusetts, PV project was released [5]. The 56 kilowatts (kW) of PV at Gardner represented a PV penetration level of 37%, and the inverters (APCC SunSines) were among the first generation of true sine wave pulse width modulation inverters. All the PV homes were placed on the end of a single phase of a 13.8 kV feeder. This was done intentionally:

> Selection of the houses comprising the Gardner Model PV Community was predicated on establishing a high saturation of inverters as may become typical on New England distribution feeders in the next century. [6]

The PV contribution to voltage distortion at Gardner was about 0.2%, which was far less than those made by many customer loads [5]. Thus, harmonics were not a problem as long as the PV inverters were well designed. This paper also mentions the potential value of PV systems being able to provide reactive power to keep the power factor of a feeder approximately constant.

A 1989 paper [7] indicates that the PV community was aware at that time of potential issues involving interactions between PV systems and automatic tap-changing transformers (load tap changing [LTC] transformers). This paper describes a computer model used to study the problem, and found that cloud-induced PV output fluctuations could cause excessive operation of LTCs, but no maximum penetration level was suggested. (See 8, pp. 2–3, for more information about this problem.)

Another cloud transient study was released in March 1990 [9]. This one used a utility in Kansas to quantify the impact of geographic distribution of PV on allowable PV penetration level at the system level. This utility was described as having only very small amounts of fast-ramping generation capacity; most of its generation was in the form of slow-responding coal-fired units. The authors concluded that under the conditions studied, the utility's load-following capability limited PV penetration to only 1.3% if the PV were in central-station mode; the limitation was caused by unscheduled tie-line flows that unacceptably harmed the utility's economics. However, the allowable penetration rose to 18% if the PV were spread over a 100-square-kilometer (km^2) area, and to 36% if the PV were scattered over a 1000-km^2 area, because of the smoothing effect of geographic diversity.

Also in March of 1990, an important Electric Power Research Institute report on the Gardner, Massachusetts, PV project was released [6]. This study looked at four areas:

- The effect on the system in steady state and during slow transients (including cloud transients)
- How the concentrated PV responded under fast transients, such as switching events, unintentional islanding, faults, and lightning surges
- How the concentrated PV affected harmonics on the system
- The overall performance of distribution systems, in which the total impact of high-penetration PV was evaluated.

This study reports a number of interesting findings:

- The authors measured the rate of sunlight change caused by cloud passages. They report measured values of 60 to 150 $W/m^2/s$.

- Spatially distributing PV systems significantly reduces the system impacts of slow transients caused by clouds, and at Gardner no unacceptable voltage regulation problems occurred as a result of cloud passages. However, the authors do note that unacceptable voltage excursions could be possible if more PV were added, and if "...the circuit is lightly loaded and var [sic] compensating capacitors are connected."
- The inverters used at Gardner used the slide-mode frequency shift (SMS) method of unintentional islanding prevention [10]. The authors of [6] report on a series of antiislanding tests that use five of the inverters at Gardner, including tests that use a 10- horsepower induction machine running in parallel with the five inverters. In none of these tests were the authors able to cause the inverters to run for more than one cycle. (These are believed to be all R-L load tests; no RLC load tests of the type required by IEEE-1547 and UL-1741 were reported.)
- The fault current provided by the inverters was limited; the maximum observed fault current was "...no more than 150% of rated converter current."

The final conclusion of this EPRI report is that the 37% penetration of PV at Gardner was achieved with no observable problems in any of the four areas studied.

The impact of high penetrations of PV on grid frequency regulation appeared in a 1996 paper from Japan [11]. This study used modeled PV systems that respond to synthetically generated short-term irradiance transients caused by clouds. The study looked at system frequency regulation and the break even cost, which accounts for fuel savings when PV is substituted for peaking or base load generation and PV cost. This paper reaches three interesting conclusions: (1) the break-even cost of PV is unacceptably high unless PV penetration reaches 10% or so; (2) the thermal generation capacity used for frequency control increases more rapidly than first thought; and (3) a 2.5% increase in frequency control capacity over the no-PV case is required when PV penetration reaches 10%. For PV penetration of 30%, the authors found that a 10% increase in frequency regulation capacity was required, and that the cost of doing this exceeds any benefit. Based on these two competing considerations, the authors conclude that the upper limit on PV penetration is 10%.

Between 1996 and 2002, a series of reports was produced by an International Energy Agency working group on Task V of the Photovoltaic Power Systems Implementing Agreement. Unintentional islanding, capacity value, certification requirements, and demonstration project results were all the subjects of reports, but the one that is of primary importance here dealt with voltage rise [12]. This report focused on three configurations of high-penetration PV in the low-voltage distribution network (all PV on one feeder, PV distributed among all feeders on a medium-voltage/low-voltage (MV/LV) transformer, and PV on all MV/LV transformers on an MV ring). This study concludes that the maximum PV penetration will be equal to whatever the minimum load is on that specific feeder. That minimum load was assumed to be 25% of the maximum load on the feeder in [13], and if the PV penetriion were 25% of the maximum load, only insignificant overvoltages occurred. Any higher PV penetration level increased the overvoltages at minimum loading conditions to an unacceptable level. This study assumed that the MV/LV transformers do not have automatic tap changers (they are assumed to have manually set taps).

In August 2003, two major studies dealing with this topic by General Electric (GE) (under contract from the National Renewable Energy Laboratory [NREL]) were released [13,

14]. The first concentrated on DGs interfaced to utilities through inverters; the second focused on larger scale system impacts and rotating DG, but still with several results on inverter-based DG. Both were simulation-based studies: the first used GE's Virtual Test Bed model and focused on a simulated distribution system; the second used positive sequence load flow and examined the entire Western Electricity Coordinating Council footprint. Key conclusions of the first study [13] include:

- For DG penetration levels of 40%, such that the system is heavily dependent on DGs to satisfy loads, voltage regulation can become a serious problem. The sudden loss of DGs, particularly as a result of false tripping during voltage or frequency events, can lead to unacceptably low voltages in parts of the system.
- The simulated distribution system was assumed to employ step voltage regulators (SVRs), which are essentially autotransformers with an automatically adjustable tap on the series winding [15]. During periods of low load but high generation and with certain distribution circuit configurations, the reverse power flow condition could cause the SVRs to malfunction. Again, voltage regulation becomes a problem.
- A voltage regulation function, implemented through reactive power control, would significantly increase the benefits of inverter-based DGs to the grid. Unfortunately, this function would interfere with most anti-islanding schemes as they are presently implemented.
- Inverter-based DGs do not contribute significantly to fault currents, and thus did not adversely affect coordination strategies for fuses and circuit breakers. The study notes that the short-duration fault current contribution of small distributed inverter-based DGs is smaller than that of distributed induction machines. However, it also points out that this might not always be true if the DG is connected at a point where the utility series impedance is unusually high. These conclusions may not remain valid if the voltage regulation controls suggested earlier are implemented.
- The inverter-based DGs did not respond adversely to high-speed transients such as those caused by capacitor switching, and thus did not degrade the system's response in such cases.
- For widely dispersed DGs, modern positive feedback-based anti-islanding appears to eliminate unintentional islands without serious impacts on system transient performance, but the complexity of the subject indicates that more study is needed.

In the second study [14], significant impacts were observed when DG penetration levels were 10% to 20%. Although this study concentrates on large DGs, which would probably not include much PV, two of the study's conclusions are relevant here.

- It echoes the sentiment that aggressive voltage and frequency trip set points will become a problem at high DG penetration levels. The study documents one case (see page 29) in which a significant transient event becomes a full-blown cascade failure because of underfrequency tripping of DGs after a major mainline generator is lost. (A similar result has been observed in the field; an official investigation committee concluded that the Italian blackout of 2003 was made significantly worse by underfrequency tripping of DGs [16].)

- It suggests that active anti-islanding, particularly involving positive feedback on frequency, has a negative but minor impact on system dynamic behavior.

Neither GE study indicates a maximum allowable DG penetration level, but the first study suggests that on the system simulated there, the maximum level is less than 40% because significant problems appeared at the 40% level.

A 2006 study [17] examined the impact of DGs on distribution system losses, as a function of penetration level and DG technology. It concluded that distribution system losses reach a minimum value at DG penetration levels of approximately 5%, but as penetration increases above that level, distribution system losses begin to increase. The reasons for this are not entirely clear, but the general result (there was a penetration level at which distribution losses were minimized) was consistent across all DG technologies. The penetration level at which minimum losses occurred was nearly doubled if voltage regulating, variable power factor inverters were used.

Yet another 2006 study focused on high penetrations of distributed generators in distribution systems [18]. This report was produced by a European consortium called Distributed Generation with High Penetration of Renewable Energy Sources (DISPOWER) that includes universities, research institutes, manufacturers, and representatives of several segments of the utility community. This report examined many types of DG in many configurations. Items in the DISPOWER report that are of specific interest here include:

- The report describes a Power Quality Management System, which is a centralized control scheme for distributed generators that has been in field tests since 2005. This system uses transport control protocol/Internet protocol and Ethernet cables as the physical communications channel. Initial field tests appear to be promising.
- One section of the report deals specifically with problems expected as DGs approach high penetration levels. The authors studied both radial and mesh/loop distribution system configurations and conclude that the mesh/loop configuration has significant advantages for mitigating the problems associated with high DG penetration. They also noted that harmonics increased slightly when the DGs were present, but never did they reach a problematic level. This study does not suggest a maximum penetration level.
- One section of the report discusses safety and protection. It notes that the practice most commonly adopted by DGs today is to disconnect at the first sign of trouble. This report, like the earlier GE reports, suggests that this approach will no longer be acceptable when penetration levels become significant, although no specific penetration level is given. Instead, ride-through must be implemented, without creating problems with unintentional islanding.

A recent study [19] reached some striking conclusions. This study examined the impact of PV penetration in the United Kingdom, where utility source series impedances are typically higher than in the United States. It examined the probability distributions of voltages in a simulated 11-kilovolt (kV) distribution system with varying levels of PV penetration, using an unbalanced load flow model. PV output was simulated using measured data with one-minute resolution. As expected, the probability density functions shown indicate that PV causes the distribution to shift toward higher voltages, but only by a small amount. The mean point of common coupling voltages increased by less than 2 V (on a 230-V nominal base). The study's conclusions include:

- If one employs a very strict reading of the applicable standard in the United Kingdom (BS EN 50160), PV penetration is limited to approximately 33% by voltage rise issues. However, at 50% penetration, the voltage rise above the allowed limits was small, and so the authors suggest that the 33% limit is somewhat arbitrary.
- Reverse power flows at the subtransmission-to-distribution substation did not occur, even at 50% PV penetration.
- Contrary to the results in [17], the authors of [19] found that at 50% penetration distribution system losses were reduced below the base-case values, largely because of reductions in transformer loading.
- Voltage dips caused by cloud transients might be an issue at 50% penetration, and the authors suggest further study of this issue.

Table 2.1 summarizes the PV penetration limits found in the literature.

Table 2.1. Summary of Maximum PV Penetration Levels Suggested in the Literature

Reference Number	Maximum PV Penetration Level	Cause of the Upper Limit
2	5%	Ramp rates of mainline generators. PV in central-station mode.
4	15%	Reverse power swings during cloud transients. PV in distributed mode.
5	No limit found	Harmonics.
6	> 37%	No problems caused by clouds, harmonics, or unacceptable responses to fast transients were found at 37% penetration. Experimental + theoretical study.
8	Varied from 1.3% to 36%	Unacceptable unscheduled tie-line flows. The variation is caused by the geographical extent of the PV (1.3% for central-station PV). Results particular to the studied utility because of the specific mix of thermal generation technologies in use.
10	10%	Frequency control versus break-even costs.
11	Equal to minimum load on feeder	Voltage rise. Assumes no LTCs in the MV/LV transformer banks.
12,13	< 40%	Primarily voltage regulation, especially unacceptably low voltages during false trips, and malfunctions of SVRs.
16	5%	This is the level at which minimum distribution system losses occurred. This level could be nearly doubled if inverters were equipped with voltage regulation
18	33% or ≥50%	Voltage rise. The lower penetration limit of 33% is imposed by a very strict reading of the voltage limits in the applicable standard, but the excursion beyond that voltage limit at 50% penetration was extremely small.

2.2. Grid-Connected with Storage

Figure 2.4 shows two basic storage architectures commonly found with grid-connected PV systems. (a) shows an architecture that many older systems have used, where a separate battery charge control device controls power collected from the PV array. This arrangement leaves the inverter to provide backup battery charge control from the utility power grid when insufficient PV power is available, but does not allow efficient extraction of excess PV power for supply to the grid when the batteries are fully charged. Figure 2.4(b) shows an architecture that is more common in modern grid-connected PV power systems that allows the PV array power to be directed optimally by the inverter to batteries or the utility power grid as appropriate.

In both cases, storage provides the opportunity to supply power to critical loads during a utility outage. This feature is not available without storage.

As with the grid-connected only configuration described previously, PV generation reduces the power taken from the utility power grid, and may in fact provide a net flow of power into the utility power grid if the interconnection rules permit. Storage has been traditionally deployed for the critical load benefit of the utility customer in the United States, but the Ota City High Penetration PV project [20] deployed local storage as an alternate destination for energy collected during low load periods to prevent voltage rise from reverse power flow in the distribution system.

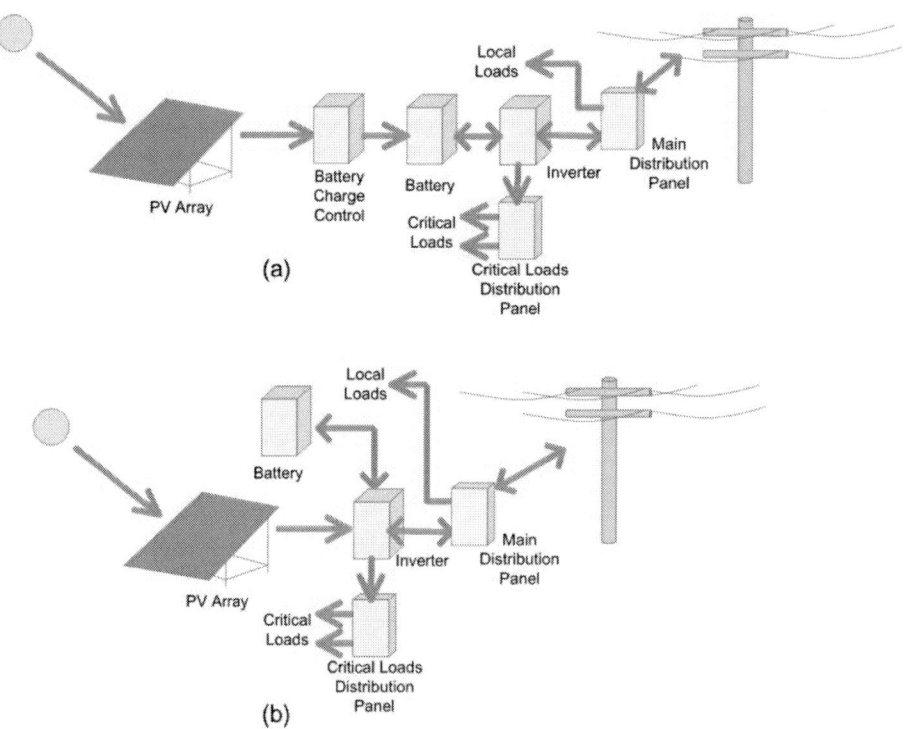

Figure 2.4. Grid-connected PV systems with storage using (a) separate PV charge control and inverter charge control, and (b) integrated charge control.

2.3. Off-Grid with Storage

Off-grid PV systems may include electricity or other storage (such as water in tanks), and other generation sources to form a hybrid system. Figure 2.5 shows the major components of an off-grid PV system with electricity storage, no additional generators, and AC loads. In a system of this type, correctly sizing the energy storage capacity is a critical factor in ensuring a low loss-of-load probability [21].

In this system configuration, the inverter acts as a voltage source, which is in contrast to the grid-tied system. The stand-alone inverter determines the voltage wave shape, amplitude, and frequency. To maintain the voltage, the inverter must supply current surges, such as those demanded by motors upon startup, and whatever reactive power is demanded by the loads.

Many stand-alone PV systems include engine-generator sets. In most cases, the generators are thought of as backup generators that are operated only during periods of low sunlight or excessive load that deplete the energy storage to some minimum allowed state of charge. The inverter senses a low battery voltage condition and then starts the generator. The generator usually produces 60-hertz (Hz) AC power directly, and thus when it starts, it powers the loads directly (the power to the loads does not pass through the inverter). The inverter operates as a rectifier and battery charger, drawing generator power to recharge the batteries. The system continues in this mode until the batteries are recharged. The generator is then stopped, and the inverter resumes regulation of the AC bus voltage, drawing power from the PV and batteries.

3. PROJECT APPROACH

As part of the work done under this task, in addition to the aforementioned literature review, a limited survey of utility engineers and a limited amount of computer modeling were carried out. Those results are presented below.

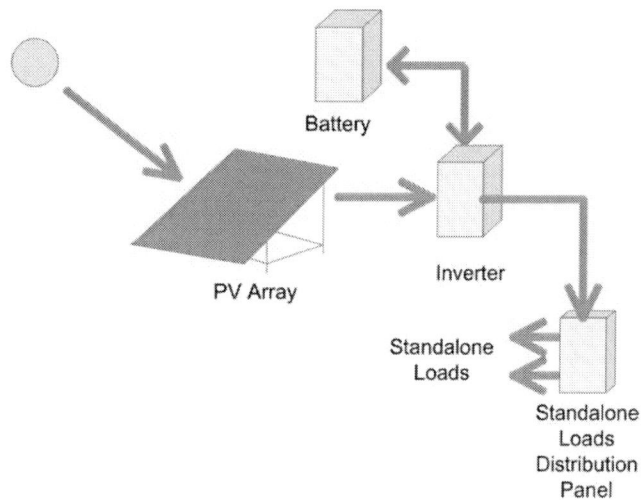

Figure 2.5. Off-grid PV system with storage.

3.1. Survey of Utility Engineers

A survey of utility engineers was conducted as a part of the RSI work. The survey developed for this work is attached as Appendix A. Survey responses were solicited from engineers at nine utilities. Eight engineers representing seven of the utilities responded, and their responses are summarized below. The utilities who submitted responses were:

- Salt River Project (Arizona)
- National Grid (Massachusetts, including the Gardner Project)
- Public Service Company of New Mexico
- Tucson Electric Power Company
- Southern California Edison
- Sacramento Municipal Utility District
- San Diego Gas and Electric.

PV penetration levels reported by the respondents varied widely. Some utilities have experienced PV penetration lower than 3% on any feeder to date, but there are high-penetration examples. For example, the Gardner, Massachusetts, project included PV at a 37% penetration level in distributed mode, and the 4.6-megawatt (MW) central-station PV plant near Springerville, Arizona, represents almost 58% penetration on its feeder. The highest system-level PV penetration reported (total PV as a fraction of total system peak demand) was about 0.2%. Three of the respondents indicated that they could not accurately state their PV penetration levels because they have only begun the process of mapping very small customer-sited PV to specific feeders. However, all three reported that they are in the process of performing such mapping.

As might be expected from the low PV penetration levels, most respondents reported having not yet seen any adverse effects from high PV penetration. Notably, the respondent from National Grid, who has in-depth knowledge of the Gardner, Massachusetts, project, reported no problems. Two problems were mentioned by other respondents as having been observed in the field: (1) voltage fluctuations during cloud passages over very large central station PV plants, caused by the slow following characteristics of the utility's thermal generation; and (2) voltage fluctuations caused by mass tripping of PV, which resulted from system events involving momentary voltage or frequency dips; some of the dips were initiated by events thousands of miles away.

Respondents were also invited to share their concerns about the possible future effects of high levels of PV penetration. Table 3.1 summarizes their concerns, and the number of respondents expressing each one.

Two respondents reported a high degree of confidence in the protection afforded them by standards like UL-1741 and IEEE-1547. These utilities expect to rely on UL-1741 and IEEE-1547 compliance to ensure that high penetrations of PV will not cause severe problems. This comment illustrates the importance of continuing the evolutionary process these standards are undergoing, to allow them to adequately deal with future PV system topologies and maintain their success with legacy systems. One respondent also mentioned that inverter testing programs, such as the one conducted by Distributed Utility Associates in California, are helpful to utilities.

Table 3.1 Summary of Utility Engineers' Concerns about Potential Future Problems Associated with High Penetrations of PV*

Potential Problem	Number of Mentions
Excessive cumulative harmonic distortion	3
False trips	3
Need for PV inverters to incorporate voltage regulation	2
Potential for unintentional islanding	2
Need to reduce impacts of cloud transients	1
Need for improved modeling tools to facilitate distribution system planning/analysis with PV	1
Need for utility control of PV inverters	1

*The number of mentions is the number of respondents (of a possible 8) mentioning each problem.

Another respondent mentioned that his utility has policies that he interprets as limiting the total allowable DG penetration on a feeder to 30%.

One respondent stated: "Appropriate planning methodologies are needed for emergencies when the PV will be disconnected from the distribution system." This could be interpreted as an anti-islanding concern, but contextual clues from elsewhere in this respondent's answers suggested that standard operating protocol required a visible lockable disconnect to isolate the PV from the utility during service. Thus, this response is not included in Table 3.1 as being related to anti-islanding.

3.2. Model Results

The modeling performed specifically for this work was limited because of the available time. One author has developed a detailed system-level model of a grid-tied PV system, and extensively experimentally verified the model with assistance from the Distributed Energy Test Laboratory at Sandia National Laboratories. This model runs in the MATLAB Simulink environment and is designed to examine issues related to islanding prevention, operation of MPP trackers, and system-level impacts resulting from tripping (or failure to trip) grid-tied PV. The harmonics produced by PV inverters are not modeled.

This model was used to study the simulated distribution system shown in Figure 3.1, which models one phase of a 13.2 $kV_{line-line,RMS}$ distribution system.

Five PV systems with their local loads are shown, but any number of PV + load blocks can be simulated. A power factor correction capacitor was included as shown and was used to compensate the reactive power demands of the distribution transformers and lines. Typical parameters for distribution system components were taken from [15] and [22]. Loads were modeled as parallel RLC circuits with a real power consumption of 5 kW (11.52 Ω at 240 V_{rms}), a Q factor of 0.5, and a resonant frequency of 60 Hz. All the PV systems and loads were single-phase. This model was used to examine the effectiveness of the Sandia Frequency Shift (SFS) and Sandia Voltage Shift (SVS) methods of islanding prevention in the multi-inverter case. First, for a single PV/load block, the generation-to-load ratio was adjusted to 1 (real power balanced) by adjusting the PV system output, and the reactive powers were balanced using the power factor correction capacitor. The run-on time (time between the tripping of the utility breaker at the left in Figure 3.1 and the deactivation of the PV system) was recorded. Then, PV/load blocks were added one at a time with interconnecting impedances as shown in Figure 3.1, and the real and reactive powers were rebalanced. This process was repeated for a utility with a low series impedance (0.5 Ω, purely resistive) and a high-impedance case (5 + j37.7 Ω). Figure 3.2 shows the results for SFS, and Figure 3.3 shows the results for SVS.

The simulation results suggest that neither SFS nor SVS loses effectiveness in the multi-inverter case; in fact, they improve slightly as inverters are added. At least for these two methods, loss of anti-islanding effectiveness at high penetration with multiple inverters is apparently not a concern. The apparent increase in run-on times from five to six inverters in Figure 3.3 in the low line impedance case is an artifact caused by numerical chattering [23].

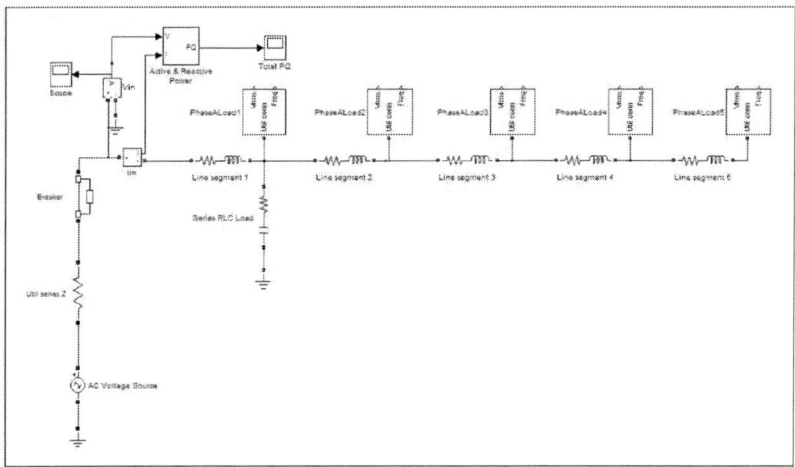

Figure 3.1. Example simulated distribution system.

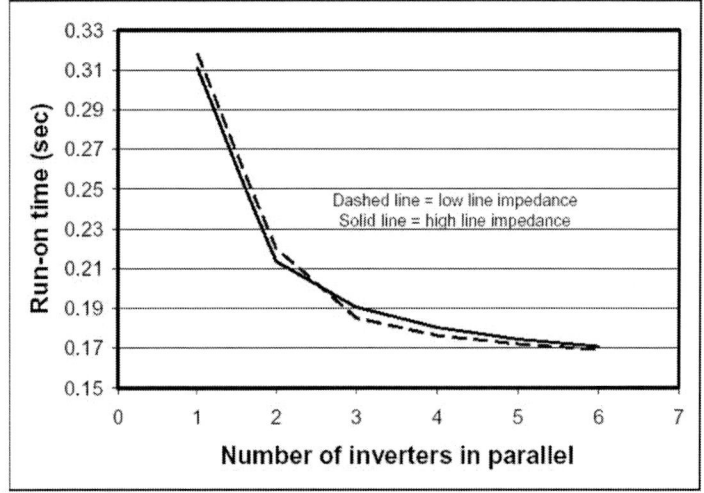

Figure 3.2. Results of simulations to test the effectiveness of the SFS active anti islanding method in the multi-inverter case.

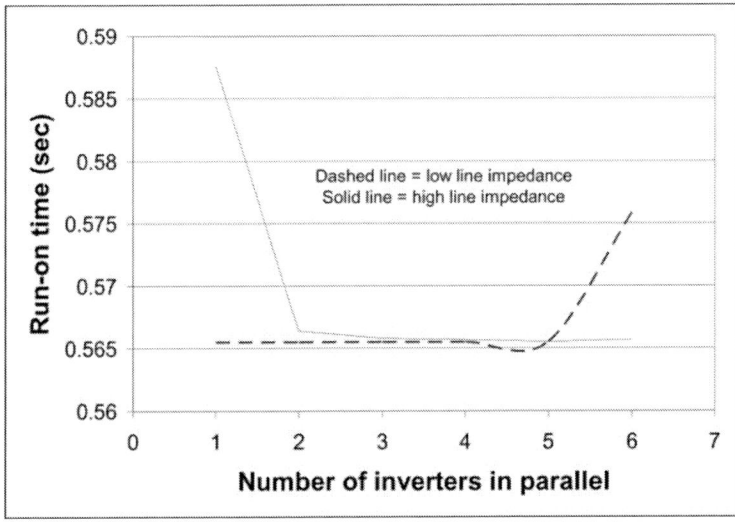

Figure 3.3. Results of simulations to test the effectiveness of the SVS active anti-islanding method in the multi-inverter case.

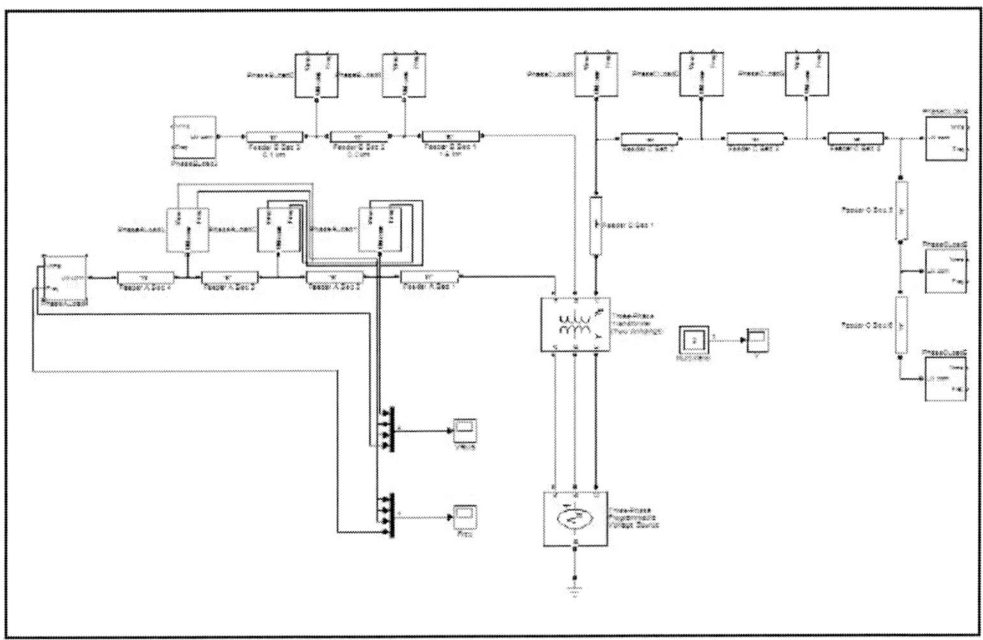

Figure 3.4. Simulated distribution system used in the modeling reported here.

The model was also used to observe the effect of a loss of PV generation during a low-voltage event on the utility. A simulated three-phase distribution system is shown in Figure 3.4. Again, "typical" distribution parameters were taken from [15] and [20], along with suggestions from experienced colleagues. The generation to load ratio in this specific case was 0.83 3, and the power factor of the aggregate load as seen from the three-phase utility source was 0.95 lagging. A simulation was performed in which the utility source dropped to 92% of nominal voltage, followed by a sag to 80% of nominal voltage (see Figure 3.5). During the first sag, the PV systems do not trip because the voltage does not fall below the UVR trip threshold, and when the sag ends the voltage returns to its pre-sag level. During the second sag, all of the PV trips almost immediately. The post-sag voltage is 2 V lower than the pre-sag voltage because of the loss of generation, and the frequency did not deviate by more than 0.05 Hz. The effect of the mass trip of the PV on the system is almost negligible because the utility source impedance is low and because the source is considered to have infinite inertia.

3.3. Description of Issues

3.3.1. Voltage Excursions

Voltage rise refers to the increase in voltage at the DG (PV) end caused by the PV system sending current back through the power system impedance. Power systems can be modeled as shown in Figure 3.6; the utility is represented by its Thevenin equivalent (voltage source and series impedance). Usually, the series impedance of the utility ("Utility_Z" in Figure 3.6) is quite small, and the voltage drop across it is not significant. However, in some areas the load current can become high enough, relative to this impedance, that the resulting voltage drop could cause the load node voltage (what utilities refer to as service voltage) to become unacceptably low. The usual solution to this problem is to increase the sending-end voltage (Utility_V in Figure 3.6) so the voltage at the load node remains within an acceptable range. Of course, this means that under light loading conditions the load node voltage can be fairly high.

If a PV system produces more power than the local loads require, the resulting reverse power will flow through the series impedance. The voltage drop across the series impedance will now be negative because of the reversal in the direction of power flow, so the voltage at the PV end becomes the utility voltage plus the voltage across the series impedance. If the utility voltage were already set fairly high, it is easy to envision a situation in which the PV array can push the load node voltage over the utility regulation limits defined in American National Standards Institute C84.1 or even the allowed overvoltage threshold [24].

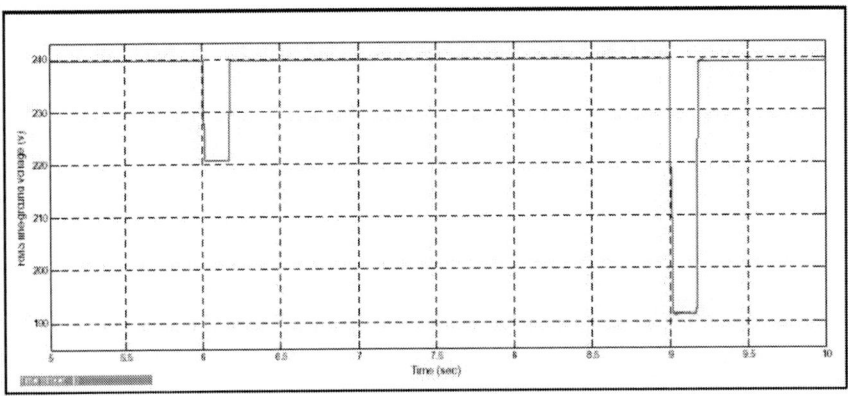

Figure 3.5. RMS voltages on Phase A during two utility voltage sags.

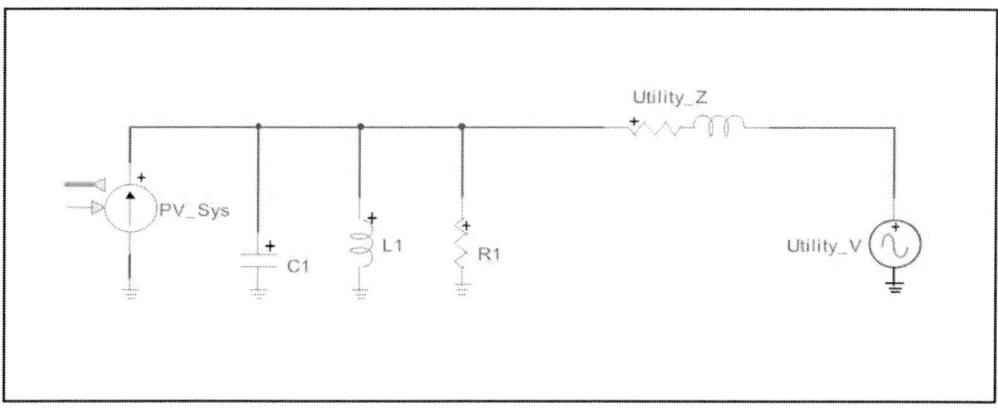

Figure 3.6. Schematic of a utility source (right) with its series impedance, feeding a node with an RLC load and a PV system.

The literature search suggests that voltage rise is a significant issue because it sets some of the lowest limits on allowable PV penetration, but there is not universal agreement on what those limits are.

Voltage increase can be lessened, and the upper limit it imposes on PV penetration eased, in the following ways:

- Decrease the utility's series impedance. Probably the most logical solution to the voltage rise problem is to design service drops and distribution systems to have very low impedances (low voltage drops), which suggests larger or multiple conductors, and larger derating factors on transformers or the use of more transformers. One obvious impact of such a redesign is increased capital cost, but another less obvious effect would be an increase in the system short-circuit current strength at the point of common coupling, and overcurrent protection would have to be modified accordingly. This approach reduces voltage drop problems in both directions and reduces losses. Theoretically, reconfiguring conductors to reduce parasitic capacitances and inductances might also help in some cases, but this is likely to be of little importance in distribution systems.
- Use energy storage. PV energy could be diverted from the utility line to a storage medium for later use when voltages are too high. The many benefits of energy storage are described elsewhere in this report.
- Use nonunity power factor operation to give PV inverters voltage control capability. Inverters can phase-shift their output to supply volt-amperes reactive (VArs) to (or draw them from) the utility. However, to significantly reduce voltage increase, the PV inverter would need to be designed for additional apparent power capability (e.g., a 5- kilowatt inverter might need a 6- to 7-kilovolt-ampere rating). For additional details and analysis, see the RSI Report titled, *Distribution System Voltage Performance Analysis for High Penetration PV*, Section 4.3. Some flexibility is allowed for utilities to make special arrangements for nonunity power factor operation, but changes to IEEE 1547 that specifically address this strategy would eventually be needed for standardization and universal adoption.

- Require customer loads to improve their power factors. Again, this would allow the utility to reduce its sending-end voltage, leaving more headroom for the PV.
- Program PV inverters to fold back power production under high voltage. This approach has been investigated in Japan, and though it can reduce voltage rise, it is undesirable because it requires the PV array to be operated off its MPP, thus decreasing PV system efficiency and energy production. It also interferes with today's positive feedback-based anti-islanding, because folding back has a voltage regulation effect and could reduce an overvoltage that might otherwise indicate islanding or another abnormal condition.
- Customers downstream of the DG or on adjacent feeders originating from the same substation bus (in other words, those who cause the low voltage condition that requires increased substation voltage) could use an energy management system (EMS) that incorporates a load-shedding scheme. Noncritical loads could be equipped with load- shedding switches activated by either a low voltage threshold or a communications signal (power line carrier or otherwise). Then, the utility's voltage setting at the substation could be reduced, leaving more headroom at the PV end. This is likely to be a very cost-effective solution, but it requires the customer to put up with the occasional loss of load caused by low voltages.
- Use diversionary or dump loads at times of high PV power production and low load. This is essentially the dual of a load-shedding regimen under low-voltage conditions; one switches in extra loads at times of very high voltage. From the grid's perspective, it provides the same effect as the storage and fold-back methods above. Facility EMS controls could also be integrated with the PV system controls to provide a more robust solution that could operate discretionary loads as needed. For example, the system could automatically start and stop a washing machine and clothes dryer during peak PV generation/peak voltage conditions. However, in many grid-connected PV applications, suitable dump loads may not be easily identified.

Looking to the distribution system of the future, additional solutions to the voltage rise issue based on power electronics may be possible. For example, power electronic transformers in substations or along distribution feeders could regulate voltage, control fault current, and improve power quality [25]. To realize this, the cost and reliability of power electronic transformers must improve. Alternatively, distributed SVCs and SVRs, centrally controlled via a communications bus, could be used to regulate distribution system voltages [26]. Another solution to the voltage rise issue that has been suggested by Japanese researchers is to consider a new distribution system architecture based on loops or meshes, instead of radial feeders [27]. The loops or meshes would be interconnected by using power electronics similar to the power electronic transformers just mentioned that would precisely control the loop power flows. These power converters could also be centrally controlled as discussed in [23]. The added degree of control provided by the power electronics and the loop architecture could effectively eliminate voltage rise.

3.3.2. Peak Load Support

Utility infrastructure costs are driven largely by the need to serve loads during high demand. To help manage loads, utility rate structures typically include charges tied to the customer's peak monthly demand. Future tariffs may be based on real-time pricing – such as hourly – and these would also reflect the cost of service during periods of peak demand.

Customers can thus incur a significant cost related to on-peak loads. These loads must be managed to lower energy costs. EMS and storage could be employed to limit customer peak loads, benefiting both the customer and the utility.

3.3.3. Distribution Outages

Most utility outages occur at the distribution level. Outages may be momentary – a few cycles or seconds – or longer term. Advanced utility automation systems may help to prevent such outages by automatically isolating sections of the line and reconfiguring sources such that the outage is confined to a small number of customers. However, the capacity of alternate sources is often limited by voltage or ampacity constraints, limiting the viability of this solution.

If a DG or storage source could be included in the utility planning, the source could be used to support a temporary, independent island until the outage could be resolved.

3.3.4. Spinning Reserve

Historically, spinning reserve has been provided by large idle power plants, kept spinning for faster response to outages by other units. The capacity of such units is therefore untapped, and the capital and operating costs of these units must be allocated to the customer base through electricity rates.

This service could be provided (or supplemented) by DG sources. When aggregated, the effective reserve provided by DG could be comparable to the plants they would displace. This would free the capacity of the larger units for ongoing energy needs.

3.3.5. Frequency Regulation (and Area Regulation)

Though the VAr control method described above is clearly a voltage regulation scheme, in this context we refer to the need for the power system to respond to rapid changes in load and PV output, and can include amplitude and frequency regulation. Rapid variances of PV output, usually caused by cloud transients, interact with the ramp rates and response times of the generating plants and voltage regulation equipment that must follow changes in load.

The literature and the survey results suggest that voltage regulation is still a concern. For central-station PV plants, voltage regulation is a particularly great concern; the lowest values of allowable PV penetration level mentioned in the literature are all for central-station PV plants, and are in the low single digits.

Voltage regulation caused by cloud transients, especially over large central station PV plants, can be improved by:

- Using fast-acting energy storage that levels the PV output during cloud transients. If one assumes that a cloud transient that cause the PV system output to dip has a linear ramp-down period, a constant low-power period, and a linear ramp-up period, the cloud transient could be leveled by an energy storage system that can produce the power profile shown in Figure 3.7. Typically, during the constant portion of the cloud transient, the PV system's output is assumed to drop to about 20% of its value under clear sky conditions. Thus, the storage system must supply the other 80%.

Assuming that the irradiance ramps at a rate of 200 W/m^2/s [28], the amount of energy storage required to level a cloud transient is shown in Figure 3-8, as a function of PV system size and duration of cloud transient (time period t_2 in Figure 3.7). This is the storage amount required to eliminate the cloud transient. The PV ramp rate could be reduced with considerably less storage.

- Integrating high time resolution cloud transient forecasting into utility dispatch and control. This is similar to wind forecasting proposals for large wind farms. Satellite imagery could be coupled with knowledge of the PV plant locations to predict PV output dips caused by cloud transients. Using this information:
  - Utility voltage regulation means could be employed preemptively.
  - The PV plant could "soften" the transient by preemptively ramping down from MPP operation just before the cloud transient, and then resuming MPP operation slowly after the cloud passes. This leads to a loss of PV energy, but that loss could be minimized.

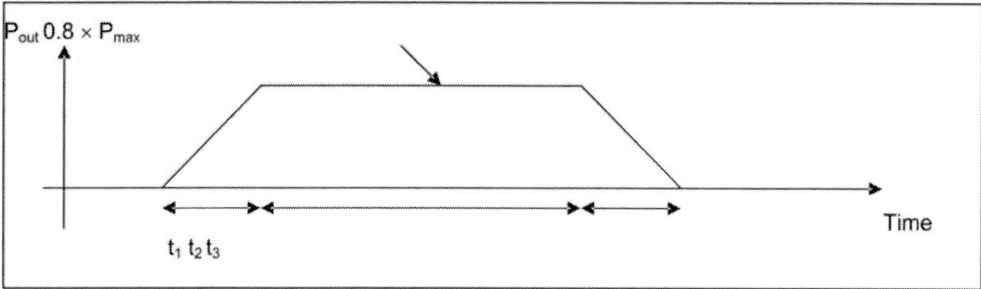

Figure 3.7. Power profile required of an energy storage unit to level a cloud transient in a PV system.

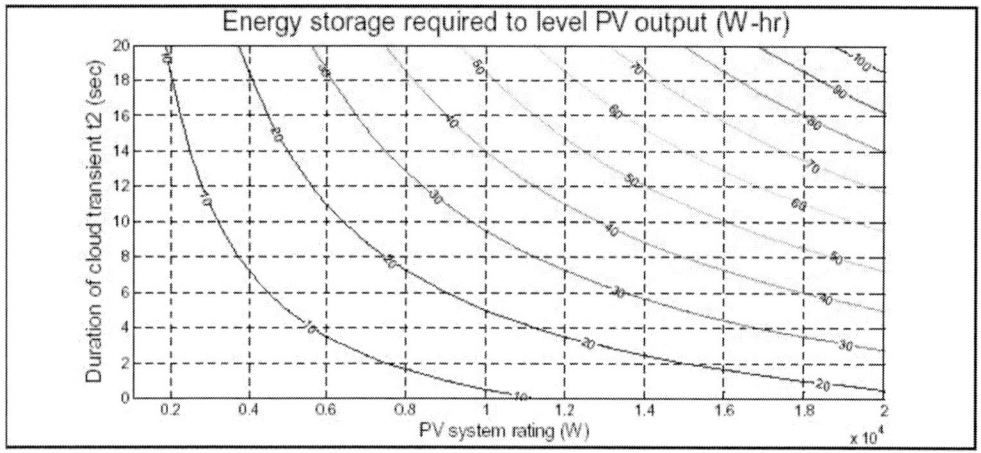

Figure 3.8. Energy storage required to provide the power profile in Figure 3.7, as a function of PV system rating and duration of the cloud transient.

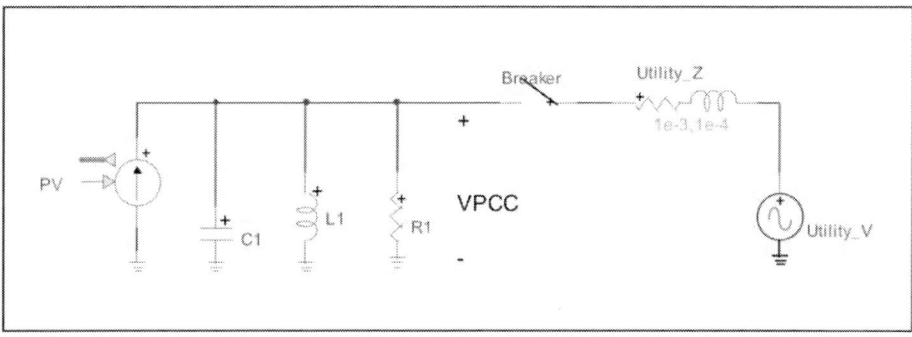

Figure 3.9. Simplified system configuration for understanding loss-of-mains detection.

3.3.6. Problems Related to Active Anti-Islanding Methods

According to today's standards, all customer-sited DG are required to incorporate a means to detect loss of mains, to ensure that inverters do not feed utility faults or open utility lines.

In general, two levels of loss of mains detection are employed in modern PV inverters. First is the traditional response to abnormal conditions affected by OV/UV and OF/UF trips. Consider the case shown in Figure 3.9, which is the same as Figure 3.6 except that a breaker was added (which could represent any current interrupting device). It is relatively straightforward to show [29, 30] that if the breaker opens at a time when the PV system's output power (real or reactive) and the RLC load's P and Q demand are not equal, there will be a detectable change in the amplitude or frequency of the point of common coupling voltage (VPCC, marked in Figure 3.9). Thus, if one sets the OV/UV and OF/UF operating windows to be very narrow, the OV/UV and OF/UF will provide effective loss-of-mains detection in most fault or open-line cases. Partly for this reason, IEEE-1547 specifies that DG should trip offline if the RMS voltage at the inverter's terminals is 10% above or 12% below the nominal value for more than two seconds (and faster at wider limits), or if the frequency is not between 59.3 Hz and 60.5 Hz.

With static inverter-based DG, there will remain a narrow range of RLC loads, called a nondetection zone (NDZ) [31], for which the OV/UV and OF/UF alone would fail to detect a loss of mains. Although the spontaneous occurrence of these tuned circuit loads and other conditions necessary for an unintentional island to form and remain stable are extremely improbable, inverter controls typically employ any of a number of active anti-islanding algorithms to further reduce the NDZ and have helped allay concerns of utility protection engineers and line workers.

The results of the literature search suggest three reasons why it is desirable to replace current active anti-islanding schemes with alternatives that facilitate the implementation of grid support functions in inverters.

- Allowing PV and other inverter-based DG to ride through voltage sags or frequency disturbances is highly desirable. This is not possible with the aggressive UV and UF tripping of PV used today. These aggressive low voltage or frequency trips can (and have been observed to) cause DGs to disconnect at a time when their continued operation would provide extremely high value to the host utility. Thus, using aggressive OV/UV and OF/UF settings to improve the detection of and response to line faults and loss of mains, has limited the ability of PV to be a "good citizen" on the grid [13]. Until a few years ago, similar trip settings were used with large scale wind farms. The advent of low-voltage ride through requirements [32, 33] signaled a change in utility perspective towards large wind, and many utilities with PV experience are suggesting that a similar change in perspective needs to follow for PV and other DG as they reach high system-level penetration.

- Most of the highly effective islanding prevention techniques used in commercially available PV inverters use some type of destabilizing positive feedback [34] to help ensure that either the amplitude or the frequency of VPCC goes beyond the OV/UV or OF/UF limits upon loss of mains. Although they prevent islanding, these types of controls require PV inverters to generally perform an anti-regulation function: they act in such a way as to attempt to make any excursion in voltage or frequency worse. The literature indicates that this type of control has a minor but negative impact on the grid.
- The literature and survey responses make clear that voltage regulation capability in PV inverters is desirable from a system-level perspective. This capability also conflicts with and reduces the effectiveness of certain anti-islanding functions.

There are other reasons why active anti-islanding as implemented today is undesirable at higher penetration levels. For example, certain active anti-islanding techniques can cause power quality problems at very high penetration levels, under certain conditions. Impedance detection can cause flicker and power system noise, if certain precautions are not taken [21, 35]. However, most real-world inverters do take those precautions. Also, almost all active anti-islanding methods require some distortion of the PV system's output current waveform [23], but that distortion can be minimized under normal operating conditions, and inverters are required to meet the harmonic limitations in [24] while their anti-islanding controls are active. The literature reports no field observations of either problem.

Today's active anti-islanding methods are also not suitable for use in microgrids. A microgrid is a collection of electrical sources and loads, along with their interconnections and associated equipment such as transformers, that can operate in parallel with the utility or in stand-alone mode as needed [36]. The defining characteristic of a microgrid is its capability to separate from its host utility and power its own loads. Some experts believe microgrids could become an important part of the utility system of the future [37, 38]. Microgrids would require a lossof-mains detection scheme that allows them to know when to switch from utility-parallel to islanded mode, but methods that rely on creating a voltage or frequency transient are not amenable to smooth transitions between these two modes. The method used in much of today's microgrid work requires the microgrid to always import power from the utility; the loss of that import power can then be used to detect the onset of islanding [39]. This restriction would prevent future microgrids from exchanging power with each other or exporting power to support the host system.

Finally, there is some concern that certain anti-islanding methods might lose effectiveness in the high-penetration case where there are large numbers of inverters. There is not yet a consensus on this issue [40].

Based on this discussion, the need for alternative loss-of-mains detection methods is clear. These alternative methods should not use destabilizing positive feedback, but rather facilitate the implementation of grid support functions, without losing islanding detection effectiveness for any combination of local loads, DGs, or system configurations. Potential solutions to this problem include:

- Use power line carrier communications (PLCC) [26, 27, 41, 42]. Any of several types of communication system could be used to replace active anti-islanding, but PLCC has a number of significant advantages for this application. If the PLCC signal meets certain criteria, such as having a continuous carrier, and if other well-known challenges to PLCC communications can be adequately solved, loss-of-mains, fault, and islanding detection could all be achieved with the PLCC signal as a continuity test of the line. If there is a fault, the PLCC signal will be lost at the PV system's end of the line. A test for the PLCC signal can then be used to detect islanding; its presence indicates that the utility is still there, and its absence indicates a condition that requires shutdown or separation from the utility. The inverter would thus "know" when it was islanding and could react appropriately, and active anti-islanding would be unnecessary. Voltage and frequency trip settings could be widened to better accommodate utility transients and provide better ride-through, or even adjusted dynamically depending on whether the inverter were in grid-tied or stand-alone mode. The PLCC receiver need not be in the inverter; the loss-of-mains detection function could be implemented at the point of common coupling, which would facilitate AC modules and microgrids.

  Almost no information content is required in the PLCC signal. It can still be used for other control functions without interfering with the loss-of-mains detection function. In addition to being continuous, the PLCC signal must be available at all endpoints, which means it must propagate well through distribution system impedances. This generally restricts the usable frequency range of the PLCC signal, and thus the available bandwidth. Subharmonic and low-frequency (< 1 kHz) systems have been successfully tested for this application, but PLCC transmitters in this frequency range tend to be expensive. Broadband over power line (BPL) might be useful in this application if issues related to propagation, generation of interference, and noise immunity can be addressed [43]. A number of commercial PLCC-based automatic meter reading systems operating in the 1-6 kHz range are also available, and some of these may be suitable. The main challenges for PLCC in this application are to develop a rugged, low-cost PLCC transmitter and identify or develop a low-cost (and preferably noninvasive) means of ensuring reliable signal availability at all endpoints.
- Integrate PV inverters into utility supervisory control and data acquisition systems or AMI systems. Inverters could be tied into utility communications systems, which would issue a warning to inverters in sections of the utility isolated from the mains. Any available channel, such as BPL, DSL, or coax, could be used. This would require that utility communications systems reach to all distribution-level endpoints, which is not presently the case. There may be other reasons to connect inverters to AMI systems, such as enabling PV systems to respond to real-time pricing signals. This would require the inverters to be connected to a high-bandwidth communications system that could, if properly configured, handle the anti-islanding function as well.

- Use other passive islanding detection techniques. One promising candidate for such a technique is harmonic signature detection [44], but at this time it has not been proven to be universally applicable in real-world power systems.

4. Project Results

The results imply that future generations of grid-tied PV inverters should incorporate a number of features, as described below. Incorporation of these features would move today's grid-tied PV system architecture toward the Solar Energy Grid Integration System (SEGIS) architecture shown in Figure 1.1.

4.1. Voltage Regulation

A PV inverter or the power conditioning systems of storage within a SEGIS could provide voltage regulation by sourcing or sinking reactive power. The literature search and utility engineer survey both indicated that this is a highly desirable feature for the SEGIS.

Implementing this feature would require modifications to the traditional PV inverter hardware design. For example, the required rating of the PV power electronics would have to be suitably oversized to support reactive needs and maintain full real power service. Also, the inverter's energy storage capacitors must be suitably sized so that excessive ripple does not reach the PV array during periods of high VAr production or absorption. The inverter's control software would also have to be suitably modified.

Technology drawn from stand-alone inverters and motor drives is sufficient for all of these requirements. However, adding this capability would increase inverter cost. The market mechanisms that would lead to acceptance of this additional cost are less clear. The problem of pricing ancillary services from DG, such as voltage regulation and VAr support, has not yet been fully solved.

Significantly enhanced communications capabilities in PV inverters must be a part of SEGIS development. These communications capabilities would allow inverters to receive and respond to market pricing signals sent from the utility, and to maintain proper coordination of their actions with those of other utility voltage regulation equipment. Communications for future PV inverters are discussed more fully below.

4.2. Backup Power (Islanding)

A utility that uses automated switching and sectionalizing could use a SEGIS to serve loads in a microgrid that operates in stand-alone (islanded) mode. In this context, the SEGIS would have to provide all the services normally provided by the utility, including load following and frequency control, and it would have to be able to resynchronize with the utility before reconnection.

SEGIS storage systems could be sized to cover momentary interruptions (one minute or shorter) or longer term, such as 15 minutes, depending on the customer's budget and required level of reliability. The system would be similar to a conventional uninterruptable power supply (UPS), except that it would be controlled by the utility and would serve multiple customers. To maximize its effectiveness, the utility could employ a parallel load management system to shed noncritical load in the island.

In general, PV improves the performance and feasibility of microgrids in two ways: (1) by reducing fuel use, thereby either extending the length of time that a microgrid can stand alone without fuel inputs from the outside, or reducing the amount of fuel that must be stored on- site; and (2) by reducing the emissions of the DG mix, which in many cases are restricted by law. Microgrids must generally be justified economically before they will be installed [45]. A full discussion of the economics of microgrids is beyond the scope of this paper. Two technical challenges associated with microgrids will be discussed here.

- Loss-of-mains detection must be provided at the point of common coupling between the microgrid and the host utility. The microgrid needs to be able to enter the standalone mode (sometimes called the intentional islanding mode) seamlessly upon loss of utility, and to reconnect automatically when the utility comes back online. The destabilizing active anti-islanding techniques used in inverters today are unsuitable for this purpose because they cannot readily be implemented at the point of common coupling, and they work by creating a voltage, phase, or frequency transient when the utility is lost. These transients work against a seamless transition between grid-parallel and stand-alone modes. Two suitable replacements, a passive method such as harmonic signature detection, or some form of communications, were identified earlier. PLCC is a preferred candidate because of the unique match between its properties and the needs of the application. Also, unintentional islanding within the microgrid would also have to be dealt with; that is, a SEGIS within a microgrid must be equipped to act appropriately if the section of the microgrid becomes isolated from the rest of the microgrid [35]. This problem could also be solved by communications.
- A SEGIS within a microgrid must be equipped with control software that enables it to operate in an environment where it must interact with other generators, possibly where no single generator has enough capacity to carry the entire load. The SEGIS and other generators must therefore work cooperatively to maintain voltage stability and power quality and meet cost goals.

Inverter control in microgrids is an active research topic today, and many questions remain unanswered [36, 46, 47]. Techniques for controlling DGs in a microgrid can be broadly grouped into centralized control and distributed control techniques, although in practice some combination of both is almost always used. Centralized control relies on high-speed communication channels between DGs and a central control computer, which may be an EMS. Distributed control schemes [48] such as agent-based controls do not have centralized control; instead, the DGs must work cooperatively to control the system, and EMS functions would be provided by this same cooperative action between sources, storage, and loads. One form of distributed control is local variable-based control, which uses only the information available in the DG's terminal voltage. Generally, either power versus frequency droop controls or active output impedance emulation are used to regulate voltage and share loads between the DGs, and the need for communications between the generators is minimized [35, 49, 50]. Other distributed control concepts go in the opposite direction, relying on high-bandwidth communications channels between the DGs, and between the DGs and other power system elements [51]. De Brabandere et al. propose a distributed control strategy that combines a low- bandwidth communication strategy based on identification of the ratio of resistance to reactance with a modified droop control applied to both frequency and voltage [52, 53].

Debate continues about whether distributed or centralized controls are preferred. In reality a combination of both will probably continue to be used. Distributed control based on local variables is robust in the face of grid disturbances, and is inherently plug-and-play in the sense that theoretically any combination of generators with the correct droop controls can be easily paralleled, and new generators can be added to the system at any point without changing the set points of the controls of the other generators. Local variable-based control is therefore ideally suited to the fast control functions of voltage regulation and maintenance of stability, and to enable generators to continue to function in the event of a failure of a communications channel. Communications-based techniques can much more easily implement financially motivated energy management functions, could eliminate the need for inverter-based active anti-islanding, and can coordinate all power system elements much more easily than local variable-based techniques. Communications standards like IEC-61 850 or LonTalk may eventually facilitate this process. These techniques are thus best suited to the slower-speed control functions of energy management and system coordination.

Additional research is needed in the area of SEGIS integration into microgrids. Agent-based controls have been successfully demonstrated in certain circumstances [54, 55, 56, 57, 58]. These early results demonstrate the great potential of agent-based control, but more broadly applicable solutions are still needed. Droop controls are being actively investigated by a number of research teams, but implementing a microgrid with droop controls seems to require a very large engineering effort; true plug-and-play functionality seems far off.

4.3. Spinning Reserve

Energy storage has been used for spinning reserve at the transmission level. Smaller DG units could likewise be used, with an aggregate capacity comparable to conventional thermal units used for this service. The storage would be sized with about 15 minutes of storage, depending on the ancillary market design. This is enough time to start up gas turbines and dispatch power.

4.4. Frequency Regulation (and Area Regulation)

Distributed storage could be used to regulate system frequency (or control area flow) by charging and discharging in response to signals sent by the system operator. Initial demonstrations using flywheel technology have been used for this purpose. Storage would have to be sized at about 15 minutes of full power to provide this.

4.5. Possible Directions for System Design Evolution

4.5.1. Communication of Price and Generation Control Signals

As described in the introduction, grid-connected PV power systems are given stringent requirements not to introduce negative impacts to the grid and are ignored for power load matching. According to work done at NREL, an economic incentive that seems to disable solar generation for limited periods appears at system penetration levels as low as 7%. [59] One possible way to adapt the control strategy for grid-connected PV system operation is to introduce communication with one or more central control or price sources. Since distribution system control communication is generally oriented toward system protection, and is usually implemented with dedicated copper circuits, conventional communication technology is not likely to scale up for application to distributed PV generation. The following sections describe the characteristics of communication systems in general, and of modern digital communication systems in particular. Its aim is to identify key features of the communications systems of the future for distributed PV.

4.5.1.1. Communication Systems

Communication systems are the means by which information is transferred between a sender and a receiver. The study of such systems is the subject of a broad academic discipline. A brief review of the capabilities of communication systems in general and the capabilities of some promising candidates for near-term implementation are presented here.

In general, communication systems all share features that will be discussed for each candidate system: latency, bandwidth, reliability, accuracy, distance limits, capital cost, and operating cost.

Latency refers to the delay between sending and receiving information. For example, with the use of satellite communications the speed of light traversing the distances involved can introduce noticeable delays during telephone conversations. Another example of latency can be found in e-mail, which may normally take only a few seconds, but if any of the e-mail relays are busy or disabled the delay to delivery may take minutes or days. Reaction times for some safety-related events are about 160 milliseconds, and real-time pricing signals may become stale after an hour.

Bandwidth refers to the rate at which data can be transferred. Returning to the satellite telephone example, a single satellite may be able to handle thousands of telephone calls simultaneously and may be upgraded to handle tens of thousands, but it cannot shorten the time-in-transit for any words spoken during any telephone call. In the area of grid-interactive DG, most data items currently considered as possible messages are very small (a few bytes each). A central management server (aggregator) will need extra data to keep each message uniquely identifiable (network overhead) and may communicate with tens to thousands of DG systems, so bandwidth at the server could be a bottleneck that limits the number of DG units that can be aggregated by one server.

Communication accuracy refers to how many messages are received in an altered form; measurement accuracy refers to how close the reported value is to the actual value. Most digital communication systems pad the data with enough information to identify unintentional alterations (errors) of a few bits in each message and depend on the sender to repeat the transmission if no acknowledgment is received. This involves time delays, which can add uncertainty to the overall time-in-transit.

Reliability indicates how frequently the communication channel will fail to transmit a message accurately. An error may be detected as such by comparison with the redundant data (checksum), so some errors that occur may be transformed into longer delays while the message is resent. A message with undetected errors is passed on as accurate, even though it is different than the one that was sent. The large bandwidths commonly available today mean that redundancy is added to the message; thus, the probability that incorrect data may be sent is reduced to vanishingly small values. However, this method of achieving reliability has a cost in increased effective latency as information is sent and resent multiple times.

Some communication methods have inherent distance limitations that are often linked with their available bandwidths. Commonly available twisted-pair 100-Mbps Ethernet has a defined distance limit of 100 m (328 ft) [60]; an asymmetric digital subscriber line (ADSL) can connect a telephone company central office to homes up to about 1.5 miles away. PLCCs also have distance limits that can vary by orders of magnitude, depending on the properties of the specific signal and power system.

Communication system capital costs will affect decisions regarding methods of communications employed. For example, dedicated copper communication conductors are commonly used for distributed utility protection systems. However, as more signals are required, the installation of large numbers of wires for communication becomes prohibitive. Wireless (usually digital radio signaling) communication options allow new wiring to be omitted entirely. Digital packet switched communication systems allow piggybacking of information over one new medium or even using extant infrastructure (for example, reprovisioning telephone service to include ADSL removes any need to install new conductors). Reuse of electrical power distribution wires for communication is appealing, but most implementations of this strategy are designed for in-home use, and there is a tradeoff between bandwidth and distance limitations.

Communication systems operating costs are primarily driven by power consumption and maintenance. A signal propagated through the variety of impedances found in utility power distribution systems will usually require more transmitting power than a dedicated closed medium such as ADSL or cable modem. One strategy for addressing power consumption is to use low-power-short-haul technologies (which typically have low bandwidth) to reach signal gateways that collect information together and retransmit all data on higher bandwidth dedicated communication media. Maintenance costs arise from communication media (conductor or insulation degradation by corrosion or mechanical means such as digging) and transmit-receive equipment.

4.5.1.2. Open Standards Institute Seven-Layer Model

To realize the benefits of communications between distributed resources, open system standards and definitions must be used. The Open Standards Institute has promoted a model of communication that separates the elements of communication systems into layers [61], where the upper layers are more conceptual and the lower layers are more physical. For example, a postal letter containing a birthday card conveys personal greetings at a very high level. The envelope provides information about the source and destination and is one level below the concept of birthday greeting. Further levels might be analogous to the bag that the envelope is placed in at the postal service, and another level could be analogous to the truck driven from one town to another that happens to carry the bag.

This analogy is useful because, just as trucks made by different manufacturers may be used to transport the birthday card with no difference in the delivery of the greeting (as long as the trucks meet their deadlines), the choice of how a particular hour-ahead price update or utility fault/disconnect now signal is delivered can involve several communication standards that apply to different aspects of the communication system. However, the latency requirements of some applications may preclude certain communication technologies. For example, satellite communications may bypass the need for dedicated communication wires, but it has too much delay to be useful for a transfer trip application.

4.5.1.3. Candidate Communication Solutions

Table 4.1 identifies basic characteristics of several communication options for communicating protective signals and price signals to the grid-connected PV power system.

Dedicated leased lines are a traditional technology for transmitting dry-contact signals for supervisory and protective functions to transmission-connected generators. These lines are often multiplexed with continuous carrier signals. However, installing leased lines into residence-sited PV systems would be prohibitive.

Ethernet wiring is commonly used for local area networks (LANs), which are becoming common for Internet sharing in residential applications. However, this technology must be connected to a wide-area-network technology such as ADSL or a cable television system, and reliability of these networks is maintained only at a convenience level, such that protective functions would be inappropriate for transmission this way.

Continuous-carrier power line communications is most often applied to automated meter reading systems. This has the advantage of being inherently coupled to the connection whose continuity is of concern for islanding. This makes using the presence of this carrier appropriate for broadcasting a transfer trip signal from the feeder circuit breaker. Unfortunately, this technology is relatively expensive and power hungry, so it has not gained momentum in the utility market. It also has a fairly low bandwidth, so it may not be appropriate for transmitting real-time pricing signal data.

BPL is a general description of several technologies that have been offered in competition with ADSL and cable TV Internet connectivity options. Unfortunately the BPL technologies tend to be sensitive interference by loads (reduced reliability) and broadcast signals that interfere with other radio spectrum users (particularly amateur radio). These concerns have so far prevented BPL from becoming widely available.

Spread-spectrum wireless radios cover various frequency bands, but to avoid interfering with other signals they usually do not provide very wide signal bandwidth. Unfortunately, these radios are not well standardized for interoperability between manufacturers, so they may not be appropriate for use in a wide variety of PV system equipment.

Bluetooth and Zigbee are normally intended for very short-range communications (the quoted distance limits are rather optimistic), and are not very stable standards. However, in so-called solar subdivisions, the communications relay feature of ZigBee may make it a practical technology for real-time price signals.

IEEE 802.11g (Wi-Fi or wireless Ethernet) has a somewhat longer track record than Bluetooth, and is becoming a common LAN implementation technology. This may make it practical to piggyback on customer Internet connections to access real-time pricing data.

Of these options, the continuous-carrier PLCC option would be technically advantageous for islanding prevention, but it is comparatively expensive to install and operate. For real-time pricing data (a function primarily in the interest of the customer and less sensitive to latency), interconnection with the customer's LAN to share an Internet connection is an attractive, low-cost option.

4.5.1.4. Signal Classes

Voltage Regulation

The line voltage can use control algorithms like droop control or output impedance synthesis to determine voltage regulation needs. These types of control can handle voltage regulation and fast (shorter than 1 s) electrical control. However, for control in slower response situations, a centralized dispatch of reactive power could provide significant advantages in some cases, and could be provided instead based on local line conditions. Communication would need to occur only every few cycles, so bandwidth requirements are minimal.

Peak Shaving (Demand Response)

Tariffs use demand charges to discourage peak loads. Under this scenario, no communications would be required from the utility.

Future systems will likely use real-time pricing schemes in which the price paid by the customer for electricity is not constant, but is determined by the market in real time. During a time of generation shortage, supply and demand would dictate that the electricity price would increase. In a real-time pricing system, the communications system would be used to send a price signal to the customer indicating that electricity rates were rising. An EMS might then operate to shut down certain noncritical loads, especially those with built-in storage such as tank water heaters, to minimize utility bills. Systems with local energy storage might switch to these local stores, depending on the relative price of energy from storage versus energy from the utility. Either scenario would reduce the peak load for the utility.

The real-time pricing signal may be generated on time scales ranging from 1 to 60 minutes, depending on utility. Communications bandwidth requirements would depend on what level of device participates in the market. In an ideal case, every electricity load and source might participate in the real-time pricing market. However, that would require communications and intelligence capabilities in every device plugged into the wall. BPL probably could not provide enough bandwidth to realize such a system. Thus, it is more likely that EMS will integrate these functions at the facility level. The EMS would then communicate with loads, sources, and storage under its control.

Backup Power (Intentional Islanding)

The presence or absence of the utility could be signaled via communications. PLCC has unique advantages, as the presence or absence of the signal could be used as a continuity test of the line. Intentional islanding must be coordinated with automated sectionalizing switches to ensure that faults are external to the island and that the DG source is internal.

Table 4.1. Communication Technology Characteristics

Technology	Latency	Bandwidth	Reliability	Accuracy	Distance Limits	Capital cost	Operating Cost
Dedicated copper wire (dry contact)	< 3 ms	200 bps	High	High	2-20 miles	High	Low
Ethernet (10BaseT, etc)	2-10 ms	10-1000 Mbps	Medium	High	100 m	Medium	Low
Continuous carrier PLCC	0.2-10 ms (loss of mains)	Low	Unknown	High (loss of mains)	< 100 miles	High	High
Broadband (BPL) PLCC	< 30ms	Medium to High	Unknown	Unknown	2000ft per hop	Medium to High	Medium
Spread- spectrum wireless	5-50 ms	10-50kbps	Variable	High	300m	Medium	Low
Bluetooth (Class 1)	50 ms	1-3Mbps	Variable	High	<100m	Medium	Medium
ZigBee (IEEE*802.15.4-2003)	> 16ms	20Kbps or 50Kbps	Unknown	High	<100m per hop (can relay)	Low	Low
IEEE 802.11 Wireless Ethernet	2-10 ms	10-54 Mbps	Medium	High	30m	Medium	Low

*Institute of Electrical and Electronics Engineers.

These sectionalizing switches must therefore have high-speed communication and controls. They also must have detection and control logic to ensure that the intentional island is synchronized with the utility before reconnection.

Furthermore, the utility could require customers to drop noncritical loads during an island protection operation. In this case, the utility would have to provide communications to the customers to indicate the intentional island.

Communications must provide instantaneous (subcycle) status and response must likewise be subcycle.

Spinning Reserve

The ramp up of spinning reserve units—triggered by a lost unit or a frequency excursion—is initiated by a signal from the system operator. The signal is sent approximately every second, but may differ between independent system operator territories. Thermal units responding to this signal normally reach full output over a few minutes. If this same signal were available to distributed EMS, they could dispatch generators or storage under their control to relieve the pressure on remaining mainline generation units, or activate load-shedding schemes. Theoretically, the real-time pricing scheme could achieve the same goal, if the market were updated often enough. The loss of generator would trigger a price spike, to which an EMS would respond by dispatching sources or storage, or by shedding load.

Frequency Regulation (and Area Regulation)

Signals from the system operator (generally calculated values derived from frequency and average frequency) control units to ramp up and down as necessary to ensure that the average system frequency over time is 60 Hz. These signals are sent every few seconds, and response time is several minutes.

In the case of DG, response time can be much faster, even on the millisecond scale. In principle, this would provide greater value than an equivalent rated thermal unit. More thermal units have to be combined to provide comparable ramp rates. Also, the faster response provided by DG would mitigate frequency excursions faster, reducing the effective capacity needed for the service.

However, DG has not historically been used and is unproven in this application. Also, the added benefit of fast DG response is not well quantified.

Control Fault Current Modes

Today's PV inverters typically do not contribute significant levels of fault current. This is often a desirable property, because it should mean that the addition of distributed PV to utility systems will not adversely affect the coordination of utility protective devices. However, the SEGIS may require more sophisticated control over its fault current contribution, and communications that allow the SEGIS to know whether it is in grid-parallel or microgrid mode would be important.

First, consider the SEGIS's grid-tied behavior. Because the SEGIS would incorporate voltage regulation capability, its fault current contribution will likely be much larger than that of today's unity pf PV inverters. The SEGIS will thus need a way to determine when it is feeding a fault, and limiting its fault current. PLCC-based loss-of-mains detection should be effective in the case of a hard (low-impedance) fault, because such a fault should lead to a loss of the PLCC carrier, signaling to the SEGIS to disconnect from the grid. High-impedance faults present a greater challenge: reliably differentiating high-impedance faults in distribution systems from poorly behaved loads and other normal distribution system conditions is a subject of ongoing research [62, 63].

However, in the intentionally islanded (microgrid) case, the utility's contribution to fault current is not available. The SEGIS could increase its fault current contribution, so that standard protective devices will reliably operate in the event of a fault. Communications could be used to determine whether the SEGIS is in grid-parallel or microgrid mode, and the inverter's surge current capability could be used to momentarily increase its fault current capability in this case. Again, reliable differentiation between a high-impedance fault and a noisy load is a key capability.

4.5.1.5. Example Command Sets To Be Sent via Communications

Some potential signals from the utility distribution control system to the PV system might be:

- Use wide voltage-frequency range. This would enable low voltage or low frequency ride-through capabilities.
- Ramp to x% power. This signal might be used in a case in which the voltage frequency in a section of the system began to rise excessively.
- Switch offline. This signal might be thought of as the utility's E-stop button on the inverter.

Some potential signals from the price aggregating clearinghouse might be:

- Hour-ahead real energy price
- Maximize real power generation
- Hour-ahead reactive power price

The International Electrotechnical Commission Technical Committee 57 Working Group 17 is developing IEC-61 850-7-420 for distributed energy resources to define relevant data and use cases describing typical uses for those data. This standard is a part of a large group of standards being developed under the overall IEC-61850 standard for power systems of the future.

4.5.2. Energy Management Systems

The SEGIS should be designed to work with an EMS that takes into consideration anticipated PV energy available, pricing signals, storage system availability and performance, and other factors. For example, PV energy could be delivered to the grid during high-priced periods or to storage during low-price periods. If prices over the coming hours are expected to increase by a factor greater than the battery efficiency losses, the energy could be stored.

4.5.2.1 Peak Shaving (Demand Response)

EMS and SEGIS can be used to lower peak customer loads and reduce demand charges. Certain non-time-critical loads, such as thermal loads, can be timed to minimize peaks. EMS controllers can be programmed so peak loads are managed without compromising customer processes. SEGIS storage would be charged during periods of low demand and dispatched during peaks by the EMS that can receive and respond to real-time pricing signals.

SEGIS storage systems would be sized based on the load profile (or net load profile), specific to each customer. Typically, the system would need to have several hours of storage, which can be kept to a minimum by smart dispatch in which the storage device follows the load in real time under the direction of the EMS. Such smart dispatch could use predictive capabilities to determine the threshold above which dispatch would occur. Under a real-time pricing scenario, the historic hourly prices could also be taken into account.

4.5.2.2. Other Energy Management System Functions

This report deals primarily with integrating the SEGIS with an EMS, but the EMS could perform a number of important grid support functions, including shedding of noncritical loads during peak demand times, fast-acting load shedding during system emergencies, and load shifting for loads with inherent thermal storage. As noted earlier in this report, realizing these capabilities would require the development of:

- High-speed communications reaching all endpoints in the distribution network
- High-speed real-time market mechanism that monetizes all aspects of power system operation, including both the market rule base and the physical communications and computational infrastructure to realize real-time market participation by all EMS and SEGIS-equipped facilities.

5. GAP ANALYSIS

5.1. Voltage Regulation Coordination

PV inverters and power conditioning systems could be used to vary reactive power, but current grid interconnection standards are not compatible with this function. The validation of voltage regulation using a large number of generators has not been demonstrated.

5.2. Distribution-Level Intentional Islanding (Microgrid)

The use of storage to provide short-term intentional islanding support on a distribution feeder serving multiple customers has not been demonstrated. Communication and control related to isolation has not been demonstrated. A demonstration of such a system on a real utility circuit would help to validate this as a distribution planning option.

5.3. Controlling Facility Demand and Export by Emergency Management System Integration

The use of storage to mitigate problems arising from the export of power from the customer facility to the grid has been demonstrated in the Ota City PV-integrated distribution system. However, effectively using storage to eliminate backfeed would require a control algorithm to be developed that could intelligently manage storage capacity. The algorithm would take into account historical PV output and load to predict optimum load dispatch set points to capture demand charge savings. It would have to apportion stored energy over the course of the week. The algorithm would have predictive capabilities based on ambient temperatures, and solar output to forecast net loads (loads less expected PV output).

Similar control strategies would need to be developed in response to demand-response or real-time pricing scenarios. Stored energy would most effectively be managed with estimates of future hourly pricing and PV output. Algorithms could be developed, for example, to forecast anticipated prices based on historical signals received by the EMS system. Storage could be charged in anticipation of needed energy during periods of high prices. The dispatch of stored energy would take into account anticipated PV output in future hours. The algorithms for forecasting pricing and PV availability could be based on statistical analysis using diurnal patterns.

An example of an integrated EMS/PV/Storage system is described in Table 5-1. The optimization problem would be to minimize the monthly cost of electricity service to the customer. This cost is a combination of demand charges and energy charges, including real-time energy charges in which future pricing is unknown.

The EMS would take into account fixed and measured parameters to control outputs. Using available PV power (in this assumed case) is always beneficial, so PV power is not a controlled output. Only loads and storage are controllable. The key to optimization would be to determine when and how to manage loads and when to charge or discharge the storage.

The problem is a combination of deterministic and stochastic effects. For example, the cost to the customer is easily calculated if the monthly loads and utility prices are known. The PV output could likewise be calculated from its design characteristics and known insolation. Such models are deterministic and readily available.

The stochastic effects, however, are more complicated. For example, charging the storage in advance of high real-time energy prices would be desirable. However, under a real-time pricing scenario, the prices are not known and must be forecast. Likewise, the availability of PV energy and thermal load requirements would have to be forecast. Historical measured data and historical forecasting errors could be used to forecast, so the respective forecasting models learn through experience.

The decision to charge or discharge storage in a given hour would therefore require a knowledge of all system states at every hour of the month, and these would be forecast based on the best available data. Forecasts might include insolation, ambient temperature, and energy prices. Deterministic models would use these forecasts to predict PV output and thermal loads. Finally, the decisions to control noncritical loads and dispatch storage would be made to minimize monthly energy cost.

Table 5.1. Example EMS Optimization Parameters

Parameters	Loads (by Circuit or Device)	PV	Storage	Pricing
Fixed Parameters	Critical versus noncritical Thermal set points (such as chil-ler water tempera-ture range)	Rated power System output versus meteor-ological values	Rated power Transient power Energy storage capacity as a function of discharge rate Turnaround efficiency	Demand charges (fixed or tiered) Energy charges (real-time and fixed tiers)
Measured Parameters (real time)	Actual loads Temperatures (air, chiller water, process heat, etc.)	Actual power Ambient temperature Insolation (from instrument or satellite data)	Actual power State of charge (measured or calculated)	Real-time price (if used) Meter read status (demand ratchet)
Forecasted Parameters (minutely or hourly)	Loads	Power		Price
Output/ Control	Noncritical loads (on/off)	(None)	Charge or discharge power	(None)

5.4. Backup Power (Intentional Islanding)

Utilities are obligated to provide nondiscriminatory pricing to all customers. The use of storage for enhanced reliability would give preferential reliability to certain customers connected to the island. It is not clear (1) how the regulatory agencies would view differential reliability; (2) whether the utilities would be willing to offer it; or (3) who would pay for this service.

Utilities can charge individual customers special facilities fees, but on circuits where some customers are willing to pay and others are not, how the regulatory rules would apply and how to recover the costs to install backup power for multiple customers are not clear.

Finally, our research has also demonstrated a need for an alternative to active anti-islanding that is compatible with microgrids and intentional islanding. Communications of several types seem to be one likely solution.

5.5. Spinning Reserve

Spinning reserve is normally performed at the transmission level. Storage at the distribution level would be effective only as spinning reserve when the distribution circuit is not taken out of service. During a major system-level outage, some distribution circuits would be curtailed in order to preserve system integrity. Circuits with substantial DG sources could be preserved for reserve service, but currently no mechanism is in place to do this. Therefore, it is not clear whether distribution sources are viable for this application.

In addition, spinning reserve is normally provided with units rated higher than 100 MW. The market for smaller units is not established.

5.6. Frequency and Area Regulation

Like spinning reserve, this service is normally provided at the transmission level. Unlike spinning reserve, which is limited to major disruptions, this service is provided continuously. During a distribution outage, the service would not be available.

Competitive markets are emerging throughout the country. Different ISOs have different minimum size requirements. Some allow systems rated at 10 MW and higher, some at 1 MW.

Energy storage or PV would provide significantly faster response times than conventional generation. Systems could respond in milliseconds (once the signal is received) relative to minutes for thermal plants. Therefore, DG provides this service more effectively than do conventional sources. This suggests that special control algorithms could be developed to take advantage of the fast response times.

Finally, energy storage or cloud transient forecasting for leveling or softening PV output during cloud transients appears to be desirable, but neither has been demonstrated.

5.7. Harmonics

Although two survey respondents mentioned cumulative harmonics as a source of concern for the future, and the topic arises in the literature occasionally, the literature, experimental results, and utility feedback indicate that harmonic pollution and excitation of power system resonances are not problems, as long as high-quality sinewave inverters are used. Even early studies [3, 5] suggest that this was not a problem with mid-1980s inverter technology, and with the introduction of IEEE-519 since then and the incorporation of those harmonic limitations into IEEE-1547, it appears safe to conclude that harmonics do NOT limit penetration levels if IEEE 519/1547-compliant inverters are used.

5.8. Effect of Distributed Generation on Coordination of Protective Relaying

The potential for PV inverters to change the conditions under which utility coordination schemes between fuses and circuit breakers are established has been raised as one possible adverse impact from high penetration levels of PV. Based on the literature search done for this work, this does not appear to be an issue with today's inverters because they contribute so little to fault currents, and thus are highly unlikely to cause fuses to melt or breakers to open out of their designed sequences. If voltage regulation controls are implemented, this issue may need to be revisited.

There is no universal agreement that PV inverters will not affect protection coordination. For example, see 8, pp. 2–10, where the authors describe a number of pathways through which high penetrations of PV could theoretically cause a protection coordination problem. Section 2 of that report details other potential issues, such as possible false tripping of protective relaying.

6. RECOMMENDATIONS FOR FUTURE RESEARCH

Future research and development should account for the following issues:

6.1. Smart Photovoltaic Systems with Energy Management Systems

Hardware and algorithms will need to be developed that incorporate communication protocols used by EMS and utility distribution systems. When hardware is available that can accept input from advanced utility distribution systems and control loads and generation, algorithms can be developed that optimize economic use of energy sources.

The physical implementation of the EMS may be incorporated within the PV system or may be a separate device, depending on market forces. Small, limited-feature smart PV systems will likely incorporate a simplified EMS function; larger and more configurable designs may choose to create a separate EMS device.

6.2. Reliability and Lifetime of Inverter/Controllers

Inverter hardware currently available has an MTBF of 5 to 10 years. Since the MTBF of the PV modules that those inverters are connected to is closer to 20 to 30 years, inverters will have to be replaced once or twice during the life of the system. Also, an inverter failure incurs a missed-opportunity cost for energy that was not generated. Thus, increasing the usable life of inverters will most likely lead to lower energy costs.

6.3. Voltage Regulation Concepts

Interconnection policies such as IEEE1547 strongly discourage voltage regulation by DG sources in the utility distribution system. However, a cohesive technical and policy approach to allowing voltage regulation by DG will need to be developed to handle projected high-penetration scenarios. Slow regulation (for managing distribution system voltage profiles or microgrid operation) and fast regulation (for addressing flicker and cloud-induced fluctuations) will both be needed in high-penetration scenarios. Demonstrations of solid technical approaches for voltage regulating DG will provide support for updated standards that will streamline commercial product development and simplify utility interconnection.

6.4. Distribution-Level Intentional Islanding (Microgrid)

Further development is needed for control strategies to manage microgrids. This area is related to the grid-connected voltage regulation needs discussed earlier, but it will most likely need to be augmented with communications to coordinate the transition between grid-connected and isolated modes of operation.

Further investigation into the regulatory issues should be conducted. For example, the customers who would benefit from the intentional island as a secondary source would have increased reliability relative to customers who would not be connected to the microgrid. Could tariffs be increased for the customers who benefit to recover the capital costs of the storage system and associated controls? Would it be preferable (and legal) to collect premium revenues only during the island operation, and how would these prices be set? How would the utility address customers connected to the island who are not willing to pay for enhanced reliability?

6.5. Energy Storage

Energy storage subsystems need to be identified that can integrate with distributed PV to enable intentional islanding or other ancillary services. Intentional islanding is used for backup power in the event of a grid power outage, and may be applied to customer-sited UPS applications or to larger microgrid applications. Stored energy may also be applied to grid ancillary services such as spinning reserve or frequency regulation if aggregation is implemented.

CONCLUSIONS AND RECOMMENDATIONS

In general, the idiosyncratic characteristics of PV as a DG have not yet caused any significant problems for utility systems. If PV penetration levels increase much more, the work conducted here suggests that the problems most likely to be encountered are voltage rise, cloud-induced voltage regulation issues, and transient problems caused by mass tripping of PV during low voltage or frequency events. Issues that are not expected to arise are power quality problems caused by active anti-islanding, excessive harmonic pollution, and major problems with coordination of protective relays and fuses.

Several short- and long-term solutions to the likely problems have been suggested in this report, but in conclusion the long-term view will be emphasized. The distribution system of the future will likely be characterized by a much greater proliferation of DGs, distributed storage, and much higher prevalence of power electronic converters, as illustrated the SEGIS concept. Major research efforts worldwide are attempting to produce power electronics transformers, new types of voltage regulators, more capable static VAr compensators, and the controls and communications required to coordinate all these power system elements. Certain control elements, such as fast electrical control, will be distributed, and economic dispatch and load control will likely be handled by a central EMS. When evolutionary steps produce this power system, most of the high-penetration PV issues discussed here will cease to be problems; the improved level of flexibility and control, coupled with the availability of distributed storage, will eliminate them.

The power converters within the SEGIS can be viewed as additional power electronics elements in this integrated system. Based on the work documented in this report, the most critical capabilities that SEGIS inverters of the future should have to be able to work within this system are ancillary-service and microgrid-ready controls, and communications capabilities to interface them to the power systems control communications bus. Hardware reliability concerns are important, but have not been specifically addressed in this report.

REFERENCES

[1] Willis, H. ed, *Distributed Power Generation: Planning and Evaluation,* Boca Raton: CRC Press, 2000.

[2] Chalmers, S; Hitt, M; Underhill, J; Anderson, P; Vogt, P; Ingersoll, R. "The Effect of Photovoltaic Power Generation on Utility Operation." *IEEE Transactions on Power Apparatus and Systems;* PAS-104, 1985, pp. 524–530.

[3] Patapoff, N; Mattijetz, D. "Utility Interconnection Experience with an Operating Central Station MW-Sized Photovoltaic Plant." *IEEE Transactions on Power Systems and Apparatus;* PAS-104, 1985, pp. 2020–2024.

[4] Jewell, W; Ramakumar, R; Hill, S. "A Study of Dispersed PV Generation on the PSO System." *IEEE Transactions on Energy Conversion;* Vol. 3, 1988, pp. 473–478.

[5] Cyganski, D; Orr, J; Chakravorti, A; Emanuel, A; Gulachenski, E; Root, C; Bellemare, R. "Current and Voltage Harmonic Measurements at the Gardner Photovoltaic Project." *IEEE Transactions on Power Delivery;* Vol. 4, 1989, pp. 800–809.

[6] EPRI report EL-6754, *Photovoltaic Generation Effects on Distribution Feeders, Volume 1*: Description of the Gardner, Massachusetts, Twenty-First Century PV Community and Research Program," 1990.

[7] Garrett, D; Jeter S. "A Photovoltaic Voltage Regulation Impact Investigation Technique: Part I—Model Development." *IEEE Transactions on Energy Conversion;* Vol. 4, 1989, pp. 47–53.

[8] Baker, P; McGranaghan, M; Ortmeyer, T; Crudele, D; Key, T; Smith, J. *Advanced Grid Planning and Operation.* NREL/SR-581-42294. Golden, CO: National Renewable Energy Laboratory, January 2008.

[9] Jewell, W; Unruh, T. "Limits on Cloud-Induced Fluctuation in Photovoltaic Generation." *IEEE Transactions on Energy Conversion;* Vol. 5, 1990, pp. 8–14.

[10] Imece et al. tests on SunSine inverter.

[11] Asano, H; Yajima, K; Kaya, Y. "Influence of Photovoltaic Power Generation on Required Capacity for Load Frequency Control." *IEEE Transactions on Energy Conversion;* Vol. 11, 1996, pp. 188–193.

[12] Povlsen, A. *International Energy Agency Report* IEA PVPS T5-10: 2002, February 2002. Available online at www.iea.org.

[13] Kroposki, B; Vaughn, A. *DG Power Quality, Protection, and Reliability Case Studies Report, NREL/SR-560-34635.* Golden, CO: National Renewable Energy Laboratory. General Electric Corporate R&D, 2003.

[14] Miller, N; Ye, Z. *Report on Distributed Generation Penetration Study*. NREL/SR-560-347 15. Golden, CO: National Renewable Energy Laboratory, 2003.

[15] Kersting, W. Distribution System Modeling and Analysis, Boca Raton: CRC Press, 2002.

[16] Union for the Coordination of Transmission of Electricity. *Final Report of the Investigation Committee on the 28 September 2003 Blackout in Italy*. 2004, p. 121.

[17] Quezada, V; Abbad, J; San Román, T. "Assessment of Energy Distribution Losses for Increasing Penetration of Distributed Generation." *IEEE Transactions on Power Systems;* Vol. 2 1(2) May 2006, p. 533-540.

[18] Dispower: Distributed Generation with High Penetration of Renewable Energy Sources, Final Public Report, 2006. Available on the DISPOWER Web site: www.dispower.org.

[19] Thomson, M; Infield, D. "Impact of Widespread Photovoltaics Generation on Distribution Systems." *IET Journal of Renewable Power Generation*; Vol. 1, 2007, pp. 33–40.

[20] Ueda, Y. et al., "*Performance Ratio and Yield Analysis of Grid-Connected Clustered PV Systems in Japan.*" In Proceedings of the 4th World Conference on Photovoltaic Energy Conversion, pp. 2296–2299.

[21] Luque, A; Hegedus, S. *Handbook of Photovoltaic Science and Engineering*, John Wiley and Sons, 2003.

[22] Short, T.A. *Electric Power Distribution Handbook. Boca Raton*: CRC Press, 2004.

[23] Watson, N; Arrillaga, J. *Power Systems Electromagnetic Transients Simulation*. Institution of Electrical Engineers, 2002.

[24] ANSI/IEEE Std 1547-2003 *IEEE Standard for Interconnecting Distributed Resources with Electric Power Systems*.

[25] Ronan, E; Sudhoff, S; Glover, S; Galloway, D. "A Power Electronic-Based Distribution Transformer." *IEEE Transactions on Power Delivery;* Vol. 17, 2002, pp. 537–543.

[26] Hatta, H; Kobayashi, H. "*A Study of Centralized Voltage Control Method for Distribution System with Distributed Generation.*" In 19th International Conference on Electricity Distribution (CIRED), May 2007, paper # 0330, 4 pages.

[27] Okada, N. "*Verification of Control Method for a Loop Distribution System Using Loop Power Flow Controller.*" In Proceedings of the 2006 IEEE Power Systems Conference and Exposition, October 29–November 1, 2006, pp. 2116–2123.

[28] Whitaker, C. *BEW Engineering, unpublished data.*

[29] Bower, W; Ropp, M. "*Evaluation of Islanding Detection Methods for Photovoltaic Utility-Interactive Power Systems.*" International Energy Agency Task V Working Group report IEA-PVPS T5-09, 2002.

[30] Ropp, M. *Design Issues for Grid-Connected Photovoltaic Systems*. PhD Dissertation. Atlanta, GA: Georgia Institute of Technology, 1998.

[31] Ropp, ME; Begovic, M; Rohatgi, A; Kern, GA; Bonn, R; Gonzalez, S. "Determining the Relative Effectiveness of Islanding Prevention Techniques Using Phase Criteria and Nondetection Zones." *IEEE Transactions on Energy Conversion;* Vol. 15, 2000, pp. 290–296.

[32] Behnke, MR; Erdman, WL. Impact of Past, Present and Future Wind Turbine Technologies on Transmission System Operation and Performance, California Energy Commission Report CEC-500-2006-050, 2006, www.bewengineering.com/docs.

[33] Erdman, WL; Behnke, MR. *Low Wind Speed Turbine Project Phase II:* The Application of Medium-Voltage Electrical Apparatus to the Class of Variable Speed Multi-Megawatt Low Wind Speed Turbines, Golden, CO: National Renewable Energy Laboratory. NREL/SR-500-3 8686, Nov 2005, www.bewengineering.com/docs.

[34] Ropp, M; Begovic, M; Rohatgi, A. "Analysis and Performance Assessment of the Active Frequency Drift Method of Islanding Prevention." *IEEE Transactions on Energy Conversion;* Vol. 14, 1999, pp. 810–816.

[35] Ropp, M; Ginn, J; Stevens, J; Bower, W; Gonzalez, S. *"Simulation and experimental study of the impedance detection anti-islanding method in the single-inverter case."* In Proceedings of the 4th World Conference on Photovoltaic Energy Conversion, May 2006, pp. 2379–23 82.

[36] Lasseter, R; Piagi, P. *Control and Design of Microgrid Components*. Power Systems Engineering Research Center, PSERC Publication 06-03, 2006.

[37] Boyes, J; & Menicucci, D. *"Energy Storage: The Emerging Nucleus of America's Energy Surety Future."* Distributed Energy, March-April 2007.

[38] Kroposki, B; Pink, C; Basso, T; DeBlasio, R. *"Microgrid Standards and Technology Development."* In Proceedings of the 2007 IEEE Power Engineering Society General Meeting, 4 pages.

[39] Piagi, P. *Microgrid Control. PhD Dissertation. Madison, WI:* University of Wisconsin-Madison, 2005.

[40] Ropp, M; Gonzalez, S. *"Development of a MATLAB/Simulink Model of a Single-Phase Grid-Connected Photovoltaic Inverter."* Submitted to IEEE Transactions on Energy Conversion, under review.

[41] Ropp, M; Larson, D; McMahon, D; Meendering, S; Ginn, J; Stevens, J; Bower, W.; Gonzalez, S; Fennell, K. *"Discussion of a Power Line Carrier Communications-Based Anti-Islanding Scheme Using a Commercial Automatic Meter Reading System."* In Proceedings of the 4th World Conference on Photovoltaic Energy Conversion, May 2006, pp. 235 1–2354.

[42] Ropp, M.E; Aaker, K; Haigh, J; Sabbah, N. *"Using Power Line Carrier Communications to Prevent Islanding."* In Proceedings of the 28th IEEE Photovoltaic Specialists Conference, September 17-22, 2000, pp. 1675–1678.

[43] Liu, W; Widmer, H; Aldis, J; Kaltenschnee, T. *"Nature of Power Line Medium and Design Aspects for Broadband PLC System."* In Proceedings of the 2000 Zurich Symposium on Broadband Communications, pp. 185-189.

[44] Liserre, M; Pigazo, A; Dell'Aquila, A; Moreno, V. "An Anti-Islanding Method for Single-Phase Inverters Based on a Grid Voltage Sensorless Control." *IEEE Transactions on Industrial Electronics;* Vol. 53, 2006, pp. 1418–1426.

[45] Willis, H. ed., *Distributed Power Generation: Planning and Evaluation*. Boca Raton: CRC Press, 2000.

[46] Soultanis, N; Hatziargyriou, N. *"Control Issues of Inverters in the Formation of L.FV. Micro-Grids."* In Proceedings of the 2007 IEEE Power Engineering Society General Meeting, 7 pages.

[47] Venkataramanan, G; Illindala, M. *"Small Signal Dynamics of Inverter Interfaced Distributed Generation in a Chain-Microgrid."* In Proceedings of the 2007 IEEE Power Engineering Society General Meeting, 6 pages.

[48] Rehtanz, C. *Autonomous Systems and Intelligent Agents in Power System Control and Operation.* Springer-Verlag, 2003.

[49] Lasseter, R; Piagi, P. *Control and Design of Microgrid Components.* Power Systems Engineering Research Center (PSERC) Publication 06-03, 2006.

[50] Lasseter, R; Piagi,P. "*Extended Microgrid Using (DER) Distributed Energy Resources.*" In Proceedings of the 2007 IEEE Power Engineering Society General Meeting, 5 pages.

[51] DISPOWER final public report, available at www.dispower.com; see Section 2.2, page 30.

[52] De Brabandere, K; Bolsens, B; Van den Keybus, J; Woyte, A; Driesen, J; Belmans, R. "*A Voltage and Frequency Droop Control Method for Parallel Inverters.*" In 35th Annual IEEE Power Electronics Specialists Conference, 2004, 7 pages.

[53] De Brabandere, K; Vanthoumout, K; Driesen, J; Deconinck, G; Belmans, R. "*Control of Microgrids.*" In Power Engineering Society General Meeting, June 2007.

[54] Ygge, F; Astor, E. "Interacting Intelligent Software Agents in Distribution Management." In Proc. Distribution Automation/Demand Side Management (DA/DSM), 1995.

[55] Ygge, F; Akkermans, H; Andersson, A; Krejic, M; Boertjes, E. "*The Homebots System and Field Tests: A Multi-Commodity Market for Predictive Load Management.*" In Proceedings of the 4th International Conference on the Practical Applications of Intelligent Agents and Multi-Agents, 1999.

[56] McArthur, S; Davidson, E; Hossack, J; McDonald, J. "*Automating Power System Fault Diagnosis Through Multi-Agent System Technology.*" In Proceedings of the 37th Hawaii International Conference on System Sciences, January 2004.

[57] McArthur, S; Davidson, E; Catterson, V. "*Building Multi-Agent Systems for Power Engineering Applications.*" In Proceedings of the 2006 IEEE Power Engineering Society General Meeting, 7 pages.

[58] Catterson, V; Davidson, E; McArthur, S. "*Issues in Integrating Existing Multi-Agent Systems for Power Engineering Applications.*" In Proceedings of the IEEE 13th International Conference on Intelligent Systems Application to Power Systems, November 2005, 6 pages.

[59] Denholm, P. Preliminary results described during presentation of "*Production Cost Modeling of Solar PV Impacts.*" Presented July 26, 2007, at the Midterm Meeting of the Renewable Systems Interconnection Distributed PV Study Project.

[60] IEEE 802.3u.

[61] ISO Standard 7498-1:1994.

[62] Benner, C; Russell, B. "Practical High-Impedance Fault Detection on Distribution Feeders." *IEEE Transactions on Industry Applications;* Vol. 33, 1997, pp. 635–640.

[63] Michalik, M; Rebizant, W; Lukowicz, M; Lee, S-J; Kang, S-H. "High-Impedance Fault Detection in Distribtion Networks with Use of Wavelet-Based Algorithm." *IEEE Transactions on Power Delivery*; Vol. 21, 2006, pp. 1793–1 802.

APPENDIX A: HIGH-PENETRATION PV SURVEY

High-Penetration PV Survey Sent to Utility Engineers

Respondent name (optional):

1. What is the highest PV penetration level you are experiencing on your system? For purposes of this survey, take the definition of penetration level to be the ratio of nameplate PV power rating (W peak) to the maximum load seen on the distribution feeder (W).
2. What is the estimated total amount of PV (watts peak) installed on your system, and what is your peak system load (MW)?
3. What adverse impacts have grid-connected PV systems had on your system? Are these impacts worse in the higher penetration portions of your system?
4. What steps have you taken to mitigate the adverse impacts of PV penetration on your system?
5. What current or future issues most concern you as the level of PV system penetration increases?
6. Please offer any further comments you would like to add. Thank you very much for your participation in this survey.

APPENDIX B: PRODUCT VENDORS

Identification of Product Vendors

A brief list of vendors that are active in the PV and storage markets

Photovoltaic Module Manufacturers

- Sharp USA: a manufacturer of poly-crystalline silicon and mono-crystalline silicon PV modules
- Kyocera Solar: a manufacturer of poly-crystalline silicon and mono-crystalline silicon PV modules
- Evergreen Solar: a manufacturer of poly-crystalline (string ribbon) silicon PV modules
- GE Energy: a manufacturer of poly-crystalline silicon and mono-crystalline silicon PV modules
- First Solar: a manufacturer of cadmium telluride thin-film PV modules
- United Solar Ovonic: a manufacturer of triple-junction amorphous silicon thin-film PV modules
- SolarWorld: a manufacturer of poly-crystalline silicon and mono-crystalline silicon PV modules

- BP Solar: a manufacturer of poly-crystalline silicon and mono-crystalline silicon PV modules
- Sanyo: a manufacturer of a broad range of industrial and consumer products, including heterojunction-with-intrinsic-thin-layer silicon modules (as well as amorphous silicon PV cells for consumer electronics applications)

Power Electronics and System Integration

- ABB Switzerland Ltd.: a manufacturer of a wide variety of electrical power generation, transmission, distribution, and consumption equipment.
- Exeltech: a manufacturer of on-grid and off-grid inverters
- Fronius AG: Austrian PV inverter manufacturer
- GridPoint: a supplier of energy management equipment designed to manage loads, storage, and renewable generation sources to optimize power flows for minimum cost
- PV Powered: a manufacturer of grid-connected PV inverters
- SatCon: a manufacturer of power system components for vehicles, machinery, and utility interactive power conversion systems.
- Sharp Electronics: manufactures a line of grid-tied PV inverters
- Siemens AG: major German manufacturer of PV inverters and other power equipment
- SMA America: subsidiary of a German manufacturer of grid-connected PV inverters
- Xantrex: a manufacturer of both grid-connected and standalone inverters

Short-Term Energy Storage

- Active Power, Inc.: a manufacturer of megawatt-scale flywheel-based energy storage equipment
- Axion Power Corporation: a manufacturer of lead-carbon batteries (an alternative battery technology)
- Beacon Power: a manufacturer of megawatt-scale flywheel-based energy storage equipment
- Electro Energy Inc.: a manufacturer of nickel-metal hydride, lithium-ion, and nickelcadmium-based battery energy storage devices
- Exide: a manufacturer of stationary lead-acid and nickel-cadmium batteries
- Gaia Power Technologies, Inc.: a supplier of multi-technology energy storage solutions
- Honda: the auto manufacturer, also manufactures a line of ultracapacitors for vehicular applications (used in the FCX)
- Maxwell Technologies: manufacturer of energy storage capacitors
- S&C Electric Company: manufacturer of electric power transmission and distribution equipment, as well as lead-acid-battery-based uninterruptible power supplies and static VAR compensators
- Saft: a manufacturer of nickel-cadmium, nickel-metal-hydride, and lithium-ion batteries

- Trojan: specializes in deep-cycle lead-acid gel-cell batteries
- Varta: a manufacturer of portable, traction, and stationary batteries in multiple chemistries

Long-Term Energy Storage

- NGK Insulators Ltd.: a manufacturer of sodium-sulfur batteries
- VRB Power Systems Inc. : manufacturer of vanadium-redox electrolyte flow battery energy storage systems
- ZBB Energy: a manufacturer of zinc-bromine electrolyte flow battery energy storage systems

Distribution

1	MS1104	Margie Tatro, 6200
1	MS1104	Rush Robinett, 6330
1	MS1033	Charlie Hanley, 6335
1	MS1110	Jeff Nelson, 6337
1	MS1124	Jose Zayas, 6333
1	MS1108	Juan Torres, 6332
1	MS 1033	Doug Blankenship, 6331
1	MS0734	Ellen Stechel, 6338
1	MS1108	Jason Stamp, 6332
1	MS1108	David Wilson, 6332
1	MS1108	Jaci Hernandez, 6332
1	MS1108	Michael Baca, 6332
1	MS0455	Shannon Spires, 6332
1	MS1108	Steven Goldsmith, 6332
1	MS1108	Jeff Carlson, 6332
1	MS 0899	Technical Library, 9536 (electronic copy)

In: Renewable Energy Grid Integration
Editor: Mitchell B. Ferguson

ISBN: 978-1-60741-325-7
© 2011 Nova Science Publishers, Inc.

Chapter 3

DISTRIBUTION SYSTEM VOLTAGE PERFORMANCE ANALYSIS FOR HIGH-PENETRATION PHOTOVOLTAICS

E. Liu and J. Bebic

ACKNOWLEDGMENTS

Reigh Walling of Power Systems Energy Consulting pointed out a significant deficiency in the earlier version of this work. His firm grounding in reality and candid criticism are gratefully acknowledged.

EXECUTIVE SUMMARY

Currently, electrical distribution systems are designed and operated based on the assumption of centralized generation, with the corollary that the power always flows from the distribution substation to the end-use customers. With the increasing penetration of residential and commercial PV, the PV power generation could not only offset the load, but could also cause reverse power flow through the distribution system. Significant reverse power flow may cause operational issues for the traditional distribution system, including:

- Over-voltage on the distribution feeder (loss of voltage regulation).
- Increased short circuit currents, potentially reaching damaging levels.
- Protection desensitization and potential breach of protection coordination.
- Incorrect operation of control equipment that may lead to an increase in the number of operations and related equipment wear, or to further aggravation of problems that affect more equipment and more customers.

Among all the potential problems that may be caused by the high penetration of PV, voltage regulation is the most likely one, because it is directly correlated to the amount of reverse power flow. This study was carried out to investigate the impact of different penetration levels of PV on the feeder voltage profile and on the equipment commonly used for feeder voltage regulation. The flow of reactive and active power on the feeder was investigated with different assumptions of inverter participation, and with various assumptions about the coordinated control of inverters and utility equipment.

A representative distribution feeder with a selection of typically used equipment was selected from a previous NREL study[4]. This feeder included commercial and residential loads. Tap-changing transformers and switched capacitors were applied at the substation and along the feeder. The model was further refined by explicitly representing the low voltage service transformers and the secondary circuits to which distributed PV generation is connected.

A series of case studies was conducted with different penetrations of PV, assuming several commonly used voltages regulation methods. The study results show:

- Reverse power flow at all studied PV penetration levels can be accommodated using traditional utility equipment with, perhaps, modified controls.
- Voltage rise on the secondary circuits is significant, and it should be included in the analysis. Establishing a communication link between service points (customer meter connections) and the utility equipment is helpful as it enables explicit control over the worst-case voltage.

In all the cases studied, PV inverters can positively contribute to feeder voltage regulation and result in an improved voltage profile. At a high enough penetration, PV inverters may be able to replace all voltage regulation equipment on a feeder. The study results show:

- At a low PV penetration level (5%), inverters do not make a significant impact on the feeder's voltage regulation during peak load.
- At a medium PV penetration level (10%), inverter voltage support can help reduce the size of the voltage support capacitors by nearly 40%.
- At high PV penetration levels (30% – 50%), PV inverters might entirely displace voltage support capacitors.

At higher penetration levels, inverter-coupled PV generation displaced some of the conventional generation. In order to match the performance of the inverters to that of conventional generators, inverters have to be able to exchange reactive power with the system. To allow for this, the following is needed:

- Current interconnection requirements need to be evolved.
- Inverter ratings need to be increased to allow for reactive power capability at all levels of power output. This can be stipulated by interconnection requirements or by carefully designed incentive programs.
- The operation of inverters has to be coordinated with the control of traditional voltage control equipment to take full advantage of the available reactive power capabilities of the inverters.

Recommendations for future research in this field are:

- Develop a set of recommended practices to reconcile existing feeder voltage control techniques with high penetrations of distributed PV. This work illustrates several such cases, but it is limited to one feeder topology. A more comprehensive coverage of feeder topologies would be beneficial.
- Develop a set of recommended practices for modeling PV inverters for load flow analysis and for other relevant planning purposes, such as short circuit current calculations. At the same time, expand the number of possible control options for traditional equipment in the analysis software. This would result in a more consistent understanding of the issues in the industry, and would simplify the test case setup so that it is possible to evaluate any specific situation.
- Create a set of benchmark cases to facilitate testing the models and the associated software. Some of the confusion and unwarranted concerns about the impact of PV generation may be a result of inconsistent and incorrect modeling.
- Develop automated screening tools that will enable evaluation of the impact of PV on the distribution system; all prospective installations could then be screened and only the ones requiring more detailed assessment would be floated up to the utilities. This would help preserve low installation costs while allowing for the more detailed assessment necessary at higher penetration levels of distributed PV.
- Develop functional requirements for communication infrastructure that will enable the coordinated operation of all equipment on the distribution feeder. The same infrastructure can be used to enable demand side management, implementation of flexible metering tariffs, and enhanced distribution system management.
- Fund demonstration opportunities that illustrate feeder operation with significant PV penetration.

1.0. INTRODUCTION

Solar photovoltaics (PV) are among the fastest growing energy sources in the world, with annual growth rates of 25-35% over the last ten years. The markets for solar PV have undergone a dramatic shift in the last five years. Prior to 1999, the primary market for PV was in off-grid applications, such as rural electrification, water pumping, and telecommunications. However, now over 78% of the global market is for grid-connected applications where the power is fed into the electrical network[1]. Furthermore, most of the new PV capacity has been installed in the distribution grid as distributed generation. As the use of solar photovoltaics continues to expand, concern about its potential impact on the stability and operation of the electricity grid grow as well. Utilities and power system operators are preparing to integrate and manage more of this renewable electricity source in their systems.

This study assesses the effects of a high penetration of distributed PV on the distribution system voltage control, and on the associated reactive power flow through the distribution system. A representative distribution system feeder that includes both residential and commercial content was selected from a previous study[4]. The selected feeder also includes voltage control equipment such as switched capacitors and on-load tap-changing transformers. The model was further expanded to represent service transformers and secondary circuits at the point of service entrance (the customer meter) – the likely connection point of the PV inverter.

The report is organized as follows. In Section 2, the current practices used for distribution voltage regulation are summarized, and the technical capabilities of inverters relevant to feeder voltage control are presented. Section 3 introduces the analysis approach, discusses modeling requirements, and presents the models used. Section 4 presents the study results; it includes establishing the baseline and analyzes the impact of the different PV penetration scenarios (i.e., 5%, 10%, 30%, and 50%) on the feeder voltage profile. It also illustrates how the reverse power flow on the distribution feeder impacts the operation of voltage regulation devices such as the on-load tap-changing transformer. Case results exploring different voltage control options are presented and discussed. Reactive power flow through the feeder is also analyzed using different assumptions for inverter participation and for the coordination of inverter and feeder control. The study is summarized in Section 5. Future research needs to explore the opportunity for optimal distribution system design for high penetration of PV are also identified.

2.0. CURRENT PRACTICE

2.1. Distribution System Voltage Control Requirements

Voltage regulation is an important subject in electrical distribution engineering, because it is the utility's responsibility to keep the customers' service voltage (the voltage at the customer's meter, or the load side of the point of common coupling (PCC)) within the acceptable range. ANSI C84.1 specifies a guideline for this range, but the utilities have the freedom to specify it differently based on their specific circumstances. ANSI C84.1 also specifies utilization voltage, which refers to the voltage at the point of use where the outlet equipment is plugged in. Furthermore, two ranges are defined: Range A is recommended for normal operating conditions, while Range B corresponds to unusual conditions, during which the occurrence has to be limited in time duration and frequency. Recommended service and utilization voltage limits according to ANSI C84.1 are shown in Table 1. Utilities are generally concerned with maintaining the service voltage within acceptable limits; the utilization voltage then follows automatically, provided that the house wiring is done according to building codes.

Irrespective of actual adopted voltage limits (by ANSI C84.1 or by the individual utility), most utilities control the service voltage indirectly by controlling the voltage on the primary circuit, the feeder. Service voltage is directly dependent on feeder voltage; when considered on the same voltage base, service voltage is equal to the feeder voltage minus the voltage drop across the service transformer and the secondary circuit connection. Consequently, it is possible to predict the service voltage based on the feeder voltage as long as the service transformer and service runs have consistent parameters for all loads. Utilities capitalize on this fact and develop internal design guidelines for sizing service transformers and for deciding the size and length of a service connection. Following the guidelines then enables them to eliminate the need to record the data related to the secondary circuits, resulting in a substantial saving in database size. (Admittedly, database size is not a significant factor nowadays, but these practices were developed more than 50 years ago when computer resources were scarce.) Hence, with design guidelines in place, service voltage is controlled indirectly by controlling the feeder voltage.

Figure 1 shows an example of voltage limits for the primary circuit, the service entrance, and utilization based on one utility's guidelines[3]. It reflects the adjustment for assumptions about additional voltage drop in the secondary circuit and allows for the necessary margin. In this study, the primary voltage and service entrance voltage limits shown in Figure 1 were used as target limits.

2.2. Voltage Regulation Methods

The voltage regulation practice applied to distribution systems is based on the radial power flow from the substation to the load. Voltage drop along the distribution feeder is inevitable and it is, in fact, required in order to move power from the substation to the customers. Typically, two (boundary) operating conditions of the feeder are considered: feeder voltage should not drop below the minimum during peak load condition, and it should not exceed the maximum during light load condition.

Table 1. ANSI C84.1 Voltage Range for 120V voltage level[2]

	Service		Utilization	
	Min	Max	Min	Max
Range A (Normal)	-5%	+5%	-8.3%	+4.2%
Range B (Emergency)	-8.3%	+5.8%	-11.7%	+5.8%

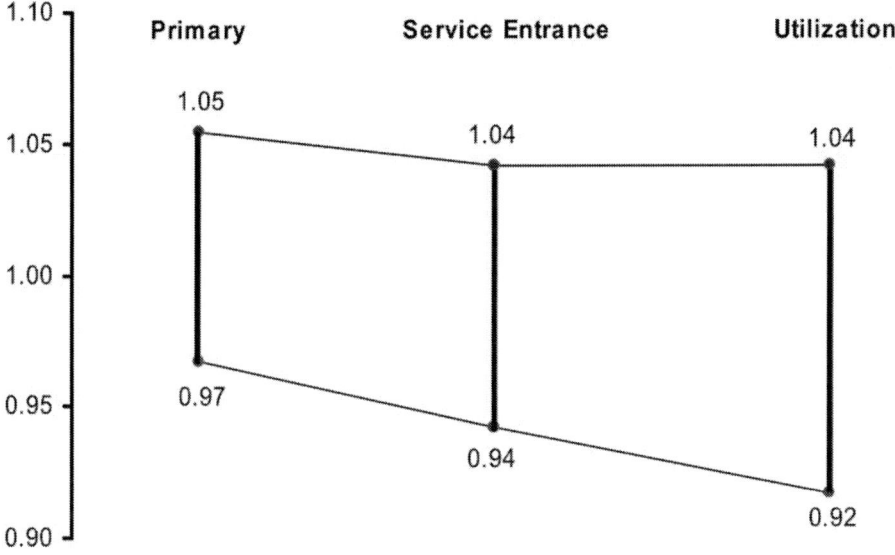

Figure 1. Voltage range limits used in this study according to Willis[3].

Voltage drop on the feeder is a consequence of current flow and the impedance (resistance and reactance) of the feeder conductor, transformer, and load. Loads require active and reactive power, and the related current that supplies the active and reactive power causes the voltage drop on feeder conductors. Feeder conductors are a given (they are selected first, based on economic considerations[*]). With conductor sizes known (i.e., their circuit parameters are fixed), there are two fundamental ways to control the voltage on the feeder: by using on-load tap-changing transformers, or by installing fixed or switched capacitors to offset the reactive power demand from the load and thus reduce the current flow through the feeder and the related voltage drop. These two methods are discussed in more detail next.

2.2.1. On-Load Tap-Changing (OLTC) Transformers / Voltage Regulators

The on-load tap-changing transformer (OLTC), or voltage regulator, is an essential part of a distribution network. Automatically adjustable OLTCs are commonly used at distribution substations to raise the starting voltage for a feeder under load, so that some point along the feeder has a desired voltage. The adjustment is proportional to the load, so this practice works well for all anticipated loading conditions. This control strategy is referred to as the "line drop compensation." The amount of permissible voltage increase is limited if there is a load (customer) near the voltage regulator, so in some cases additional voltage regulators along the feeder run might be necessary.

Voltage regulators, or OLTCs, are typically constructed as autotransformers with automatically adjusting taps. The controls measure the voltage and load current, estimate the voltage at the remote (controlled voltage) point, and trigger the tap change when the estimated voltage is out of bounds. Multiple tap change actions may be performed until the voltage is brought within bounds. The taps typically provide a range of ±10% of transformer rated voltage with 32 steps. Each step of voltage is therefore 0.625% of the rated voltage.[2]

2.2.2. Switched Capacitor

It was already explained that loads require active and reactive power, and that the related current causes the voltage drop on feeder conductors. The load's reactive power demand can either be supplied from the substation or by inserting capacitor banks along the feeder. The reactive power supplied by the capacitor banks offsets the reactive power of the load and consequently reduces the amount that needs to come from the substation and the associated voltage drop. The capacitor banks can be fixed (permanently connected) or switched (connected when needed), so that their supplied reactive power matches the need of the load. In practical installations this matching is seldom perfect, because the load and its reactive power demand vary continuously while the capacitor banks are switched in chunks. Moreover, the reactive power from capacitors varies with voltage squared, and drops off at low voltages when it is most needed. Overcompensation of the feeder (associated with too much capacitance) leads to voltage rise on the feeder, and it might require the voltage regulator in the substation to take action to lower the voltage to accommodate the rise due to overcompensation by the capacitors.

The controls used for switching capacitor banks can be based on: a time clock (load is correlated with time of day); the temperature (heavy load such as air-conditioning is correlated with ambient temperature); the voltage (low feeder voltage is an indication of the heavy load); the reactive power flow (to balance the reactive power actually drawn by the load); or the feeder current (similar to reactive power control, but less expensive to implement)[2].

2.3 Inverters' Reactive Power Support

Currently, standards such as IEEE 1547[†] and UL1741 state that the PV inverter "shall not actively regulate the voltage at the PCC." Therefore, PV systems are designed to operate at unity power factor (i.e., they provide only active power) because this condition will produce the most real power and energy. This limitation is a matter of agreement, not a technical one; many inverters have the capability of providing reactive power to the grid in addition to the active power generated by their PV cells. This is illustrated in Figure 2. The inverter's ratings are represented by a vector with magnitude S; the semicircle with radius S denotes the boundary of the inverter's feasible operating range in PQ space. Assuming that the power produced by PV array is P_{pv}, the feasible operating space reduces to the red line denoted by Ppv or, more precisely, to the segment of the red line delimited by its intersection points with the semicircle. Reactive power (Q) limits are then found by projecting the end points of the segment down to the Q axis; the values are labeled $-Q_{limit}$ and Q_{limit}. It follows that the inverter can supply positive and negative reactive power, that is, it can behave as both an inductor and a capacitor. The advantage of an inverter relative to a fixed capacitor is that it can vary the supplied reactive power continuously. The amount of Q available from the inverter depends on its ratings (S) and the active power supplied by the PV array. Consequently, the inverter can use its entire rating to supply Q if P_{pv} equals zero (there is no sun), and at the other extreme, it has no Q capability if P_{pv} equals S. Some Q capability can always be retained by over-sizing the inverter; this will be discussed in a later section. Note that this is for a unidirectional grid-connected PV inverter. Inverters connected to energy storage may allow for full four-quadrant charging and discharging of real and reactive power. In addition to the continuous reactive power support, inverters can operate very fast (milliseconds to microseconds with high switching frequency inverters) in comparison with capacitors, which can cause switching transients.

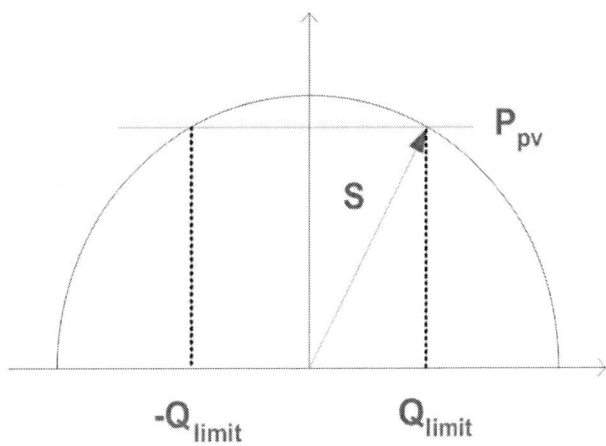

Figure 2. Determining the inverter's reactive power limits.

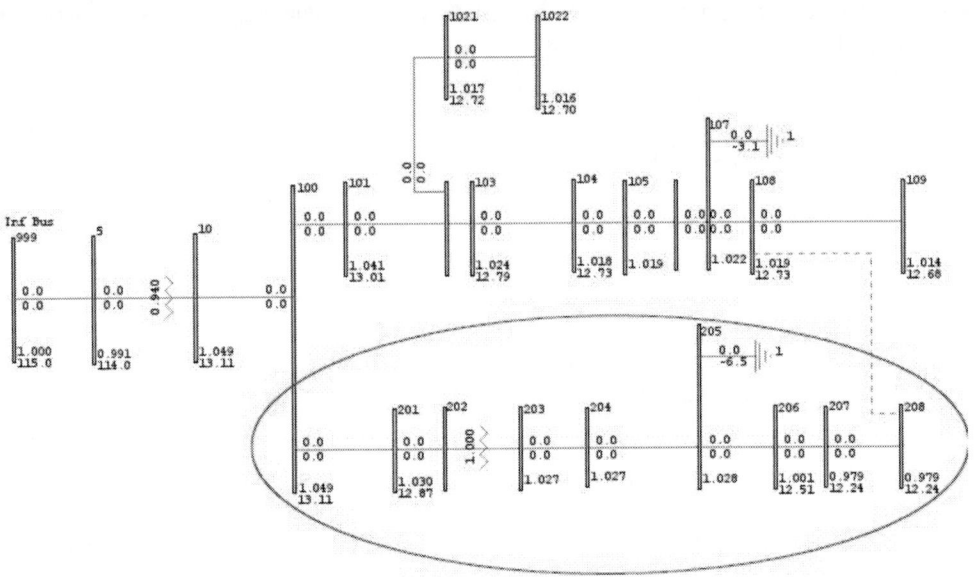

Figure 3. PSLF single line diagram of the distribution system model.

A number of publications are available that address the benefits of using inverter-based distributed generation for voltage support, their challenges, and their potential solutions. [8,9]

3.0. PROJECT APPROACH

3.1. Analysis Approach

The approach used in this study involves the following procedure:

1. Select and model a representative distribution feeder with various typical voltage support equipment.
2. Refine the model by adding a low-voltage service transformer and a secondary circuit.
3. Conduct load flow simulation of various scenarios, including peak load and various PV penetrations with maximum reverse power flow.
4. Review feeder voltage profile, the active and reactive power flows, and the impact of various control configurations.

3.2. Model Development

A representative distribution system was based on a previous distributed generation study conducted by General Electric under the contract with NREL[4]. The selected system includes the most commonly used distribution system components that are important for investigation of voltage regulation; there are voltage regulators at the substation and along the feeder, switched capacitors, and distribution transformers. In order to explicitly investigate the voltage at the customer service entrance service transformers and secondary circuits were added to the original model. The model is suitable for examination of equipment interaction and the impact on the feeder voltage profile at different PV penetration levels. The details of the feeder and the component models are introduced in the following sections.

3.2.1. Distribution Feeder Model

The selected distribution system was modeled in PSLF[10]. A one-line diagram of the feeder as modeled in PSLF is shown in Figure 3. The assumptions are:

- The considered distribution system is radial and supplied by a medium-voltage transformer equipped with a tap changer;
- The distribution system includes two main feeders with laterals and distributed loads;
- Distribution substation protections allow active and reactive power back feed;
- Each load bus has a PV connected to it, with the size relative to the size of the load on the same bus;
- All the load buses are modeled as PQ buses;
- The system base is 10 MVA;
- Bus 999 represents the infinite bus and is the slack bus in the power flow model;
- A loop, shown as a dashed connection between buses 108 and 208, can be closed to link the two feeders. However, for this study, the loop is not connected.

Although the entire system was modeled and analyzed, only the results from Feeder 2 (circled in red in Figure 3), which includes all the interested voltage regulation devices, are illustrated in this report. Feeder 2 is described below.

- It is about six miles in length.
- Seven aggregated loads represent a mixture of residential load and commercial loads ranging from 0.3 MW to 5 MW. The total load is 11 MVA.

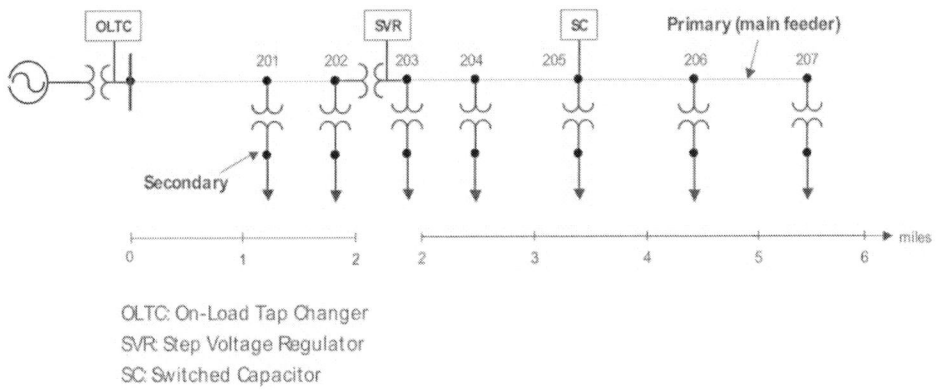

OLTC: On-Load Tap Changer
SVR: Step Voltage Regulator
SC: Switched Capacitor

Figure 4. Illustration of Feeder 2 in the distribution system model.

- The primary feeder voltage is 12.5 kV. The secondary voltages are 240 V for residential loads and 600 V for commercial loads.
- A switched capacitor is located at about 4.6 miles from the substation, at bus 205.
- Two voltage regulators are employed – one in the substation and another at 2.6 miles from the substation, between buses 202 and 203. These two devices have similar characteristics, but they are purposely labeled differently to simplify notation in the following figures and text. The voltage regulator at the substation is referred to as the *on-load tap changer*, abbreviated as OLTC, and the one along the feeder is called the *step voltage regulator*, abbreviated as SVR.

Figure 4 shows an illustrative one-line diagram of Feeder 2.

3.2.2. Component Models

In this section, the models of the major components used in the study feeder are introduced.

Table 2. Feeder Line Impedance

From	To	Length (mile)	Conductor	X/R
100	201	1.3	Z1	2
201	202	0.65	Z1	2
203	204	0.65	Z1	2
204	205	0.97	Z1	2
205	206	1.1	Z2	2
206	207	1.1	Z2	2

Table 3. Feeder Conductor Selection

Z1	0.648 Ohm/mile	ACSR 556.6 18/1
Z2	0.768 Ohm/mile	ACSR 266.8 18/2

3.2.2.1 Primary Circuit

Table 4. Feeder Voltage Equipment

Type	From	To	Rating
Transformer	5	100	20 MVA
SVR	202	203	10 MVA
SVR	202	203	10 MVA
switched capacitor (baseline 1)	205		6.2 MVAr
switched capacitor (baseline 2)	205		2.5 MVAr

3.2.2.2. Load

Loads were modeled as a combination of 40% constant power (active and reactive) load, and 60% of constant impedance load[3]. The loads in the study feeder have the power factor of 0.92, which is representative of the mixture of residential and commercial loads.

3.2.2.3. Secondary Circuit

In order to investigate the voltage performance at the customer service entrance the secondary circuit, including the low voltage service transformers and the service feeder, were added to the original feeder model.

The service transformers are rated at 1.5 pu relative to the load they serve, with an impedance of 2.5% [2] and an X/R ratio of 1.5. The service feeders are typically 50 feet to 600 feet in length[3]. For this study, an average 200-foot feeder length was selected. The impedance for the secondary feeder is calculated based on the conductor with 200 amps of thermal capacity. To simplify the model the impedance of the service transformer and the secondary feeder are added together and represented by the transformer impedance.

3.2.2.4. Photovoltaics

Grid-connected PV systems are designed to inject all of the real power produced by PV modules; they control the power precisely regardless of the voltage level, so they are best represented as negative constant power loads. The size of the negative PV load is defined proportional to the actual load connected at the same bus, based on the penetration level.

3.2.2.5. PV inverter Reactive Power (VAR) Support

The PV inverter model is not readily available in PSLF. To represent the reactive power capability of the inverters additional devices called Static VAR Devices (SVD), which are available in the PSLF standard library, were used. In PSLF, SVDs can be configured as switched or continuously controlled shunt elements whose admittance is adjusted in order to regulate the voltage at a specified bus. For the purpose of this analysis, the continuously controlled SVD model connected at each PV bus was used to simulate the VAR support feature of the PV inverter.

Each SVD is given the reactive power limits to represent the VAR support capability of the PV inverter at a studied condition. As was discussed in reference to Figure 2, the reactive power capability is decided by the size of the PV inverter S, and the active power output P_{pv}.

3.2.2.6. OLTC Transformer and SVR

The OLTC transformer and the SVR are modeled as tap-changing transformers that monitor the voltage at a remote bus and regulate the voltage to the defined limits by changing the turn ratio of the transformer. The tap range is ±10% of rated and the step voltage is 0.625%. The controlled bus depends on the case studied and will be defined in the specific case discussion.

4.0. PROJECT RESULTS

In this section, the results obtained for different steady state conditions on the feeder are presented. Cases include different PV penetration scenarios combined with different control configurations. First is a review of the baseline conditions – a feeder under the peak load.

4.1. Baseline

In normal operating conditions the feeder voltage decreases as the distance from the distribution substation increases, and it may become lower than the voltage specified by the utility's guidelines. Two feeder configurations are reviewed for peak load, which were chosen because of the different Q (reactive power) profiles on the distribution feeder, with different voltage regulation devices and locations.

4.1.1. First Baseline Configuration

In the first baseline configuration, the OLTC regulates the service voltage of the last bus on the feeder to the voltage limits. A capacitor bank is used during peak load conditions to inject capacitive reactive power and to boost the voltage along the feeder. As shown in Figure 5, the combined action of the OLTC and the switched capacitor ensures that the voltages of all the buses at the primary feeder and the service entrances are within the specified limits. OLTC action shifts the entire curve up or down, while the switched capacitor raises the voltage at its bus.

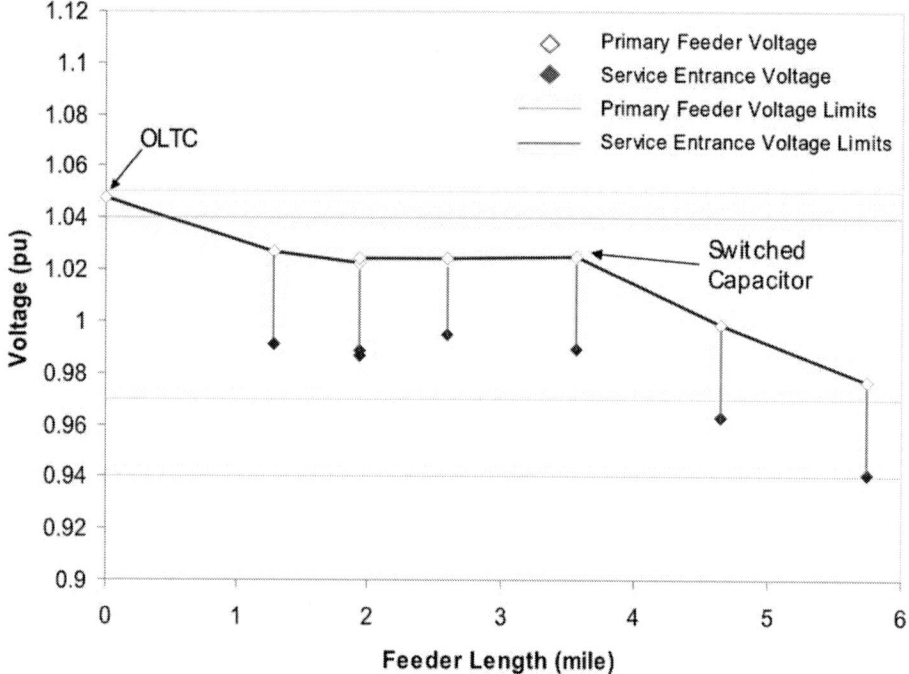

Figure 5. Baseline 1: voltage profile at peak load with the switched capacitor capacitor.

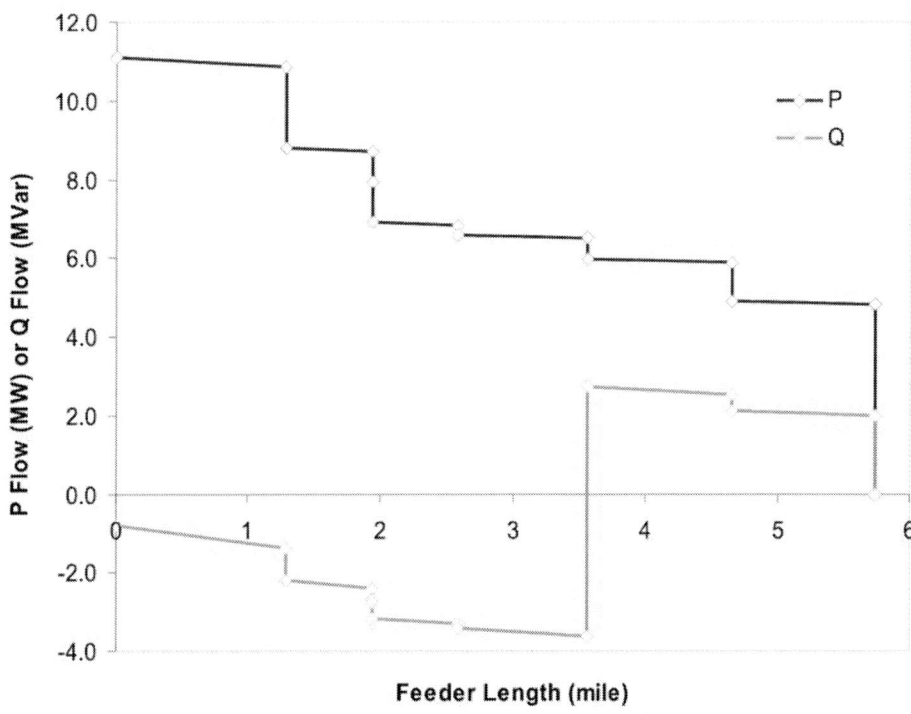

Figure 6. Baseline 1: power flow at peak load with the switched capacitor.

The corresponding P and Q flow on the feeder is shown in Figure 6. P is fed from the substation to the load; the step curve begins at approximately 11 MW, then steps down to ~9 MW at the first load bus, indicating that the load at this bus is ~2 MW. The P curve is characterized by a soft negative derivative between steps; this is representative of losses on the corresponding feeder segment. The Q curve, on the other hand, starts negative at the substation, indicating that the feeder is overcompensated by the switched capacitor; capacitive reactive power is delivered to the grid from this feeder. This is not an indication of a problem. It is a peculiar feature of this operating point. The steps in the Q curve are representative of reactive power consumption by the loads, which are generally inductive, resulting in negative steps. The sudden jump in Q at the switched capacitor bus shows the contribution of the switched capacitor. It overpowers the inductive consumption of the load at the switched capacitor bus and results in a net Q increase of more than 5 MVAr. Past this point, Q is consumed at load nodes; approximately 2 MVAr are drawn by the load at the end of the feeder.

4.1.2. Second Baseline Configuration

The second baseline configuration demonstrates the operation of SVR. The voltage regulation capability of SVR is constrained by the autotransformer tap limits; ± 10% was used. These limits may result in the need to combine SVR with other equipment, which was the case in this example. Hence, SVR is used along with the OLTC and the switched capacitor as follows. The OLTC regulates the primary side of the SVR, the SVR regulates the service voltage at the last load bus, and the switched capacitor is switched in to offset the reactive power consumption of the load along the feeder. The corresponding voltage profile is shown in Figure 7. At the discontinuity, where a step change in voltage occurs at the node with the SVR, the voltage is "stepped up." The contribution of the switched capacitor is analogous to the one in the first baseline configuration, but the capacitor is smaller, as will be discussed next.

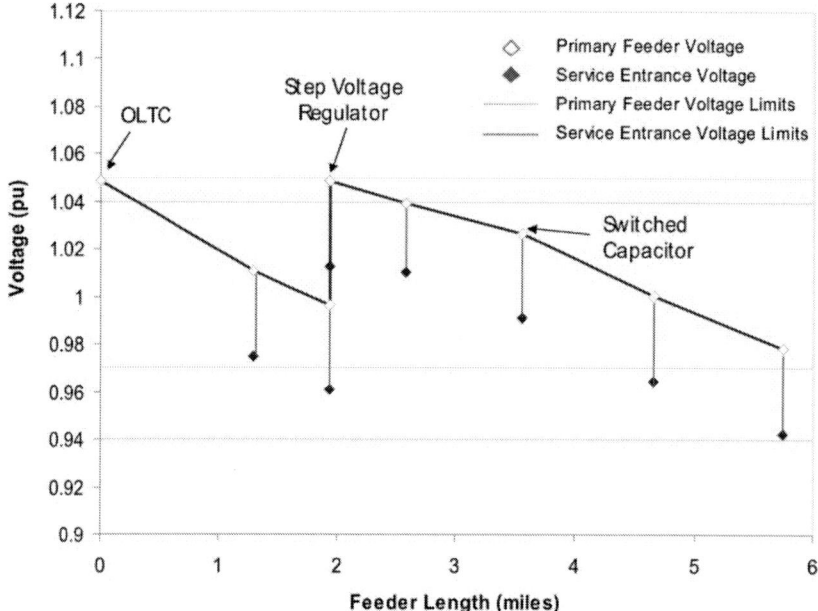

Figure 7. Baseline 2: voltage profile at peak load with SVR.

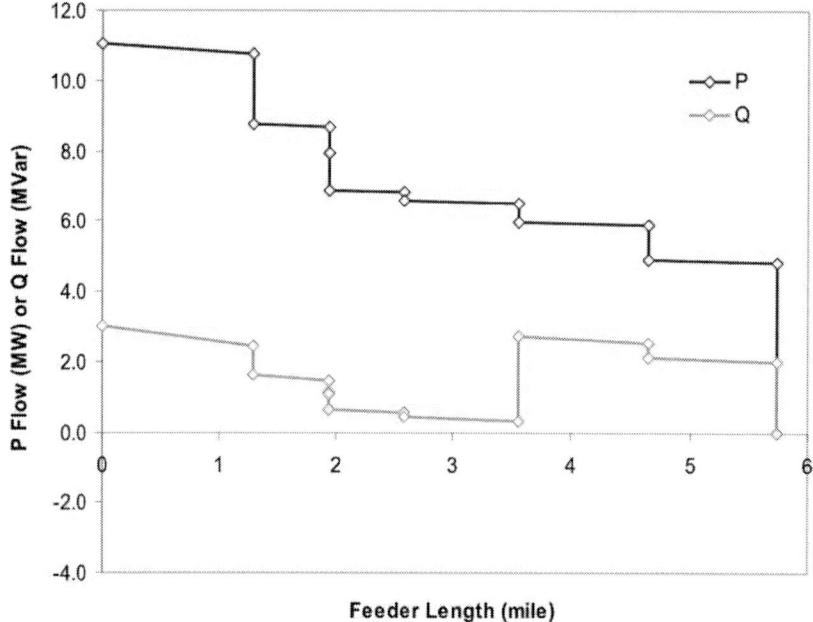

Figure 8. Baseline 2: power flow at peak load with SVR.

The P and Q flows corresponding to the second configuration are shown in Figure 8. The P flow matches the one before, so it does not need additional discussion. However, the Q flow is markedly different. In this configuration, reactive power flows from the substation into the feeder, and it gets used up gradually at each load point. There is also a noticeable Q loss across the SVR. The SVR is the transformer, and reactive power is consumed on its leakage impedance. Reactive power injection by the switched capacitor is evident from the Q jump at the switched capacitor bus, but in this case net Q injection is less than 3 MVAr, compared with over 5 MVAr in the first configuration.

Both baseline configurations result in the feeder and service voltages being within the limits, but the underlying circuit behavior is quite different as is apparent from the above discussion.

4.2. Description of the Issue

Introducing PV at the load side reduces the load demand and in turn leads to reduced losses and improved voltage profiles on the feeder. This is a fair observation as long as the PV generation coincides with the substantial load demand so that the net power flow remains from the substation to the load. As the penetration levels of PV rise, there may be time periods during the day when the net power flow is from the load (distributed PV) towards the substation – a situation not normally anticipated in the distribution system design. To illustrate this condition, consider a predominantly residential area with a significant penetration of PV. PV production is generally at its optimum at approximately 11 a.m., which is generally a period of light load condition, so power export through the distribution feeder and the substation back to the system becomes possible.

The associated reverse power flow tends to raise the voltage on the distribution feeder. The two presented base cases demonstrate that voltage drop drives power flow, so if the power flow reverses, the voltage slope will reverse as well. The obvious questions then are: will this situation cause problems, and how will the distribution circuit behave under these conditions?

These issues are discussed at various levels of PV penetration, combined with different assumptions about feeder equipment and different roles for inverters in feeder voltage control. To reduce data requirements the assumption was made that feeder load is zero during PV generation. This may appear unrealistic, but it is a matter of defining penetration. Assuming that 5% penetration causes 5% power export is equivalent to assuming that 10% penetration, combined with 5% load, results in a net export of 5%. Although estimating feeder load during optimal insolation (peak PV production) would require significant data gathering, this assumption, while arguably incorrect, is pessimistic relative to the study of the feeder voltage profile. It results in worse voltage conditions.

4.3. Results of the Research

As indicated in Table 5, there are many possible variations of options for analyzing feeder voltage control. The full set of variations includes 192 cases, but not all combinations are meaningful. For example, it does not make sense to combine reverse power flow with reactive compensation using the switched capacitor, as it would lead to voltage problems. The number of options was carefully reduced to include "only" 56 cases that were then set up and evaluated. However, presenting all these results does not add value, as the conclusions often repeat between similar cases. The report was therefore reduced to the following cases.

Peak load cases are presented first under assumptions of different PV penetration using the inverters for reactive power support – beyond their current role prescribed by IEEE 1547. After that, reverse power flow on the feeder for 50% PV penetration is presented by combining two options of voltage regulators with four options of inverter participation.

Since the role of the inverter is crucial for this discussion, assumptions about inverter capabilities are defined first in the next section.

4.3.1. Assumptions About PV Inverter Capabilities

As discussed in Section 2.3, inverters have the capability to supply inductive and capacitive reactive power to the grid, and this ability is limited only by their ratings[‡]. At the present time, inverters are not allowed to provide reactive power, so the manufacturers select their ratings to be equal to the maximum production of the connected PV modules. In terms of our previous discussion, $S = P_{PVmax}$. On the other hand, when $P_{PV} = 0$ (no sun) the entire inverter capacity can be dedicated to reactive power support at essentially no extra cost, so it is often argued that the standards should evolve to allow reactive power support. Evolution of standards will be required at higher levels of PV penetration, so it is logical to assume that inverters could be allowed to provide reactive power to regulate voltage. Other questions also emerge. Most notably, how should this capability be used during maximum power production, and what is the reasonable increase in ratings to provide reactive power support?

To provide reactive power injection while supplying maximum active power from PV modules, it is necessary to increase the inverter size. Figure 9 illustrates the active and reactive power capability of the inverter versus the size. As shown in the figure, by increasing the inverter size by 10%, making $S = 1.1 P_{PVmax}$, the reactive power capability can be increased from zero to nearly 46% in the maximum PV power generation condition. This will give the power factor range of unity to 0.91 leading/lagging. The Q capacity during no sun condition is then 110%.

In all the studied cases, inverters with 10% increased ratings were used, to allow for ample reactive power capability.

Table 5. Variational Space for Case Studies

Variable	Range	Number of options
PV penetration	5, 10, 30, 50%	4
Feeder load	0, 100%	2
PV inverter participation	IEEE 1547, voltage control, max Q, PF control	4
Voltage regulators	OLTC, OLTC + SVR,	2
Switched capacitor	0, base line 1, base line 2	3

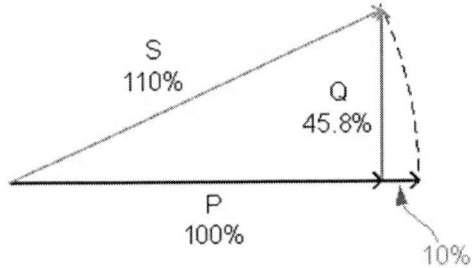

Figure 9. Relationship between inverter size and its reactive power capability.

4.3.2. Peak load, 5%, 10%, 30% and 50% Penetration, OLTC + SVR, Inverters Supplying Reactive Power

The reactive power capabilities of PV inverters can be used to offset the reactive load. This reduces the reactive power flow on the distribution feeder, and in turn reduces the voltage drop along the feeder. Inverters are configured to supply reactive power, and their output capacity Q is equal to their ratings. As in baseline configuration 2, OLTC controls the primary side of SVR, and SVR controls the service voltage at the last customer. The feeder's voltage profile and the associated active and reactive power flows for penetration levels from 5% to 50% are shown from Figure 10 through Figure 17.

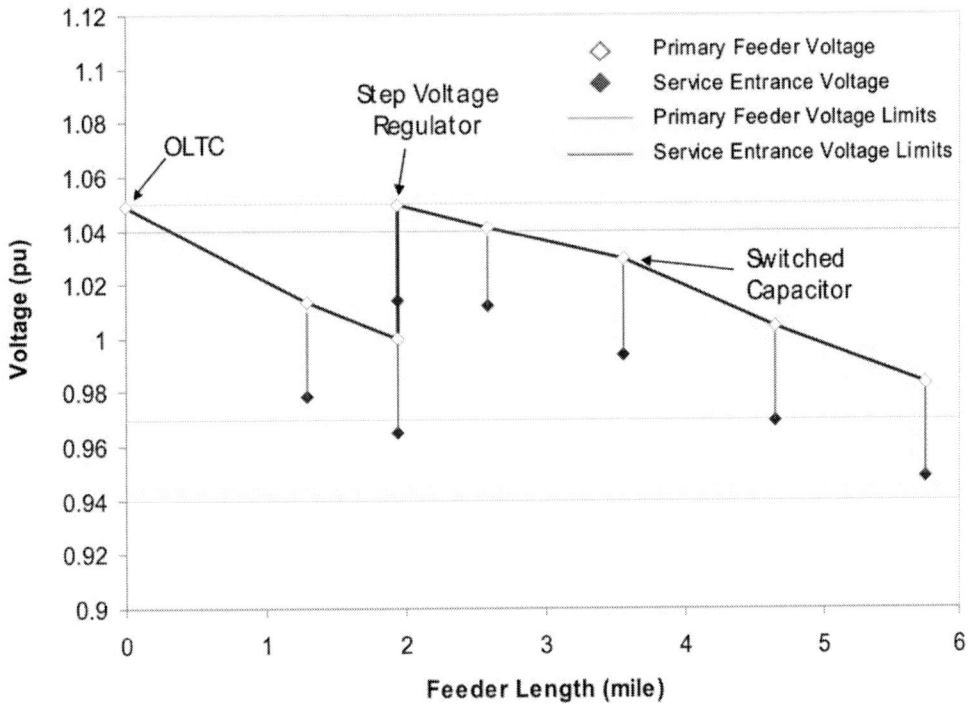

Figure 10. 5% PV penetration voltage profile: peak load with SVR.

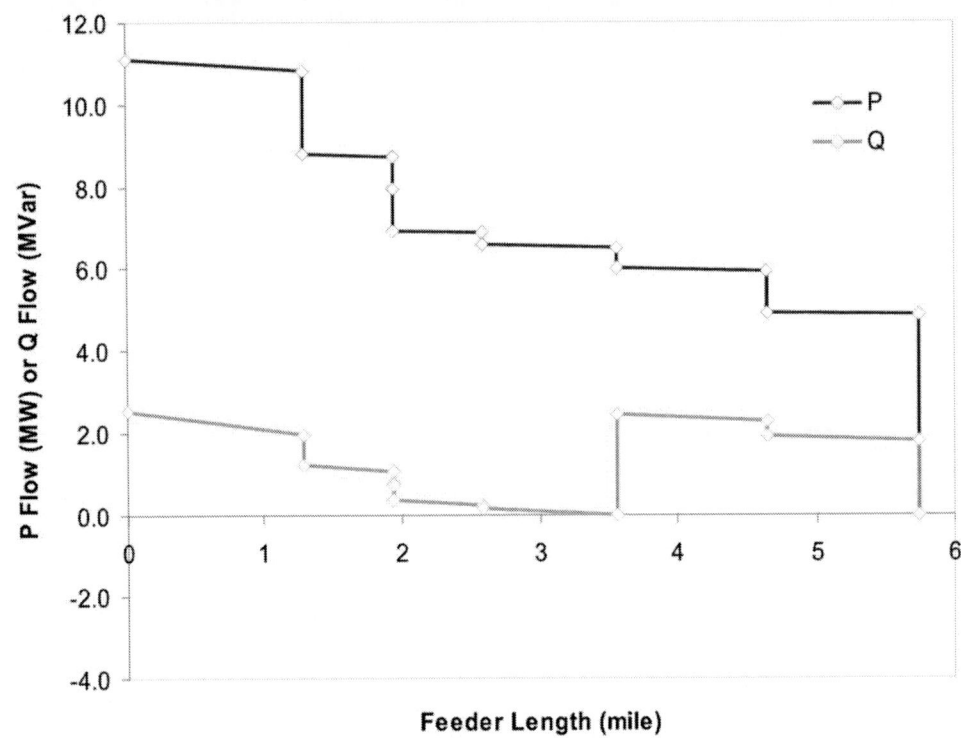

Figure 11. 5% PV penetration power flow: peak load with SVR.

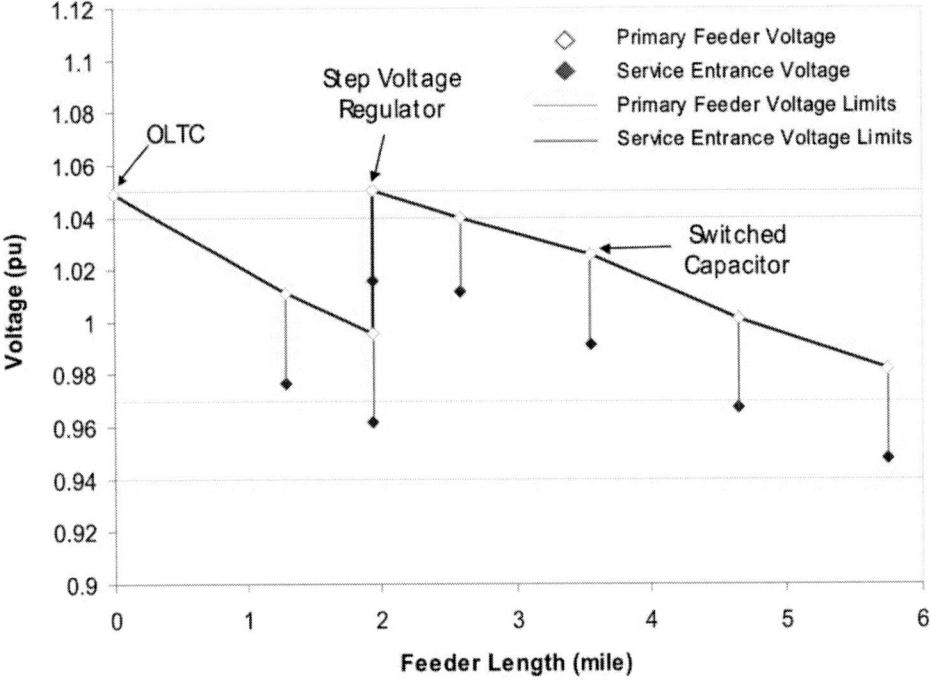

Figure 12. 10% PV penetration voltage profile: peak load with SVR.

The performance is similar to the performance discussed for baseline configuration 2, so only the differences will be highlighted. Notice that the amount of required reactive compensation by the switched capacitor progressively reduces as the penetration level increases. At 30% penetration, reactive power injected by the switched capacitor almost completely diminishes, and none is necessary for 50% PV penetration. The amount of reactive power supplied by the substation is also progressively lower with increased levels of PV penetration; this results in lower current through the feeder and, consequently, in lower I^2R and I^2X power losses. Compared with baseline configuration 2, losses are reduced as summarized in Table 6. Note that since the inverter losses were neglected, these results are somewhat optimistic. A more detailed evaluation of losses will be a topic of future research.

Table 6. Reduction in Feeder Losses Due to Inverter Q Support Relative to Baseline Configuration 2

PV penetration [%]	Loss reduction relative to baseline configuration 2 [%]
5	2
10	2.5
30	3.5
50	7

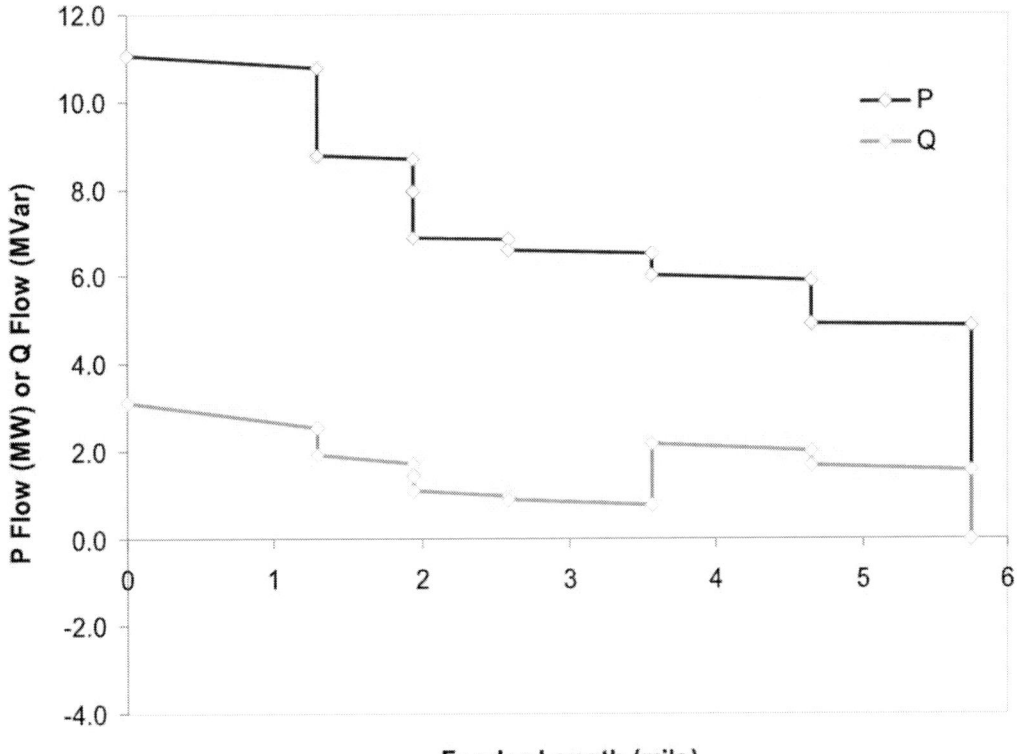

Figure 13. 10% PV penetration power flow: peak load with SVR.

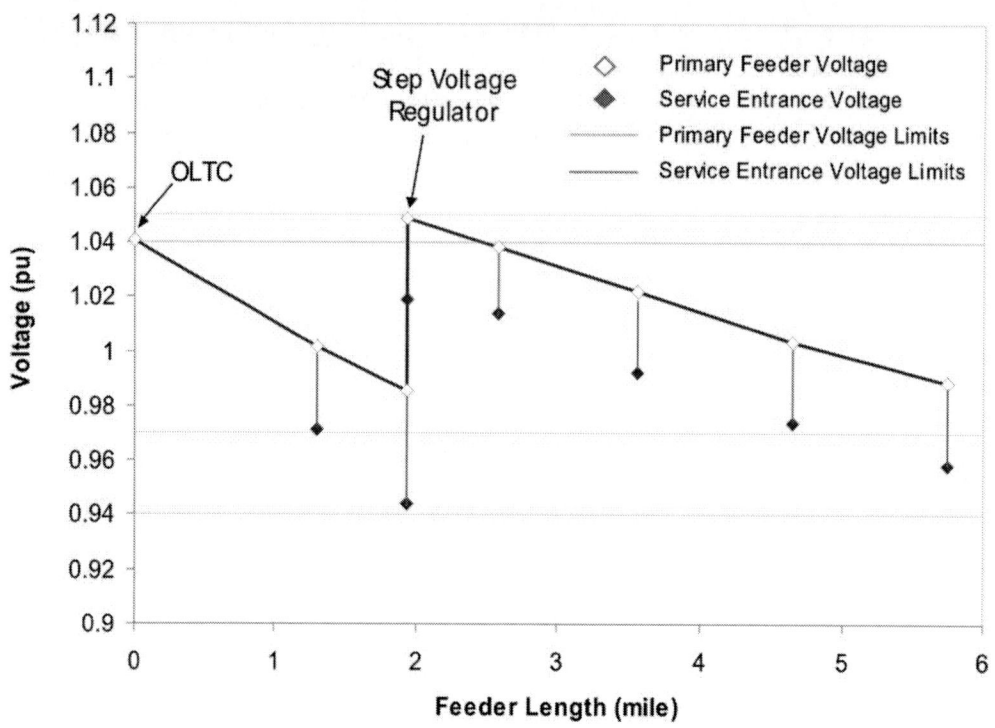

Figure 14. 30% PV penetration voltage profile: peak load with SVR.

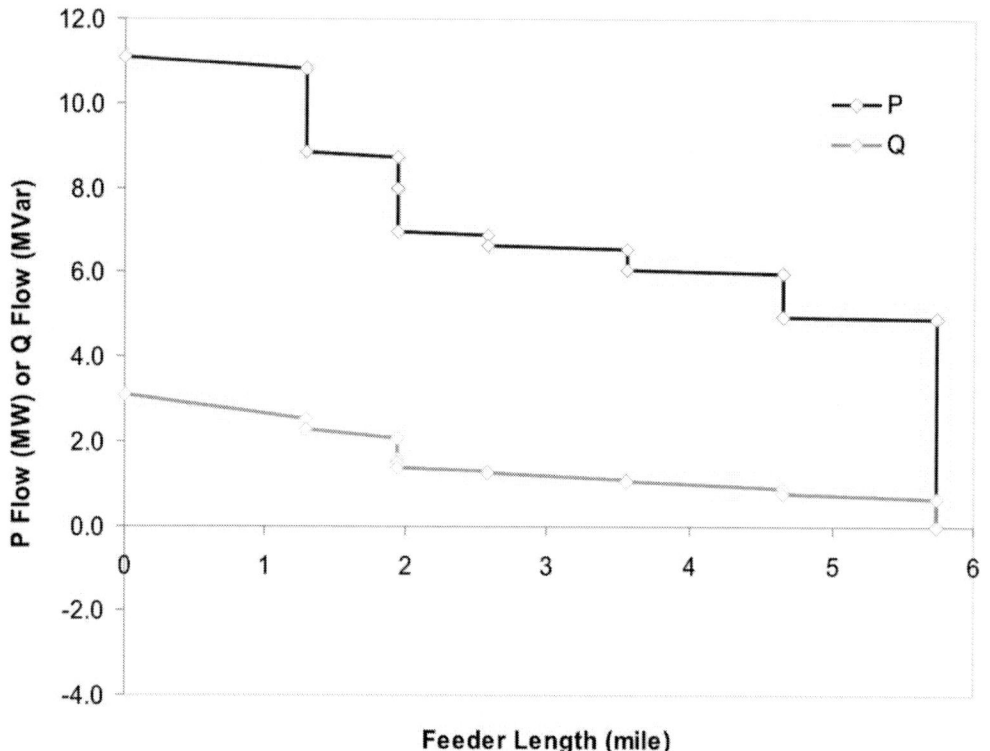

Figure 15. 30% PV penetration power flow: peak load with SVR.

4.3.3 Peak Load, 50% Penetration, OLTC, Inverters Supplying Reactive Power

Similar performance is observed when the inverters are used to supply reactive power in the circuit of baseline configuration 1. For brevity, only the 50% penetration case is presented. Here too, the reactive power flow through the feeder is reduced significantly, minimizing power losses on the feeder. In this case, the primary circuit voltage is slightly below the limit at the last load – this is not important since the service voltage is still maintained within the desired range, and the primary circuit limits are only a guideline.

The voltage profile on the feeder is show in Figure 18 and the corresponding P and Q flows are shown in Figure 19.

4.3.4 Power Export, 50% Penetration, OLTC, IEEE 1547 Inverters

This is the second part of the case study, and deals with power export through the feeder. Work begins with an inverter that is compliant with IEEE 1547, and combined with an OLTC that is controlling the voltage at the service connection of the last load. The voltage profile is shown in Figure 20, and the corresponding active and reactive power flows are shown in Figure 21.

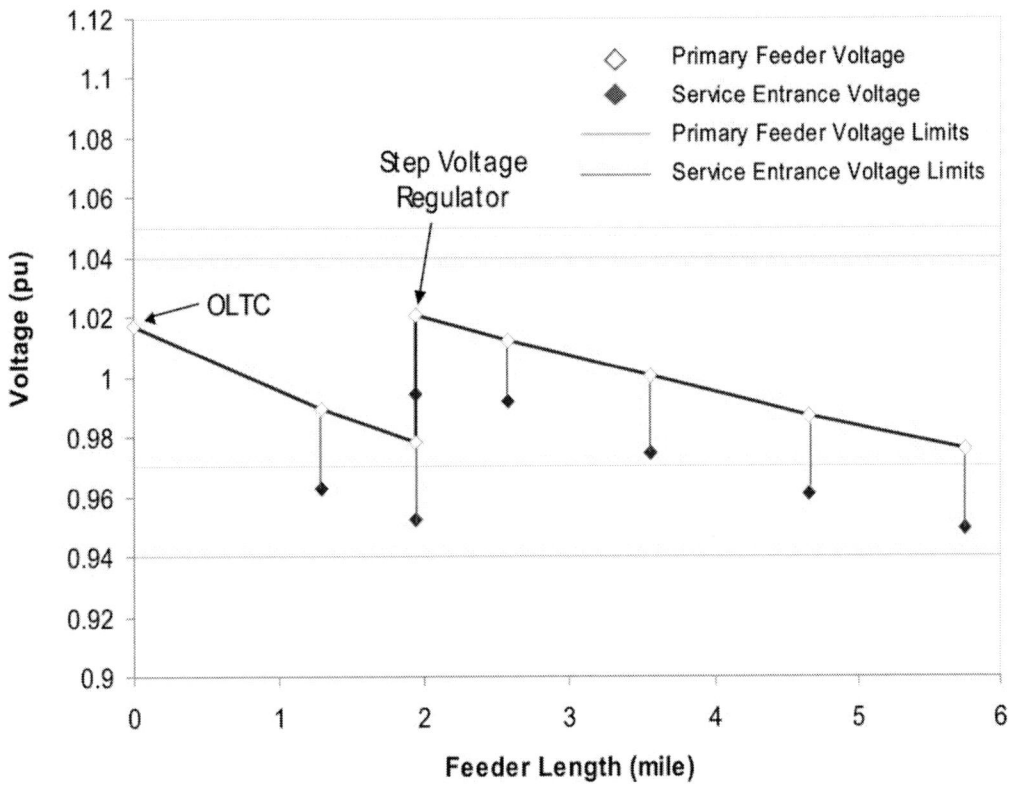

Figure 16. 50% PV penetration voltage profile: peak load with SV.

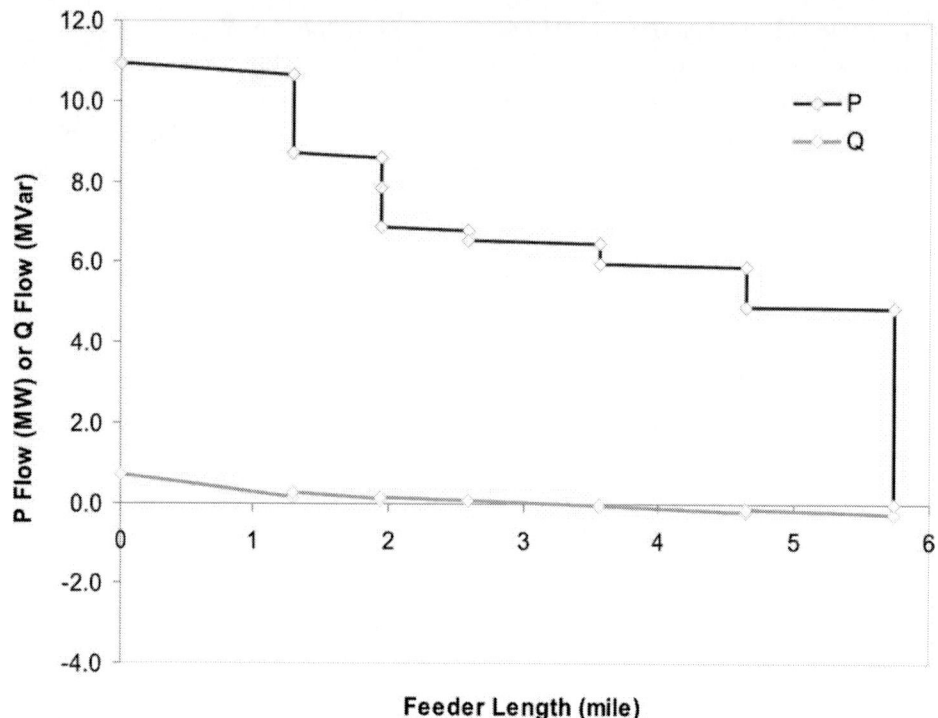

Figure 17. 50% PV penetration power flow: peak load with SVR.

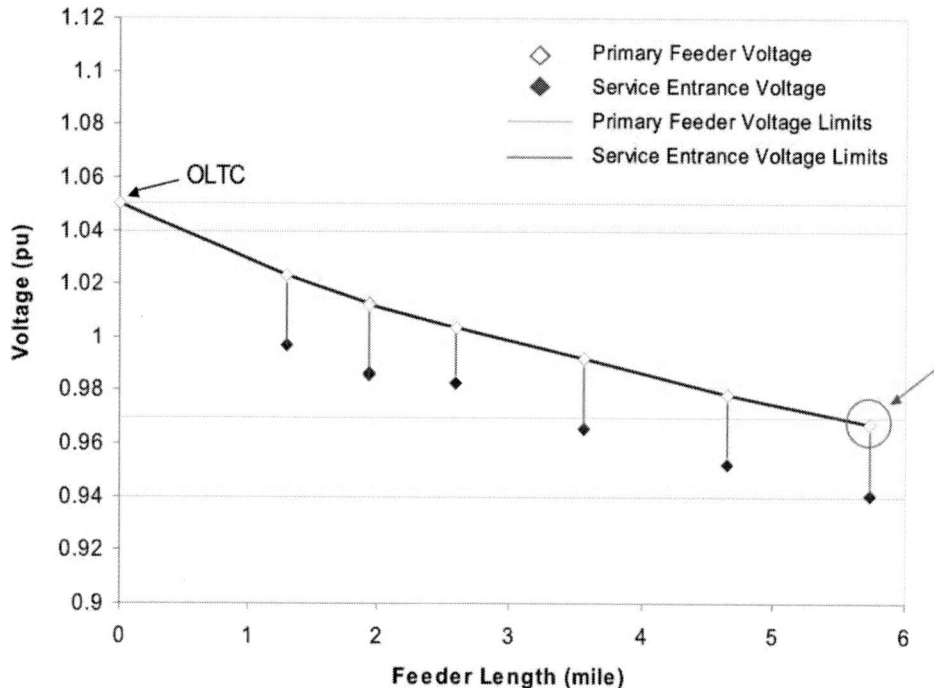

Figure 18. 50% PV penetration voltage profile: peak load with inverter only.

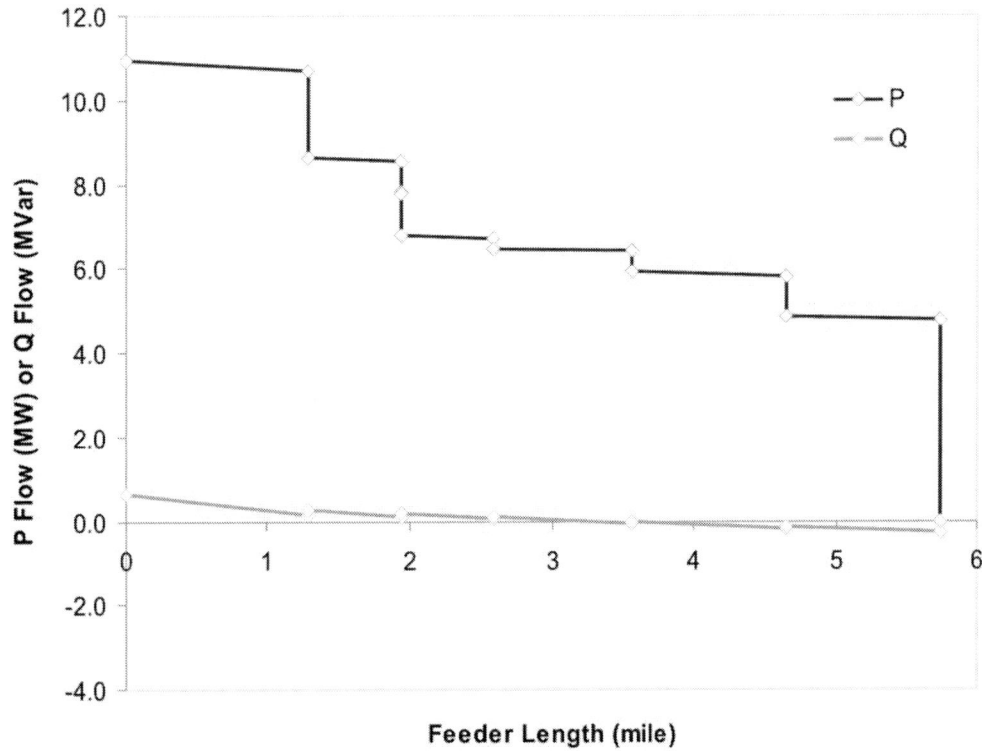

Figure 19. 50% PV penetration power flow: peak load with inverters only.

Relative to baseline configuration 1, a switched capacitor is not used, and the OLTC lowers the voltage at the substation to allow for the rise due to reverse power flow. As expected, voltage rises with increased distance from the substation, since the power flow is now towards the substation not away from it. This negative power flow is also indicated in Figure 21. Starting from the feeder's end, power flow is progressively more negative as generation is collected by distributed PV sources. Notice that the slope of the power curve between the nodes continues to be negative, because losses are still in the "same direction," as in the baseline cases.

The resulting Q flow is relatively low; Q supplied from the substation end covers only feeder Q losses, since the inverters operate at unity power factor.

This is a simple and effective measure, and the performance of the feeder is acceptable.

4.3.5 Power Export, 50% Penetration, OLTC, Inverters Controlling Feeder Voltage

In the following cases, the role of inverters with capabilities beyond the IEEE 1547 requirements is reviewed and compared with cases studied in previous sections.

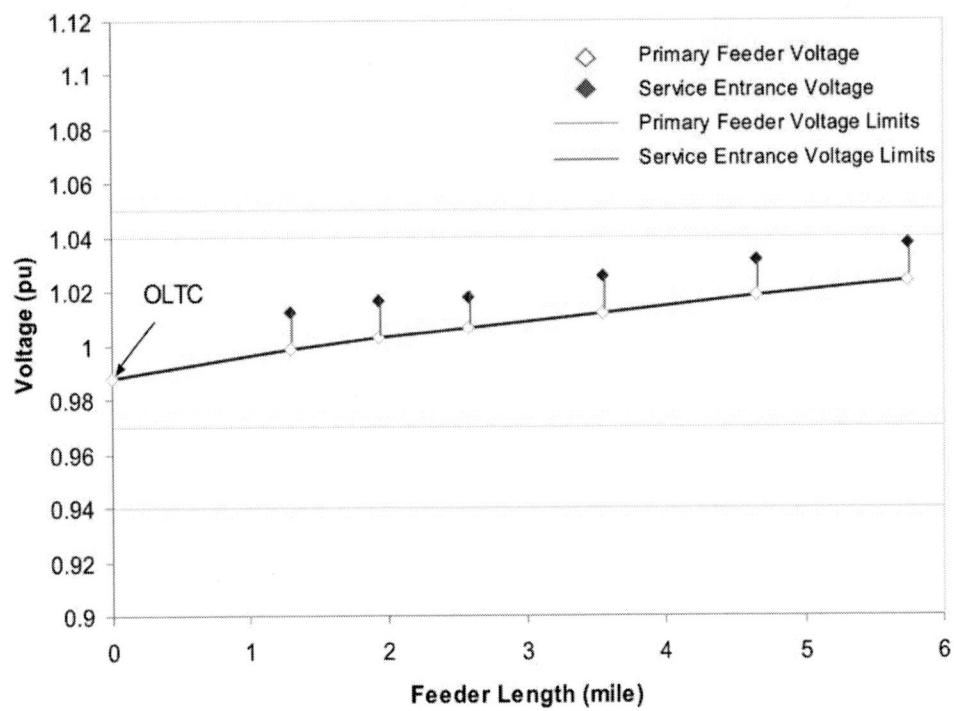

Figure 20. 50% PV penetration: voltage profile at max reverse power flow with OLTC.

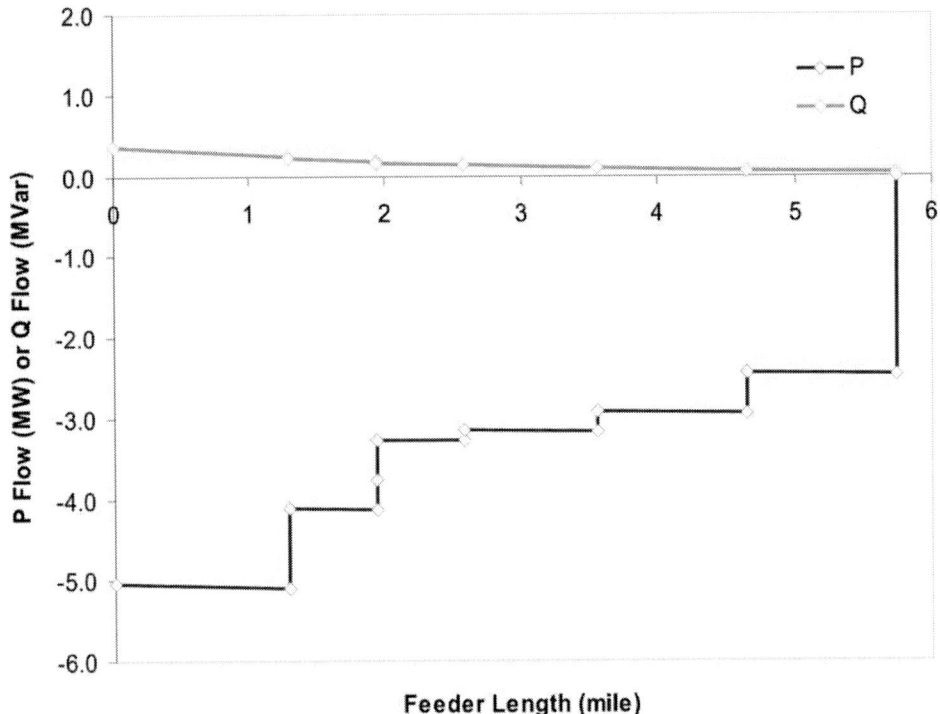

Figure 21. 50% PV penetration: power flow at max reverse power flow with OLTC.

If the requirement for the unity power factor operation of inverters were removed, the instinctive reaction would be to allow inverters to control the voltage at their terminals. This case is representative of such a strategy. The voltage profile is shown in Figure 22 and the active and reactive power flow is shown in Figure 23. The resulting voltage profile is nearly flat, as was desired, and active power flow is as it was in the previous case. Reactive power flow is, however, interesting relative to the previous case. The inverters are trying to reduce the voltage, and they accomplish this by absorbing reactive power, i.e., by operating as inductors. The resulting inductive loading of the feeder is substantial, as indicated by the approximately 1.5 MVAr supplied from the substation.

This is detrimental in two ways: first, feeder losses are increased due to unnecessary reactive power flow, and second, the reactive power demand on the transmission system is increased because of the use of distributed PV. Most traditional generators are based on synchronous machines, and they normally supply reactive power to the system, and do not absorb it. It would be beneficial to change the control strategy for the inverter and combine it with feeder voltage control to allow for net export of reactive power. This is discussed in the next section.

4.3.6 Power Export, 50% Penetration, OLTC, Inverters Supplying Capacitive Reactive Power

In this case, the control of the inverters is combined with the control of the OLTC in order to allow for Q export from the feeder.

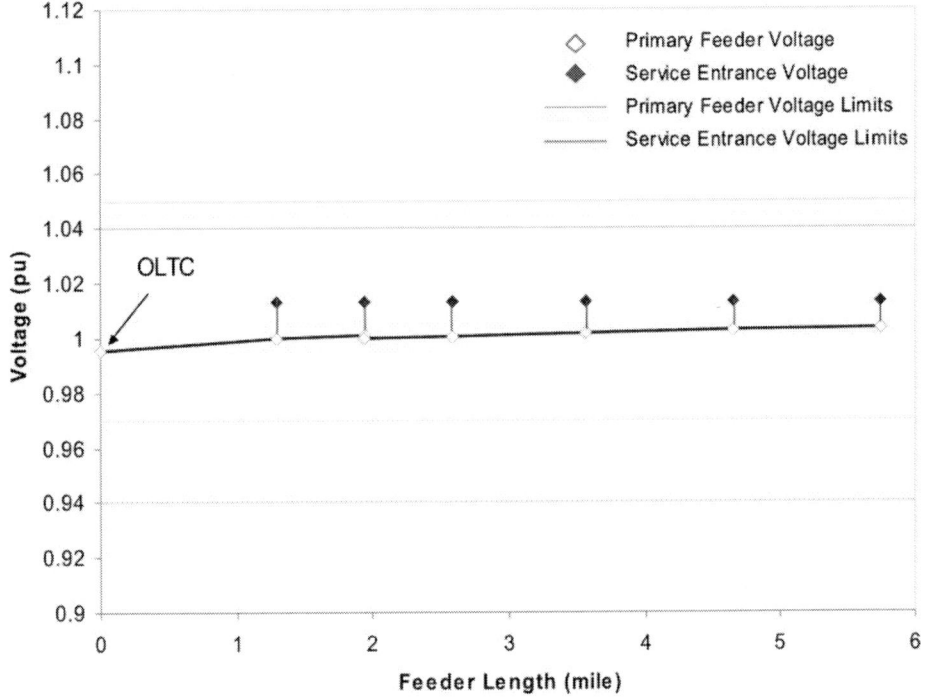

Figure 22. 50% PV penetration voltage profile: inverter voltage support.

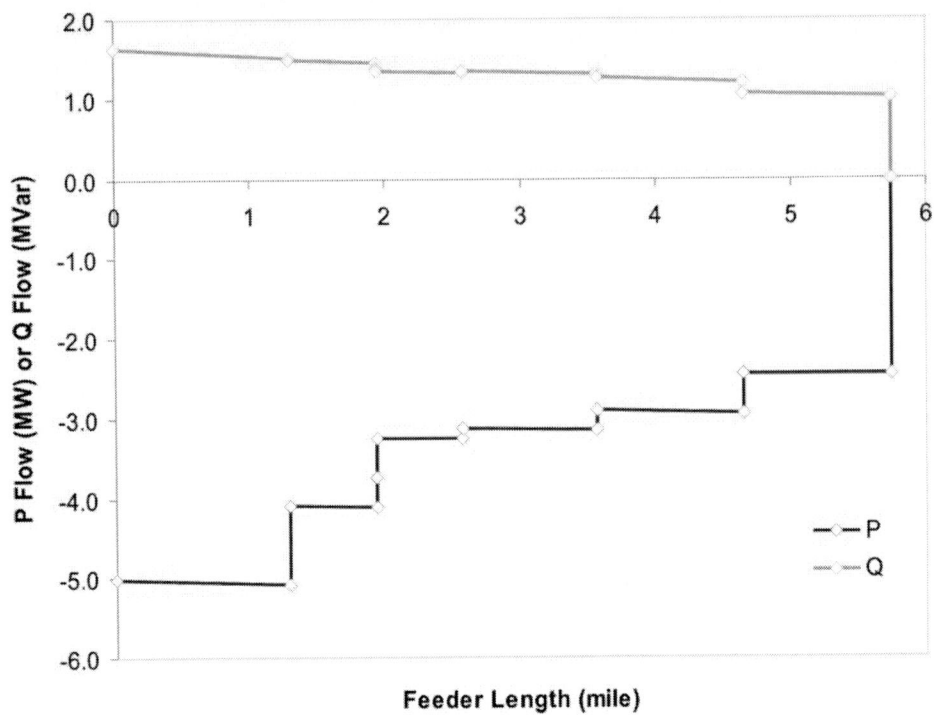

Figure 23. 50% PV penetration power flow: inverter voltage support.

Distributed PV generation is envisioned to displace some of the conventional generators in the near future. An important feature of conventional generators is their ability to provide reactive power to the system. Conventional generators can generate or absorb reactive power within the limits of the exciter, and they are normally configured to supply reactive power. The PV inverters have a similar reactive power capability. PV inverters generally have analogous capabilities, but since they are connected to the distribution system their operation has to be coordinated with the feeder voltage control in order to allow net reactive power export.

This case illustrates a control strategy that allows net export; the OLTC now reduces the output of the substation to the lower limit and the inverters raise the service voltage to the upper limit. The resulting voltage profile is shown in Figure 24, and the active and reactive power flows are shown in Figure 25.

As shown in Figure 25, net Q export is accomplished, but it is hampered by the voltage constraints on the feeder. Specifically, the Q capability of the equivalent inverter at the last node is above 1 MVAr and it supplies less than 0.5 MVAr. This is a consequence of voltage limitations and feeder topology, and given the availability of the voltage control equipment, nothing can be done to increase Q supply from the last node. An important conclusion is that over-sizing the inverters to provide Q support does not always help, as it can be restrained by the inverter placement on the feeder and available voltage control equipment.

In the next case, a way to remedy this limitation is illustrated. SVR is used to increase Q flow through the feeder.

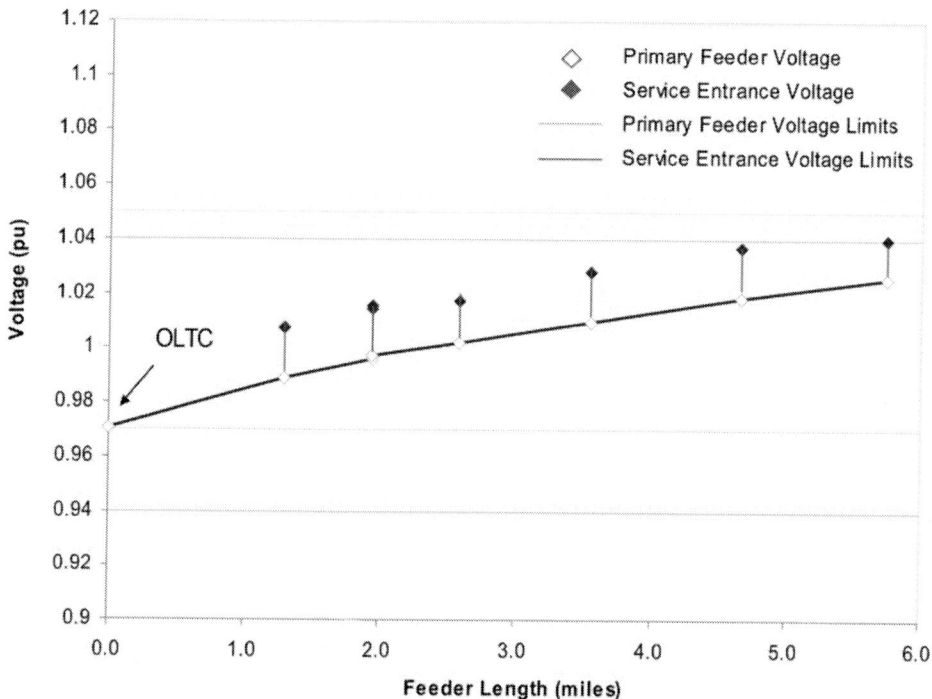

Figure 24. 50% PV penetration voltage profile: inverter VAR support.

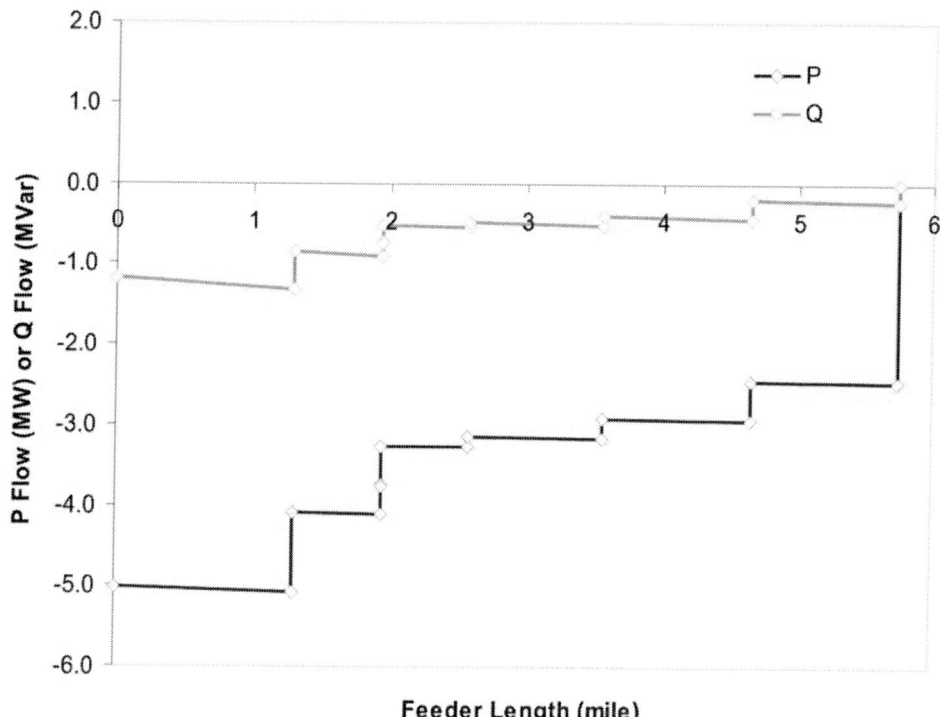

Figure 25. 50% PV penetration power flow: inverter VAR support.

4.3.7. Power Export, 50% Penetration, OLTC + SVR, Inverters Supplying Capacitive Reactive Power

In this case, the coordinated action of the OLTC, the SVR, and the inverters is used to maximize Q export from the feeder and use up all the available Q capacity from the inverters. The resulting voltage profile is shown in Figure 26, and the corresponding P and Q flows are shown in Figure 27.

The Q export to the feeder is now maximized. It is slightly over 2 MVAr at the substation, and the feeder and service voltages still remain within the desired limits. Conceptually, this is a simple modification of the circuit, and it can be generalized to maximize export of Q on any feeder regardless of length. Using such a control strategy enables the full utilization of the inverters' Q capabilities, and enables distributed PV to displace conventional generation without compromising performance.

4.3.8. Power Export, 50% Penetration, OLTC, Inverters Controlling Total Service Power Factor

Finally, the inverters are used to control the power factor at their total service connection, making the connection slightly capacitive in order to cover for the reactive drop through the service transformer and the corresponding feeder impedance.

Figure 28 and Figure 29 show the resulting voltage profile and the P Q flow of the study feeder. This case demonstrates that the inverters can achieve approximately unity PF at the feeder connection to the substation – a slightly better performance relative to the IEEE 1547 case, where reactive power flow was supplied from the substation.

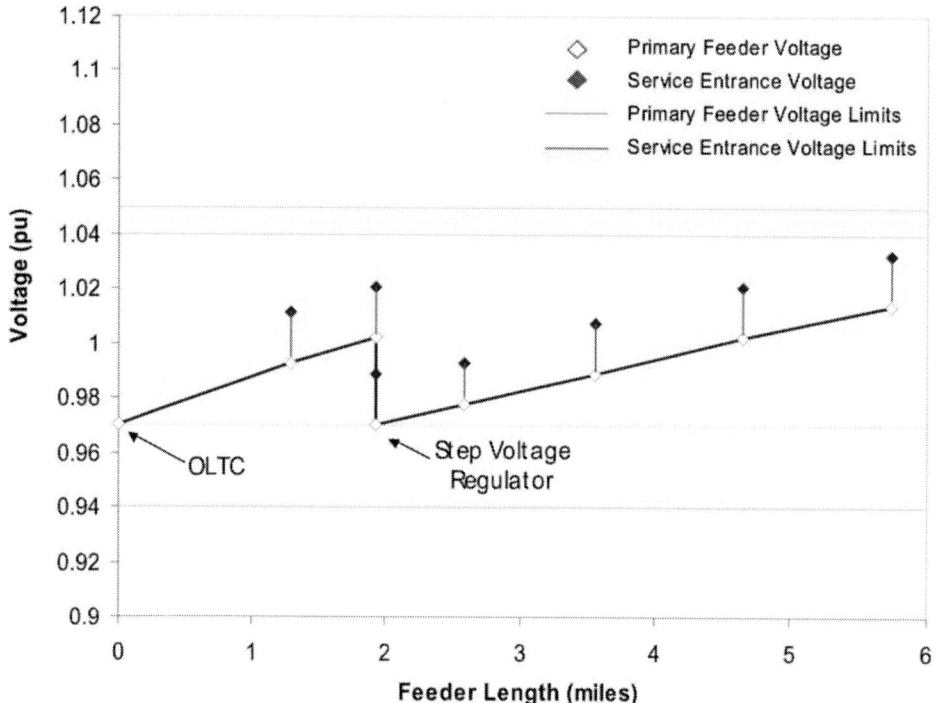

Figure 26. 50% PV penetration voltage profile: inverter VAR support with SVR.

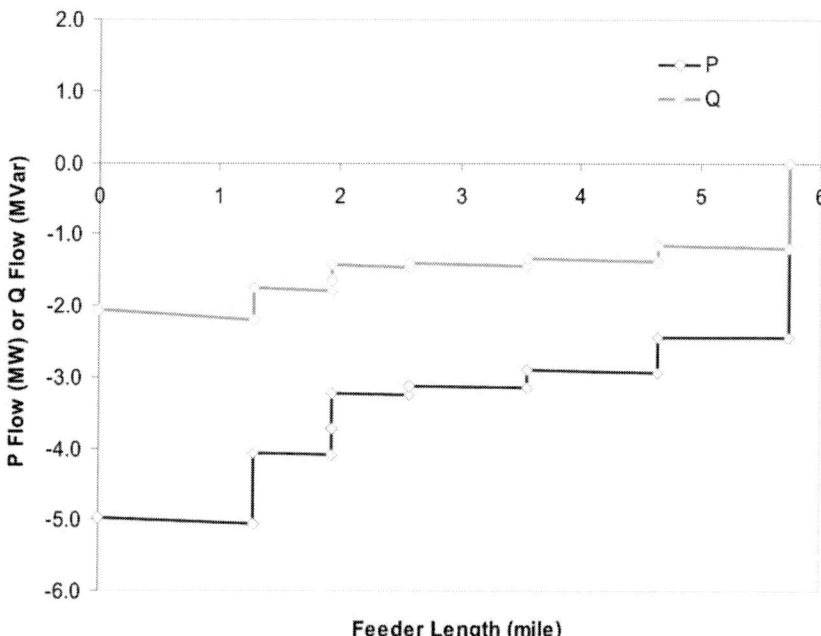

Figure 27. 50% PV penetration power flow: inverter VAR support with SVR.

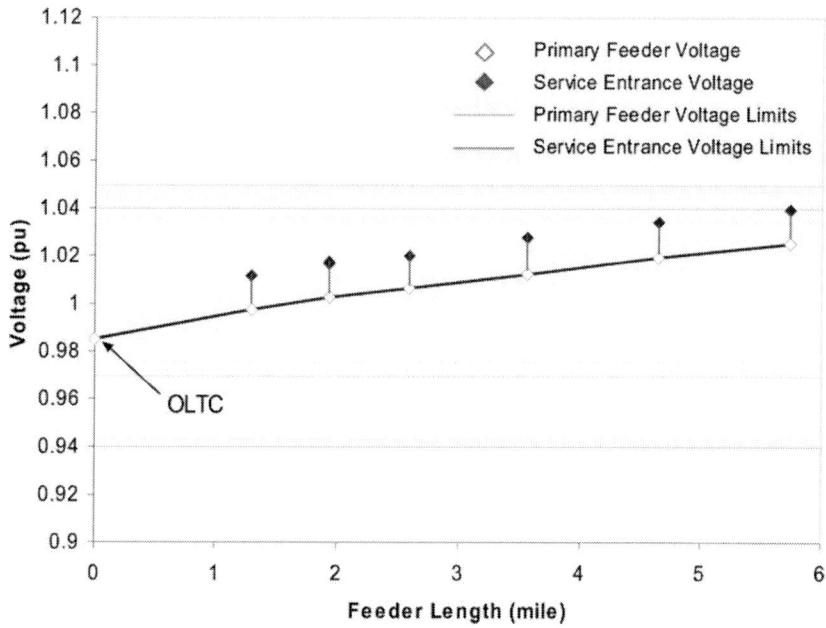

Figure 28. 50% PV penetration voltage profile: inverter power factor correction.

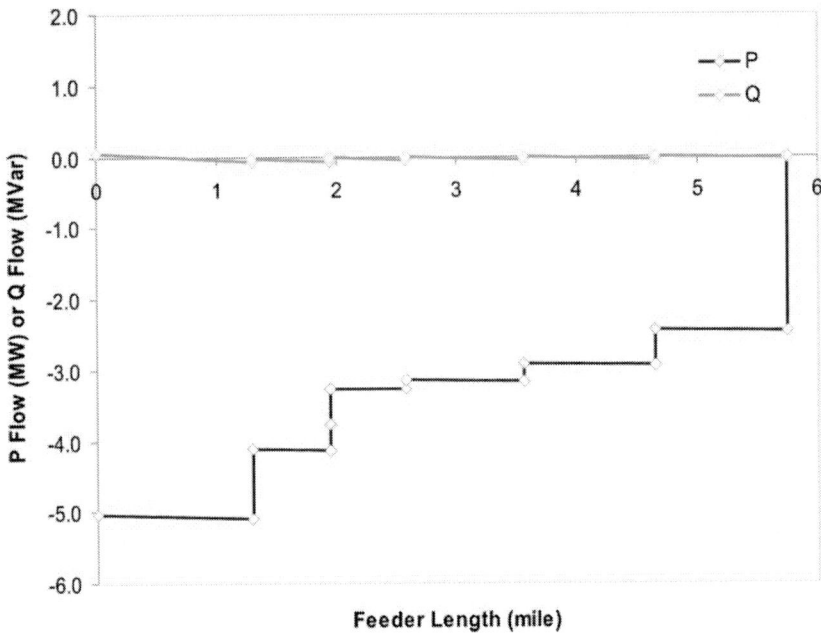

Figure 29. 50% PV penetration power flow: inverter power factor correction.

Another very important characteristic of this control strategy, relative to most of the others presented, is that it can be implemented using local measurements only. To illustrate some of the control schemes, voltage magnitude was used at the service entrance of the last load to control the OLTC. Such setup requires a point-to-point communication link, or if a more general setup is desired, a communication link should exist between the OLTC and all of the service voltage connections to allow for control based on the worst-case voltage. Much work is needed to come up with an optimal communication infrastructure to ensure effectiveness without excessive costs.

CONCLUSIONS AND RECOMMENDATIONS

In this study, a representative distribution feeder was selected for analyzing the impact of different penetrations of PV on the feeder voltage regulation. Commonly used voltage regulation equipment, such as a step voltage regulator and a switched capacitor, were applied individually and in combinations to create different feeder configurations. Two extreme scenarios, peak load and maximum reverse power flow, were simulated in order to study the feeder's voltage performance under different conditions.

A series of case studies was conducted with different penetrations of PV, each assuming several commonly used voltage regulation methods. The study results show:

- Reverse power flow at all studied PV penetration levels can be accommodated using traditional utility equipment with perhaps modified controls.

- Voltage rise on the secondary circuits is significant and it should be included in the analysis. Establishing a communication link between service points (customer meter connections) and the utility equipment is helpful as it enables explicit control over the worst-case voltage.

In all the cases studied, PV inverters can positively contribute to the feeder voltage regulation and result in an improved voltage profile. At a high enough penetration, PV inverters may be able to provide feeder voltage support. The study results show:

- At a low PV penetration level (5%), inverters do not make a significant impact on the feeder's voltage regulation during peak load.
- At medium a PV penetration level (10%), inverter voltage support can help reduce the size of the conventional voltage support capacitors by nearly 40%.
- At high PV penetration levels (30% – 50%), PV inverters might be sufficient to provide all of the feeder voltage support.

As the inverter-coupled PV sources displace conventional generation, they will also have to match most of their performance characteristics. With respect to reactive power supply to the system, PV inverters are disadvantaged because their reactive power injection may be limited by the feeder voltage limits. This can be resolved by coordinated control of utility equipment and inverters, and in some cases additional utility equipment might be needed to take full advantage of the inverters' reactive power capabilities.

Recommendations for future research in this field are:

- Develop a set of recommended practices for reconciling existing feeder voltage control techniques with high penetration of distributed PV. This work illustrates several such cases, but it is limited to one feeder topology. A more comprehensive coverage of feeder topologies would be beneficial.
- Develop a set of recommended practices for modeling PV inverters for load flow analysis and for other relevant planning purposes, such as short circuit current calculations. At the same time, expand the number of possible control options for traditional equipment in the analysis software. This would result in a more consistent understanding of the issues across the industry and would simplify the test case setup to evaluate any specific situation.
- Create a set of benchmark cases to facilitate testing the models and the associated software. Some of the confusion and unwarranted concerns about the impact of PV generation may be a result of inconsistent and incorrect modeling.
- Develop automated screening tools that will enable evaluation of the impact of PV on the distribution system; all prospective installations could then be screened and only the ones requiring more detailed assessment would be floated up to the utilities. This would help preserve low installation costs while allowing for the more detailed assessment that would be necessary at higher penetration levels of distributed PV.
- Develop functional requirements for a communication infrastructure that will enable coordinated operation of all equipment on the distribution feeder. The same

infrastructure can be used to enable demand-side management, the implementation of flexible metering tariffs, and enhanced distribution system management.
- Fund demonstration opportunities that illustrate feeder operation with significant PV penetration.

REFERENCES

[4] *"The Potential of Solar PV in Ontario"*, Rob McMonagle, The Canadian Solar Industries Association, January 30, 2006
[5] *"Electric Power Distribution Handbook"*, Tom Short
[6] *"Power Distribution Planning Reference Book"*, H. Lee Willis
[7] "Reliable, Low Cost Distributed Generator/Utility System Interconnect, 2001 Annual Report", *GE Corporate Research and Development*, August 2003, NREL/SR-560-
[8] "Study of the Impact of PV Generation on Voltage Profile in LV Distribution Networks", S. Conti, et. al., *IEEE Porto Power Tech Conference*, September, 2001, Porto, Portugal
[9] *"IEEE Standard for Interconnecting Distributed Resources with Electric Power Systems"*, IEEE 1547 Standard, 2003.
[10] "IEEE-1547 Comments", *Idaho Power Company*, Oregon PUC Technical Workshop, Oct. 2006, http://www.puc.state
[11] "Voltage regulation - Tapping Distributed Energy Resources", John D. Kueck, et. al., Public Utilities Fortnightly, Sept. 2004
[12] *"Dynamic Voltage Regulation Using Distributed Energy Resources"*, Yan Xu, et. al., 19th International Conference on Electrical Distribution (CIRED), May 07
[13] "PSLF - Power System Analysis Software", http://www.gepower.com/prod_serv/products/utility_software/en/downloads/pslf05.pdf

GLOSSARY

PV	Photovoltaic	
ANSI	American National Standard Institution	
TC	Tap Changer	
SVR	Step Voltage Regulator	
OLTC	On-Load Tap Changer	
VAR	Reactive Power	
S	Apparent Power (also known as Complex Power)	
P	Active Power (also known as Real Power)	
Q	Reactive Power (also known as Imaginary Power)	

\multicolumn{2}{REPORT DOCUMENTATION PAGE}		Form Approved OMB No. 0704-0188

REPORT DOCUMENTATION PAGE	Form Approved OMB No. 0704-0188
The public reporting burden for this collection of information is estimated to average 1 hour per response, including the time for reviewing instructions, searching existing data sources, gathering and maintaining the data needed, and completing and reviewing the collection of information. Send comments regarding this burden estimate or any other aspect of this collection of information, including suggestions for reducing the burden, to Department of Defense, Executive Services and Communications Directorate (0704-0188). Respondents should be aware that notwithstanding any other provision of law, no person shall be subject to any penalty for failing to comply with a collection of information if it does not display a currently valid OMB control number. **PLEASE DO NOT RETURN YOUR FORM TO THE ABOVE ORGANIZATION.**	

1. REPORT DATE *(DD-MM-YYYY)* February 2008	2. REPORT TYPE Subcontract report	3. DATES COVERED *(From - To)*
4. TITLE AND SUBTITLE Distribution System Voltage Performance Analysis for High-Penetration Photovoltaics		5a. CONTRACT NUMBER DE-AC36-99-GO10337
		5b. GRANT NUMBER
		5c. PROGRAM ELEMENT NUMBER
6. AUTHOR(S) E. Liu and J. Bebic		5d. PROJECT NUMBER NREL/SR-581-42298
		5e. TASK NUMBER PVB7.6401
		5f. WORK UNIT NUMBER
7. PERFORMING ORGANIZATION NAME(S) AND ADDRESS(ES) GE Global Research 1 Research Circle Niskayuna, NY 12309		8. PERFORMING ORGANIZATION REPORT NUMBER ADC-7-77032-01
9. SPONSORING/MONITORING AGENCY NAME(S) AND ADDRESS(ES) National Renewable Energy Laboratory 1617 Cole Blvd. Golden, CO 80401-3393		10. SPONSOR/MONITOR'S ACRONYM(S) NREL
		11. SPONSORING/MONITORING AGENCY REPORT NUMBER NREL/SR-581-42298
12. DISTRIBUTION AVAILABILITY STATEMENT National Technical Information Service U.S. Department of Commerce 5285 Port Royal Road Springfield, VA 22161		
13. SUPPLEMENTARY NOTES NREL Technical Monitor: Ben Kroposki		
14. ABSTRACT *(Maximum 200 Words)* This report examines the performance of commonly used distribution voltage regulation methods under reverse power flow.		
15. SUBJECT TERMS distribution system; photovoltaics; PV; voltage regulation; inverters; renewable systems interconnection; GE Global Research; National Renewable Energy Laboratory; NREL		

16. SECURITY CLASSIFICATION OF:			17. LIMITATION OF ABSTRACT	18. NUMBER OF PAGES	19a. NAME OF RESPONSIBLE PERSON
a. REPORT Unclassified	b. ABSTRACT Unclassified	c. THIS PAGE Unclassified	UL		19b. TELEPHONE NUMBER *(Include area code)*

Standard Form 298 (Rev. 8/98)
Prescribed by ANSI Std. Z39.18

End Notes

[*] Larger conductors have lower voltage drop and lower power losses, but they also cost more, so there is a tradeoff between savings due to lowered losses and increased costs of the conductors. Utilities commonly have design guidelines that are based on underlying economic consideration, practicality of carrying varying sizes of conductors in stock, and other considerations such as feeder pick up from an adjacent substation for increased reliability.

[†] IEEE 1547 Standard for Interconnecting Distributed Resources with Electric Power Systems, 2003.

[‡] Some inverters are highly optimized for efficiency, and are not capable of Q injection, but this beyond the scope of this discussion.

In: Renewable Energy Grid Integration…
Editor: Mitchell B. Ferguson

ISBN: 978-1-60741-325-7
© 2011 Nova Science Publishers, Inc.

Chapter 4

POWER SYSTEM PLANNING: EMERGING PRACTICES SUITABLE FOR EVALUATING THE IMPACT OF HIGH-PENETRATION PHOTOVOLTAICS

J. Bebic

ACKNOWLEDGMENTS

The help and support from Power Systems Energy Consulting of GE Energy are greatly appreciated. Nicholas Miller provided insights into transmission system planning and operating practices; Gary Jordan helped address the interaction of high-penetration PV with generation planning, production scheduling, and power markets; and Reigh Walling helped review the impact of high penetration of solar PV on the distribution system. It has been a pleasure and a source of inspiration to work with these experts.

EXECUTIVE SUMMARY

This report explores the impact of high-penetration renewable generation on electric power system planning methodologies, and outlines how these methodologies are evolving to enable effective integration of variable-output renewable generation sources. All three areas of system planning are considered—generation, transmission, and distribution—and the impact of high penetration of solar PV analyzed relative to each.

Generation planning is shifting from planning for peak load towards planning for system energy. This shift is centered on using net load as a basis for capacity planning and this creates a set of requirements for reliable and comprehensive renewable resource data. Furthermore, a new dimension is being introduced into generation planning—the need for explicit evaluation of generation flexibility relative to the variability of net load at the time scale of load following. Increased penetration of intermittent renewable generation means that the operational flexibility of the balance of generation portfolio will become strategically important—the lack of flexibility inevitably will result in curtailment of renewable generation. To avoid this, more flexibility must be provided. Such flexibility can be achieved in three essential ways: balancing the generation portfolio, load control, and energy storage. This process can be accelerated by targeted R&D investment, and by creation of efficient markets to address future load-following needs. Quantifying the variability to determine required flexibility also requires correlated historic load and resource data at the time scales that currently are not being collected. Integration of renewable-resource data into generation planning is an important area of future work.

Transmission planning practices can readily include renewable generation, but significant effort is required to develop models that adequately represent distributed solar PV generation at the time scales of interest for transmission planning. Standardized modeling guidelines and test cases are required to facilitate harmonization of various software tools, and to prevent confusion and unwarranted concerns that will arise as a result of inconsistent—and possibly inaccurate—modeling.

Distribution planning and engineering practices already incorporate processes that allow connection of distributed generation. These processes were developed for integrating co-generation and are not optimized for integration of small, distributed sources of power, such as solar PV. Currently, this results in unnecessary administrative and engineering hurdles that could be eliminated by dedicated, comprehensive, and coordinated treatment of solar PV installation in all relevant codes and standards. Remaining technical hurdles are possible to predict through careful analysis (simulation), but the analysis software should be harmonized with respect to the representation of PV inverters, the impact inverters have on feeder voltage, and their contribution to fault currents. Developing a set of test cases and modeling guidelines to enable benchmarking the software and the models can accelerate this process. Funding field installations and showcasing simple and effective solutions also would help build confidence within the industry.

When the penetration levels increase to a point of becoming a significant source of energy in the electric power system, communications links between system control centers and distributed PV sources will be helpful—and even necessary. This communication infrastructure, assuming that it has ample capacity, can be leveraged in many different ways. It also would create opportunities for more-effective distribution system management, more-flexible system configurations, faster restoration, more- selective protection, efficient deployment of demand response, efficient implementation of real-time metering, introduction of flexible and creative electricity tariffs, and likely for many other applications. The main impediments to deployment of this communication infrastructure are its significant cost and the uncertainty that the benefits it provides will justify the required capital investment. With high-speed communications already bringing Internet service to many homes, sizable pilot programs relying on Internet infrastructure can be created to help evaluate the benefits and to guide design decisions for creating a dedicated infrastructure.

1.0. INTRODUCTION

Recent cost reductions and the increases in production of solar photovoltaics (PV) are driving dramatic growth in domestic PV system installations. Programs such as Solar

America Initiative are setting out to make solar energy cost-competitive with central generation by the year 2015. As the costs decline, distributed PV becomes an increasingly significant source of power generation and, at some point, its further growth might be limited by the challenges of its integration into the power grid.

To prevent these integration challenges from limiting the growth of solar PV installations and to maximize the overall system benefit, it is necessary to consider solar PV in all areas of power system planning, and to evolve the planning practices to better accommodate increased energy supply from solar PV.

This report reviews the entire power system planning process, including generation, transmission, and distribution. It discusses how the planning practices are changing to accommodate variable renewable generation, with a focus on future changes required to accommodate high penetration levels of solar PV and how to maximize the positive impact of other technologies such as load control and energy storage. The report also proposes several areas for future research that will help evolve planning methodologies and enable easier and more-effective integration of solar PV.

Electricity produced by solar PV currently is not cost-competitive with electricity generated by central stations, consequently solar PV has limited penetration in grid-connected applications. As the technology develops and solar PV becomes more competitive, it is expected that it will start supplying residential and commercial loads at the customer's side of the meter. This area of the power system has the highest cost of electricity, therefore it is where cost-competitiveness will be achieved first.

It is due to this assumption that solar PV commonly is regarded as a form of distributed generation and is being developed in accord with codes and standards that govern distributed generation, such as IEEE 1547[1], and UL 1741.[2] These are modern standards (in active development) and as such they provide ample support and guidance for current and near-term applications of distributed solar PV. The standards, however, are being developed on the important implicit assumption of low total penetration of distributed generation. Essentially, the envisioned purpose of distributed generation is to offset the consumption of its adjacent load, and it is not expected to ship much power back to the system. In contrast, this work is centered on the assumption of high penetration of distributed solar PV, and on analyzing what impact such a development will have on the power system.

Understandably, a sharp increase in the use of any one source of generation is likely to present integration challenges, but this especially is the case with the distributed solar PV for the following reasons.

- Solar PV is a variable source of generation—its power output depends on insolation and it is subject to potentially abrupt changes due to cloud coverage.
- Solar PV will evolve as a distributed source of generation first used to offset the connected load. As the penetration levels increase even further, two options are possible. Energy storage could be used to ensure that no power is returned to the system, and the power could be sent to other loads in the system to avoid capital investment for dedicated storage. The second option necessitates shipping power "backwards" through a part of the electricity delivery network—the distribution system—and backwards power flow is not a design feature of present-day distribution systems.

- The codes and standards that guide the integration of solar PV are focused on simplifying installations and prescribe grid interconnection requirements that cause minimal interaction with the grid. When solar PV becomes a significant overall source of generation in the power system, some of the present interconnection requirements likely will be counterproductive.

These challenges will be best addressed by concerted efforts of power utilities and solar PV technology developers, and will greatly benefit from carefully designed incentives and policies. Furthermore, continued collaboration of industry, utilities, and government in developing and evolving relevant codes and standards is seen as a key factor in ensuring graceful developments in this field.

In the field of power systems, *planning* is the activity with the most strategic impact, and it is a key to enabling adoption of any new technology. It therefore is of utmost importance to ensure that planning practices are ready to consider the new technology, and that such consideration is as convenient as possible.

2.0. TRADITIONAL PRACTICES IN POWER SYSTEM PLANNING

Traditional electric power systems are designed on the premise of power production in central generating stations and its delivery to the points of end use via transmission and distribution systems.

The role of generating stations is clear—they produce electric power or, more precisely, convert energy from another source into electric energy. The roles of transmission and distribution systems are more interrelated; both are concerned with power delivery, so additional clarification might be helpful. The role of transmission systems is to interconnect many generators and loads across entire regions and over state and country boundaries. Transmission systems enable the transfer of power over long distances, and thus facilitate economic and system benefits. They are designed and operated to optimize the use of the generation portfolio. They make it possible to supply loads from the most economical sources of power and to operate generating stations flexibly, allowing for optimization of their maintenance schedules and improved overall system reliability. Conversely, distribution systems are the part of electric delivery infrastructure that brings the power to the loads; they "touch" the load. The interface point between the transmission and a distribution system is a (distribution) substation. A distribution system usually includes the substation and all other infrastructure between the substation and the load, including primary circuits (feeders and laterals), service transformers, secondary circuits, and customers' meters. Generally, distribution systems are designed for unidirectional power flow from the substation to end-use loads, and it is implicitly assumed that there is a sufficient supply of power from the transmission system (at the high-voltage side of the substation).

Traditional system planning activities follow this functional division, and commonly are segregated into generation planning, transmission planning, and distribution planning. Traditional planning practices are discussed in more detail in the following sections.

2.1. Generation Planning

The electric power industry is one of the oldest and well-developed industries in the United States. Consequently, all power generation planning is performed in the context of modifications to the existing system. The process begins with electricity load demand forecasting, which is followed by reliability evaluation to determine if and when additional generation is needed. Finally, optimal capacity expansions are selected based on economic considerations. These processes are reviewed briefly in the following sections.

2.1.1. Load Forecasting
Total system load generally is well known and a wealth of historic data is available. In the short term, load can be forecast with great accuracy, and this is performed daily to determine generation units' commitment. Load forecasting for the purpose of generation planning, however, requires a substantially longer time horizon, because system expansion projects require long lead times, often between 2 and 10 years.

The outputs from a load forecast are a forecast of annual energy sales (in kilowatt-hours), and the annual peak demand (in kilowatts). There are two widely used methods in energy sales forecasting: econometric regression analysis, and end-use electricity models.

The usefulness of each method depends on data availability, customer segmentation, and the degree of detail required. Generally, the accuracy of predictions depends on the accuracy of assumptions, and the predictions can't be made with absolute certainty. For more details on econometric regression analysis, interested readers are referred to Pindyck and Rubinfeld (2000). End-use electricity models are physical, engineering- based methods that often are used in forecasting the residential load, and sometimes for commercial and industrial loads. Additional information and literature sources can be found in Stoll (1989).

Forecasting the peak demand is done based on forecasted energy sales by multiplying forecasted energy with an empirically determined load factor coefficient. Peak load is extremely sensitive to weather, and both the historic data and the forecast must be adjusted consistently to normalize them relative to the weather. After this baseline prediction is made it is adjusted based on the sensitivity to weather and the peak load is then predicted with the desired degree of confidence (Stoll 1989). To illustrate the consideration of weather effects, suppose that a baseline prediction is made that a system will have a future peak load demand of 10 gigawatts (GW) for an expected daily high (temperature) of 77°F. Let us further suppose that the daily high conforms to a normal distribution with a standard deviation of 3°F, and that the historically observed correlation between temperature and peak load is 300 megawatts (MW)/°F. It can then be concluded with 95% confidence that the peak load will be below 11.8 GW; 95% confidence corresponds to two standard deviations away from the mean, and this further corresponds to 6°F and 1800 MW of additional load. Note that this example is intentionally oversimplified; several other factors influence peak load, including wet bulb temperature (to account for humidity), wind speed, solar intensity, weather conditions over the past two days (thermal buildup effect), time of day, and time of year.

Peak load forecasting is important because it directly influences the required generation capacity—on every day of the year there must be enough available generation to feed the peak load. This is discussed below.

2.1.2. Relationship Between Capacity Reserves and Reliability

Generating stations require regular maintenance, which means that during some periods of the year they are not available to serve the load. The stations also can be out of service due to unforeseen equipment failures; these outages, called forced outages, also contribute to reduced availability. Assuming that maintenance requirements are known, and that forced outages can be characterized by probability, a natural question arising is, what is the appropriate capacity of generation for a given load forecast. Appropriate in this context is directly tied to reliability of service, and it then follows that we need to find a mapping between capacity and service reliability or, more precisely, between capacity margins and service reliability. Capacity margin is a better measure of reliability because it represents the difference between capacity and peak load (capacity alone is meaningless).

Required capacity reserves commonly are determined using a probabilistic approach that examines the probabilities of simultaneous outages of generating units and compares the resulting remaining capacity with the peak system load. A number of days per year with capacity shortages thus can be determined and this measure, termed loss-of-load-probability (LOLP) index, provides a consistent and sensitive measure of generation system reliability (Stoll 1989).

To determine LOLP index, both scheduled and forced outages are evaluated. Scheduled outages are representative of the downtime required for regular maintenance, and these outages are scheduled deterministically to avoid periods of high peak load. The forced outages are determined probabilistically, and the LOLP index is computed based on a large number of probabilistic experiments. Using a probabilistic method is advantageous in implementation as it allows for convenient inclusion of other factors, such as transmission limitations between interconnected systems, and for simulation of a large number of units. LOLP calculations commonly are performed for an entire interconnected system, as this properly evaluates the benefits of shared generation reserves. A common target value for the LOLP index is 1 day per 10 years, which is equivalent to 0.1 day per year.

Therefore, given a system and the outage characteristics of the units, planners can determine whether it satisfies the desired LOLP index. The converse however is not true; it is not possible to go from a desired LOLP to the optimal system expansion. Planning the expansion to meet the desired LOLP (i.e., reliability), and do so at a minimal cost is discussed next.

2.1.3. Capacity Resource Planning

The question of what type of generating station (hydroelectric, nuclear, coal, gas turbine, or other) would be the most economical addition to the system is answered by combining a production cost analysis with an investment cost analysis.

The process is illustrated in Figure 1. The evaluation begins by preparing a set of expansion scenarios. An expansion scenario includes additions of multiple units and the planners are required to hypothesize the type and the number of units that should be considered. Hypothesizing on the type of units can be aided by using *technology-screening curves*—more details on this can be found in Stoll (1989). The planners also make assumptions on unit additions over time (e.g., a 200-MW gas turbine by 2011, a 400-MW coal-fired plant by 2015); and they also consider unit retirements. Deciding which scenarios to evaluate is a subjective process, and it depends on the planners' preferences and experience.

The scenarios then are evaluated one at a time, beginning with a multiyear reliability simulation to determine the LOLP index for each year of study. If the reliability requirements are not met they often can be improved either by advancing the installation dates of some units, or by delaying retirement dates of others. The corrected scenario then is reevaluated and possibly refined again until the reliability targets are met. Note that these iterations eventually might fail to give acceptable reliability; this possibility should not be regarded as a deficiency of the process, but rather as an indication of an inadequate scenario. If found, then inadequate scenarios are removed, and the process continues to consider the scenarios that meet the reliability target.

The next step in evaluating scenarios is to run a multiyear production simulation for each. Production simulation determines the dispatch of every unit and its associated running costs—such as costs of fuel and maintenance. Cumulative fuel costs of a unit depend on the unit's dispatch—how often it runs and at what operating point. Production simulation determines the dispatch and associated costs for all units in the system, and these costs are recorded for each year of the multiyear study. This is shown symbolically as the data output to the right of the "multiyear production simulation" processing block. Multiple data outputs are shown (stacked)—each corresponds to one expansion scenario.

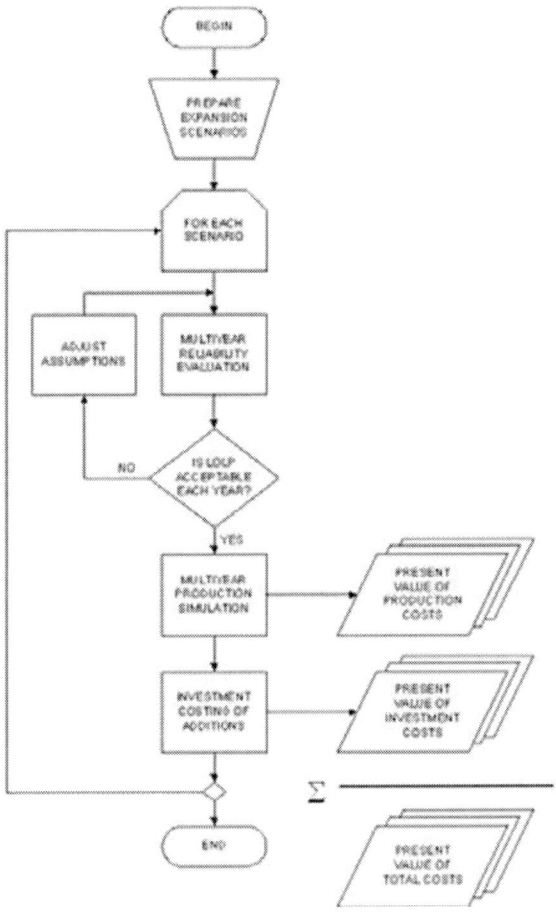

Figure 1. Least-cost generation planning (adopted from Stoll 1989).

Of course, each expansion scenario also has associated construction costs. This is shown as an "investment costing of additions" block—it outputs yearly expenditures for each scenario.

The cost data from production simulation and from investment costing are expressed on a basis of present value to account for time value of money. The total costs then can be computed, and the least-cost scenario can be selected by simple inspection.

Note that this process is centered on cost, and as such it is best suited for use by vertically integrated utilities. When deregulation of electric power industry occurred, generation companies became independent of other utility businesses and generation capacity development became a result of market forces. In the deregulated environments, separate markets exist for energy and capacity. It is the capacity market that responds to the system-reliability requirements, but this discussion is outside of scope of this study. Integration of high penetration solar PV into generation planning is discussed later. Traditional transmission planning methodologies are discussed below.

2.2. Transmission Planning

As noted, the chief role of a transmission system is to optimize the use of a generation portfolio; a transmission system makes it possible to supply loads from the most economical sources of power, and operate generating stations flexibly and thus improve overall system reliability.

Transmission planning therefore ensures that the transmission infrastructure can deliver power from the generators to the loads, and that all the equipment will remain within its operating limits in both normal operation and during system contingencies. Contingencies in this context mean unexpected failures of any system element; for example a generator or a transmission line could have an unexpected outage, which would force the remainder of the system to transition to a new operating point. Studying these transitions and ensuring that a stable operating point can be reached after any contingency is an essential part of transmission system planning.

Transmission system planning is closely interrelated with generation planning. To understand this, it is helpful to note that power flows through a transmission system are a direct result of generation dispatch; a transmission system itself has very limited ability to control the line flows. Therefore, to study the power flow through a transmission system, it is necessary to know the corresponding generation dispatch; to determine the (optimal) generation dispatch, however, the parameters and flow limitations imposed by the transmission system must be known. This "loop" is not always easy to resolve, and it might require complicated iterations between the two planning processes.

Generation dispatch and the associated power flows change many times throughout the day and often follow rather different seasonal schedules. The transmission system therefore can exist in many diverse operating states, and in each one it must be able to cope with the loss of any single element. Transmission planning is tasked with evaluation of all these operating states and their associated contingencies and determining the stability of the system for the set of worst-case conditions. Selecting a set of worst-case conditions is not straightforward and it is most often based on historical system performance and planners' experience and judgment.

Fundamentally, evaluating power system stability is equivalent to evaluating its dynamic performance following system events. This commonly is done using specialized computer programs that include a variety of component models—generators, excitation systems, governor-turbine systems, loads, and other components are all represented, and their dynamic performance (time domain response to disturbances) is simulated and evaluated. It is a common practice to explicitly model the dynamic behavior of generators, excitations systems, governor-turbine systems, and loads by differential equations, and to represent the network elements—transmission lines and transformers— by algebraic equations (Grigsby 2007).

Stability typically is evaluated in three categories: rotor angle stability, voltage stability, and frequency stability (Grigsby 2007), discussed below.

2.2.1. Rotor-Angle Stability

Rotor-angle stability is the ability of generators in the interconnected power system to remain synchronous after a system disturbance. As discussed in the introduction above, generating stations convert energy from some other source of energy into electric energy. Traditionally the interface between the two is a generator and, under steady-state conditions, the electrical torque balances the mechanical torque that is driving the generator, so that the generator operates at a constant speed. This balance can be disturbed at any time, leading to excursion of rotor angles and corresponding electromechanical oscillation. Based on the type of disturbance, rotor-angle stability consideration can be further classified into small signal (or steady-state) stability, and large disturbance (or transient) stability.

2.2.1.1. Small Signal Stability

Small signal stability refers to disturbances sufficiently small to warrant analysis by inearization of system equations around the operating point. Consequently, they can be analyzed in the context of linear systems theory. Small signal stability is evaluated relative to the following physical phenomena:

- Local modes—oscillations of a small group of machines (often in the same power station) relative to the power system
- Inter-area modes—oscillations of a group of machines in one part of the system against another group of machines in another part of the system
- Control modes—oscillations brought on by control interactions between system elements
- Torsional modes—commonly associated with the interaction between a turbine-generator shaft system and another system element, usually a line compensated by a series capacitor.

2.1.1.2. Transient Stability

Transient stability deals with large disturbances and evaluates the ability of a system to maintain synchronism when subjected to a severe disturbance. The resulting system response involves large excursions of generator rotor angles and the governing equations are nonlinear. The analysis typically is done by time domain simulations that include models of generator prime mover dynamics, excitation dynamics, and load dynamics.

2.2.2. *Voltage Stability*

The essential cause of voltage instability is the voltage drop that occurs on the inductive reactances associated with the transmission network. In a heavily loaded system, voltage to the load reduces due to these voltage drops, and this increases current draw from the load, so the positive feedback leading to instability can be established easily. The situation becomes progressively worse as some of the generators reach the reactive power capability limit (essentially the current limit), and the end result is voltage collapse at the load. These events can be precipitated both by loss of generation and loss of transmission, and typically are evaluated by time domain simulations that include voltage-sensitive models of load, and the responses of generator excitation systems.

2.2.3. *Frequency Stability*

Frequency stability studies determine the system's ability to maintain steady frequency within a nominal range following a severe system disturbance that results in a significant imbalance between generation and load. A system's response to frequency stability includes block load shedding and other special protection schemes that typically are not considered in simulations that deal with rotor stability and voltage stability.

In theory, if the generation capacity is correctly planned, then the system should not be exposed to transients associated with frequency stability. Unforeseen circumstances can arise in operations, however, and planners try to be prepared to deal with them. Furthermore, generation capacity is planned with a very long time horizon and construction delays or other events can cause unplanned capacity shortages.

2.3. Distribution System Planning

Distribution systems are the part of electricity delivery infrastructure that serves the load. Traditionally distribution systems are optimized for the lowest cost that meets the desired reliability of service, and reliability is carefully tracked and reported. This has profound implications on planning practices, because reliability is explicitly engineered into the system, and is used as an important metric in evaluating planning options.

2.3.1. *Load Forecasting*

Load forecasting is critically important in distribution system planning and, arguably, distribution utilities are in the best position to make accurate load forecasts. Distribution utilities directly meter their customers and therefore have access to the exact data needed. They also are notified of development projects in their service territory early in the process and, through that mechanism, have a good insight into prospective load growth. Other than that, load forecasting generally follows the procedures discussed in section 0 above. Given their proximity to the load, distribution utilities have the necessary data to successfully employ end-use electricity models.

Table 1. Typically Reported Distribution Reliability Indices (IEEE 1366[3])

Name	Acronym
System Average Interruption Frequency Index	SAIFI
System Average Interruption Duration Index	SAIDI
Customer Average Interruption Duration Index	CAIDI

2.3.2. Planning for Reliability

Reliability in distribution planning is defined and evaluated quite differently compared with reliability evaluation in generation planning. Typical reliability indices used in distribution planning are listed in Table 1. Evaluation of reliability is not absolute (as is the case in generation planning via computing of an LOLP index) but incremental. Reinforcement and planning options are considered relative to their impact on reliability. One example of such a process, termed Cost-Effective Reliability Improvement (CERI) is described in Willis (2004). It begins with known baseline reliability, and then evaluates many possible improvement options relative to their impact on customer reliability. Options are ranked based on their cost-benefit ratio, and the best ones are implemented.

2.3.3. Distribution System Engineering

Of course, there is much more to distribution system design than load forecasting and planning for reliability. Important design choices include distribution substation siting and sizing and feeder layout (including choosing a number and placement of reclosers and sectionalizers). Additionally, studies that address feeder voltage control, feeder protection, and motor starting also are required. These activities are often classified as distribution system engineering and are discussed below, in the context of their interaction with high penetrations of distributed PV.

3.0. PROJECT APPROACH

The review presented in this report is based on recent developments in electric power industry that were triggered by the adoption of Renewable Portfolio Standards and by the related move within the industry towards integration of renewable sources of generation, primarily towards transmission-connected wind generation.

The knowledge gained in integrating wind generation can and should be leveraged for integrating solar PV, and this report capitalizes on the similarities between the two. At the same time, solar PV also is saliently different from transmission-connected wind because it is installed as a distributed resource and has no inherent inertia. The impact of these specific features is evaluated based on their envisioned effect on system planning practices; this process is inevitably subjective but it is expected that the most important aspects are covered in this report.

4.0. IMPACT OF HIGH-PENETRATION SOLAR PV ON POWER SYSTEM PLANNING

To set the context of the discussion that follows, it is helpful to define *high penetration*. There are two fundamental ways to define penetration, either by a metric of energy or by a metric of peak power.

Defining penetration by energy quantifies energy supplied to the system from renewable sources of interest, and such a definition relates directly to displaced fossil generation and the associated savings in fuel consumption and lowered emissions. The energy-based definition is very useful in consideration of large systems and is used in many Renewable Portfolio Standards. The inherent complication of using this definition is that it implicitly depends on the quality of a resource. To achieve equal penetration, more equipment is needed in regions with lower insolation, so the same level of penetration can result in different underlying circuit behavior when evaluated in different regions, depending purely on the quality of the resource.

A power-based definition provides for a more consistent relationship between penetration and circuits' problems—it is defined as nameplate capacity of intermittent generation (installed in a circuit or system) divided by the peak load (of that circuit or system). This report deals primarily with circuits' problems, therefore a power-based definition of penetration is preferred.

This report considers high penetration to be levels up to 50%, but the absolute percentage is of limited value unless it is considered with respect to some other aspect of the system. In general, it is more appropriate to examine the sensitivity of studied phenomena to the level of penetration rather than the penetration level itself.

4.1. Impact of Variable Renewable Energy Generation

The variability of renewable energy sources is a key challenge associated with their integration into the power system. Generation planners think in terms of peak load and generation capacity—at any time, they must have enough available capacity to serve the peak load. To illustrate the notion of availability, compare a 200-MW thermal power plant with a 200-MW wind farm. Assuming a 6% outage rate, a thermal power plant generally can provide its full 200 MW during 94% of considered hours, whereas a 200- MW wind farm might be anywhere between zero and 200 MW depending on the available wind.

The uncertainty associated with renewable generation variability adds complexity to the planning process, and generally results in more demanding operation of the balance of generation portfolio. Non-renewable generators now have to maneuver more in order to accommodate the variability of renewable sources. This increases the operating costs per unit of energy from thermal generation but it also results in lower overall thermal generation and, thus, lower cumulative fuel usage, lower cumulative fuel costs, and lower emissions. These beneficial effects are the exact reasons for the industry's move towards using renewable energy, but this is of little value to the owners of thermal plants whose operating costs per unit of produced energy become higher. These incremental costs aretermed "integration costs," and their fair allocation has been—and still is—a subject generating strong interest in the industry.

4.2. Implications for Generation Planning

One way to effectively include intermittent renewable generation in the capacity-planning process is to plan for system energy, not for peak load. This is discussed below.

4.2.1. Capacity

Process flows of traditional and emerging capacity-planning practices are compared in Figure 2.

The traditional planning process is not designed to consider variable generation, therefore the initial response of the industry was to simply exclude it from capacity planning. The overall process is recapped here to illustrate the specifics of dealing with renewable generation. The process starts with the forecasting of the load energy growth, and this is immediately followed by the associated forecast of the peak load. The generation and transmission capacity then are planned to match the forecasted peak load. Renewable generation is taken in "as available" during system operation, and the output from committed thermal units is reduced to enable the intake of energy supplied by the renewable sources. The end result is sub-optimal system operation; on average, thermal units run below their rated power point, resulting in lower efficiency, higher emissions, and greater operating costs.

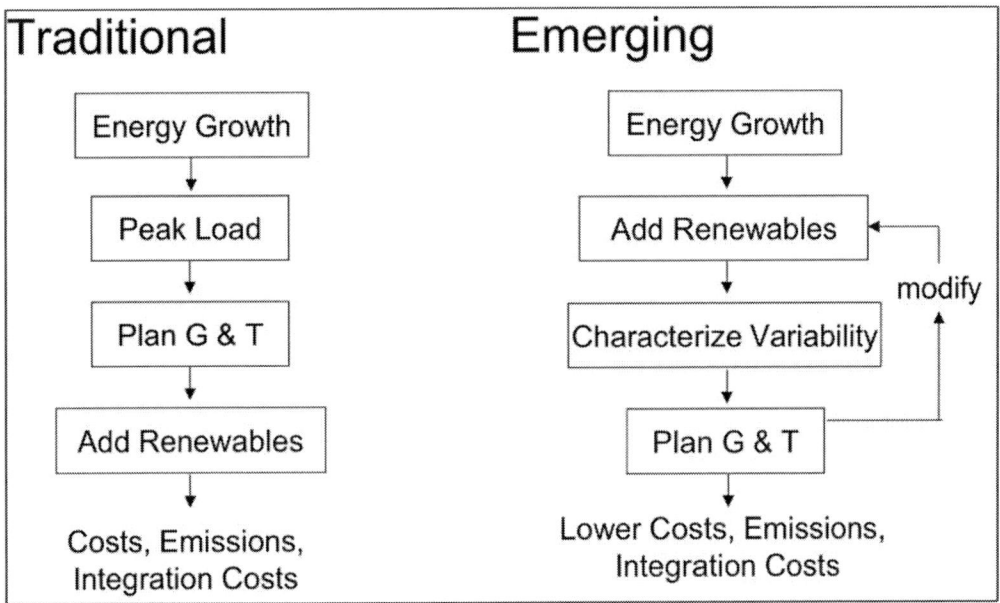

Figure 2. Traditional and emerging practice in capacity planning.

The emerging practice is to include renewable energy supply early in the planning process and consider it during energy growth forecast. This allows for full integration of renewable generation into the planning process; the key is in viewing variable renewable generation as a part of the load. The planning process is thus based on the net load—the system load is reduced to account for contribution from renewable generation. The amount of applicable load reduction is estimated based on historical renewable resource data that is scaled to predict renewable generation from existing and planned new installations. The resulting net load has increased variability compared with the original system's load, and at high penetration this variability must be explicitly characterized to ensure that the balance of generation portfolio has enough flexibility to cope with increased variability. Generation and transmission then are planned relative to net load, and with sufficient flexibility to meet the net load requirements. This evaluation of flexibility is a fundamentally important step, as it has a direct impact on the system's operating costs. Namely, a dispatch order might have to be changed to accommodate net load variability, which will result in different operating costs as compared with supplying the equivalent amount of "pure load." For example, it might happen that a lowest-cost power plant does not have sufficient flexibility to match the variability of a net load. The original plant would be replaced in the dispatch by a more expensive but flexible plant.

Another way to manage "flexibility shortage" is to curtail production from variable renewable sources and thus reduce the net load variability to a manageable level. This is acceptable if the flexibility shortage occurs infrequently enough to not warrant a change in dispatch order, but if curtailments become extensive it will make renewable resources unattractive and also impede their development. Future options might include providing flexibility by introducing direct load control or by using energy storage to cancel out the variability of renewable sources.

One attractive option is to reconsider the siting of renewable generation. An initial plan might concentrate all of the planned additions to the region with the best renewable resource. This makes sense, but if it ultimately leads to transmission congestion then it could prove cost effective to avoid transmission upgrades and simply develop part of the renewable resources at an alternative site.

The evaluation process should be iterative, denoted by the "modify" return step in the flow chart. Evaluating various options in the context of generation planning enables meaningful comparison of the costs versus the benefits they provide.

Implications of high-penetration solar PV on the balance of generation portfolio are discussed by reviewing its impact on net load and on required generation flexibility. This is the examined the sections below.

4.2.2. Characterizing the Net Load

System load is variable; it varies with the time of day, the day of the week, and with different seasons. Renewable generation also varies; it follows the variations of a renewable resource which, in turn, follows its own daily and seasonal patterns. An illustrative example of these variations is depicted in Figure 3, where the hourly loads of California ISO for July 2007 are compared with solar generation based on actual resource data scaled to represent system-wide 30% penetration of PV. The resulting net load also is shown. Dots on the chart represent actual data points, and the thick lines are the computed averages.

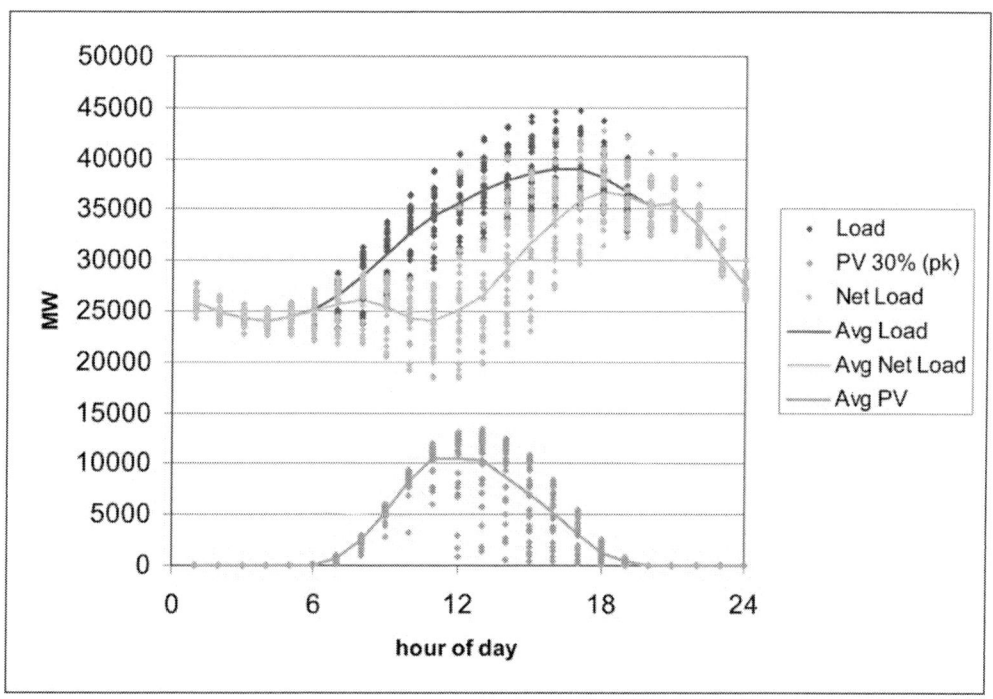

Figure 3. System load, PV generation, and net load (CAISO July 2007).

The significant shift in net load relative to a system load pattern can be observed. It is apparent that ~30% penetration of solar PV shifts the timing of minimum net load to around 11 a.m., and that for many samples this minimum reaches below the minimum original load. This can cause significant shifts in generation scheduling, and it has to be further evaluated for its impact on daily commitment and dispatch, and the associated generation costs. A detailed study requires production simulation, but a rudimentary insight can be obtained by studying load duration curves.

Load duration curves are created by sorting load samples from highest to lowest; the resulting plot provides a powerful insight into the relationship between peak and light load, and into the overall shape of the load profile. Load duration curves for an original system load and net loads with increasing solar PV are shown in Figure 4.[4]

Several important observations can be made. Peak load is reduced by the addition of solar PV; this generally lowers the system operating costs because the expensive peaking units do not run as often as before. Additionally, a penetration level of up to 10% reduces system load during high demand and makes no appreciable impact on the minimum load. This means that penetration of solar PV of up to 10% would have an insignificant impact on scheduling of other generation, and consequently would have low integration costs. Penetration levels greater than 30% cause a reduction in the minimum system load, and in an extreme case of 50% penetration, cause a significant reduction during approximately 100 operating hours over a period of one month (13.4% of the total number of hours). As shown in Figure 3, this minimum load occurs around 11 a.m., and can cause significant integration challenges.

The process of integrating renewable sources is explained as iterative via the "modify" step in the flow chart. One effective and simple way to lessen the integration challenges brought by 50% solar PV is to change the orientation of some of the panels. Instead of all of the panels facing south and producing their maximum output at about 11 a.m., a proportion of panels can be oriented more towards the west, which would shift their maximum power production to a later time. Note that offering higher electricity rates in mid afternoon would result in a change of orientation of all panels, which would just shift the problem from 11 a.m. to another time. Optimally, a spread of orientations would be achieved, and some creativity in structuring rates to drive such behavior will be necessary. This is a simple illustration of the need for active participation of the utilities and system operators in managing high penetration of solar PV.

4.2.3. Characterizing the Impact on Fuel Mix

Net load curves also can be used to gain rudimentary insight into which type of generation is displaced by the intermittent renewable generation. Each system has an associated generation portfolio, and this portfolio is dispatched daily to serve the load. Generated electrical energy then can be linked back to the type of fuel used to produce it, and this establishes the understanding of the fuel mix of the system. The daily dispatch is done to minimize the total cost. Each type of generation has an associated cost, and load duration curves also can be used to predict the dispatch order and the costs of electricity associated with different levels of load. This is illustrated in Figure 5, where the net load duration curve corresponding to 30% solar PV is "filled" by a generation fuel mix representative of California. Starting from the bottom of the graphic, the load is shown to be served by nuclear, then hydro,[5] followed by renewable (wind),[6] then coal and petroleum, and finally natural gas. It is clear from this figure that solar PV displaces only gas-fired generation so the wholesale market price for electricity generated by solar PV in this example would be equivalent to the price of gas-fired generation.

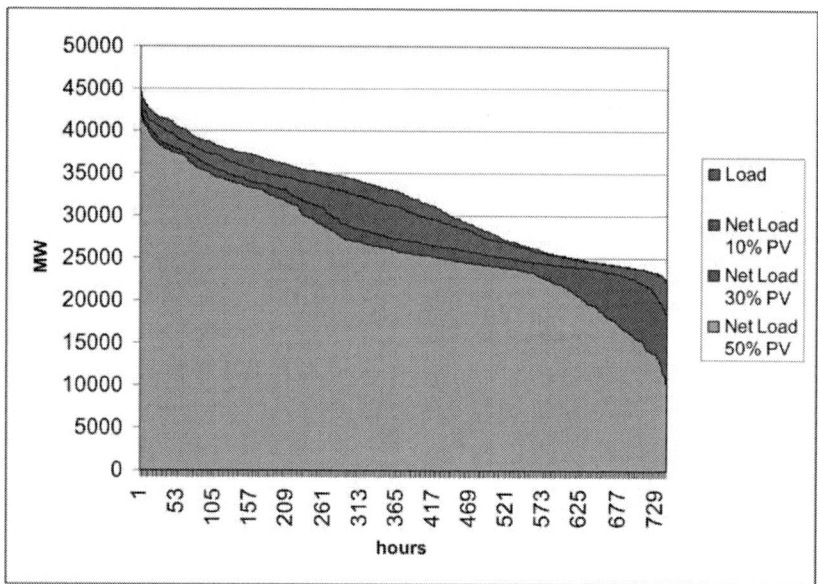

Figure 4. Load duration curves, load and net load with 10%, 30%, and 50% PV penetration (CAISO July 2007).

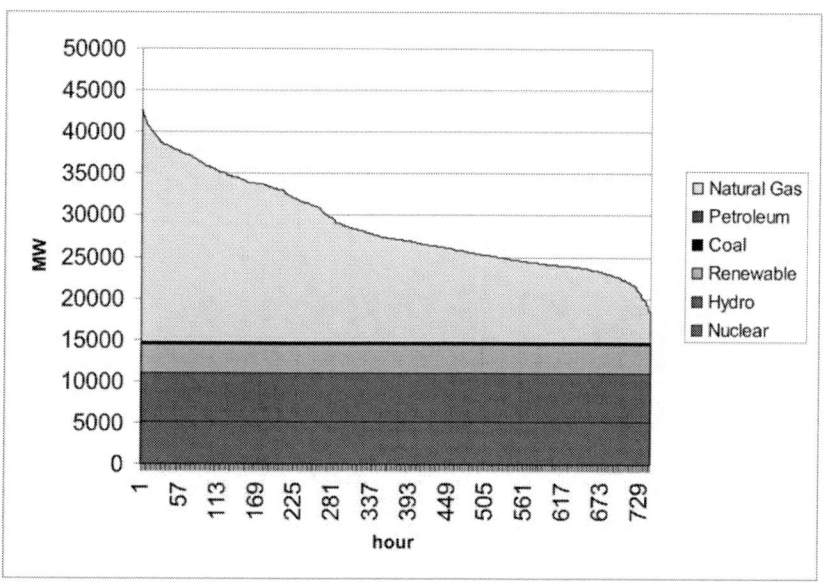

Figure 5. Dispatch order for 30% PV penetration, assuming California fuel mix (CAISO July 2007).

For comparison, Figure 6 shows a dispatch order for the same net load but assuming the U.S. fuel mix. Again, most of the displaced generation is gas-fired, but there also is some from coal-fired plants. This could be an indication of possible integration challenges, because coal plants typically are cycled only once per day. They are taken off-line during the night and brought back on at some point during the morning load rise. The minimum load caused by solar PV occurs around 11 a.m., this means that coal-fired plants would have to be kept off-line longer then before, and that the morning load rise would need to be served by more flexible gas-fired plants. Such dispatch order would result in increased system operating costs, but it would also reduce the number of running hours for coal plants, possibly forcing some of the less efficient ones into retirement. This is an example of a situation where running a full production simulation is required in order to accurately predict actual dispatch order.

The requirements for increased flexibility of the balance of generation portfolio are discussed below.

4.2.4. Generation Flexibility

The need for generation flexibility comes from the need to control the system frequency, so it is helpful to briefly review frequency control in the power system. Present-day power systems rely on rotating generators for most of their energy and, accordingly, the system frequency is directly proportional to the rotating speed of the generators. Furthermore, the rotating speed of the generators depends on the balance of generation and load. If the load is greater than available generation the system slows down, and if the generation is greater than the load, the system accelerates. The load changes continuously, so the generation must be adjusted continuously to control the frequency to its rated value. How these adjustments occur is discussed below.

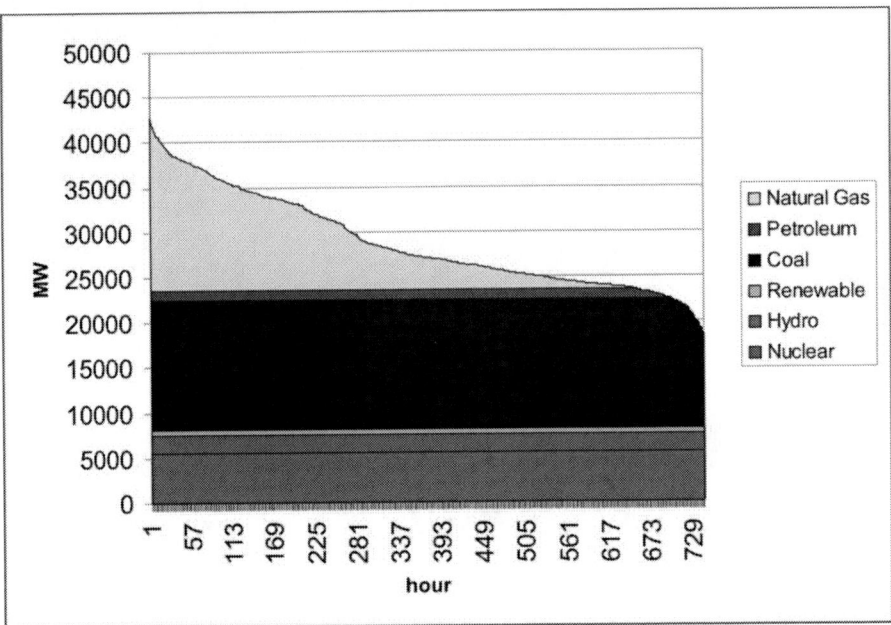

Figure 6. Dispatch order for 30% PV penetration, assuming U.S. fuel mix (CAISO July 2007).

Consider the sketch shown in Figure 7; a frequency control area[7] is represented by a closed curve and it includes the total generation "G" and the total load "L." The area also can exchange power with adjacent areas—two importing interfaces are shown on the top, and an exporting one at the bottom left. The speed of the area's "equivalent generator" is determined by the differential equation:

$$J\omega \frac{d\omega}{dt} = P_{gen} - P_{load} + P_{imports} - P_{exports}$$

where J stands for the equivalent moment of inertia of all generators, $\dot{u}$ is the equivalent angular velocity (proportional to frequency), and the power terms have expected meanings.

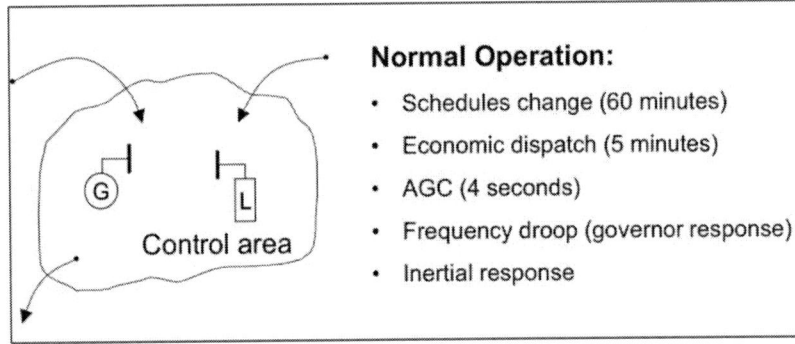

Figure 7. Control area and layers of frequency control (values typical).

In normal operation, the control area follows a predetermined schedule of power exchange with its neighboring areas. Schedules are prepared based on the forecasted load and are negotiated well in advance; they are an input to system operation in real time. Schedules between areas typically change on the hour and therefore the generation is re- dispatched every hour. Then, because the load is continuously varying, units are re-dispatched in economic order every five minutes—the new set points are communicated from the control center to all generators. The residual generation/load imbalance is handled by the automatic generation control (AGC) and also is done centrally. Every four seconds the updated set points are communicated to the subset of units (those participating in AGC). The remaining two actions—the frequency droop and inertial response are done locally at each generator—generators are configured to lower their output if the system frequency rises and to increase it if the frequency drops. This has a stabilizing effect on the power system and ultimately limits the frequency drop/rise following system disturbances (such as an unplanned outage of a generating unit or loss of a large load). Finally, inertial response of generators is the response with fastest dynamics—if the system frequency drops, then a generator connected to the system must slow down, and to do so it must deliver the energy from its rotating mass to the system.

Planning for generation flexibility deals with the first two aspects of frequency control: economic re-dispatch of units every five minutes (called load following), and automatic generation control (called regulation). Both aspects should be evaluated relative to the net load. Understanding the load-following and regulation capabilities of the system is important in determining the system's response to load changes and in evaluating its ability to maintain the frequency within the desired control range (NERC's CPS2 defines a performance standard for frequency control).

4.2.4.1. Load Following

Load-following requirements for the system can be determined based on statistical analysis of net load *data. Net* load variability must be carefully evaluated at various levels of system load and then compared to the flexibility of committed balance of generation portfolio. Correlation between intermittent renewable resource and system load plays an important role in this; generally solar resource is better correlated to the load than wind. This has two important consequences.

- Solar generation is easier to integrate into the system than wind generation because of its lesser impact on incremental variability of net load compared to variability of load alone.
- Solar generation has higher value to the system than wind because of its availability during higher load demand—compared to wind, solar PV displaces more expensive generation.

As the penetration of renewable generation increases, net load variability could eventually become higher than the available load-following flexibility of the balance of generation portfolio. It is important to understand that this will not be an abrupt change; the shortage of load-following capabilities first will appear during a few hours over the course of the year, and the total time in which shortages are noticeable gradually increases as the penetration levels increase. Forcing the burden of variability management on other generation might then become uneconomical, and other options to manage net load variability should be considered. There are a variety of options that can be used to reduce net load variability at the time scale of load following. Spatial diversity of the resource, flexible conventional generation, grid operations and control areas, limited curtailment for extreme events, load management, and, at high penetrations, energy storage, all can be used to reduce net load variability at the time scale of load following. Evaluating these mitigation options (based on their cost-benefit ratio relative to the flexibility they provide to the system) is an important area of future study. Design of efficient markets dedicated to load following is another example of needed future work.

4.2.4.2. Regulation

As in the case of load-following requirements, regulation requirements are determined based on statistical analysis of net load data, but at the finer time scale (recall that load following relates to five-minute updates, while regulation relates to four-second updates). After the regulation requirements are determined, establishing appropriate regulation capabilities is done based on past operating experience—the appropriate regulation for known variability is extrapolated to determine the required regulation for expected variability.

Present deregulated power systems in the United States operate markets for regulation service and it is reasonable to expect that these markets will correctly respond to increased needs for regulation associated with increased penetration of intermittent renewable generation. As with load following, regulation services need not be provided only by the balance of generation portfolio, other technology options are available as well. Emergence of new technologies and their participation in regulation service markets will be an interesting future development.

4.3. Implications for Transmission Planning

Integration of renewable generation does not require strategic changes to the process of transmission planning. Renewable generation must be modeled accurately, however, and this accuracy becomes critically important as the penetration levels increase.

When building the models, it is not reasonable to represent each PV source individually; they have to be organized into aggregates and connected through aggregate equivalent impedances. Equivalent impedances should represent the parameters of the distribution feeder and at least two levels of voltage transformation that exist between transmission- level voltage and distributed PV.

Furthermore, distributed PV generation is connected to the system through the inverters, and modeling the performance characteristics of the inverters in the time scale of interest for transmission system dynamics requires consideration. The general characteristics of PV inverters are reviewed and the relevant modeling requirements are discussed below.

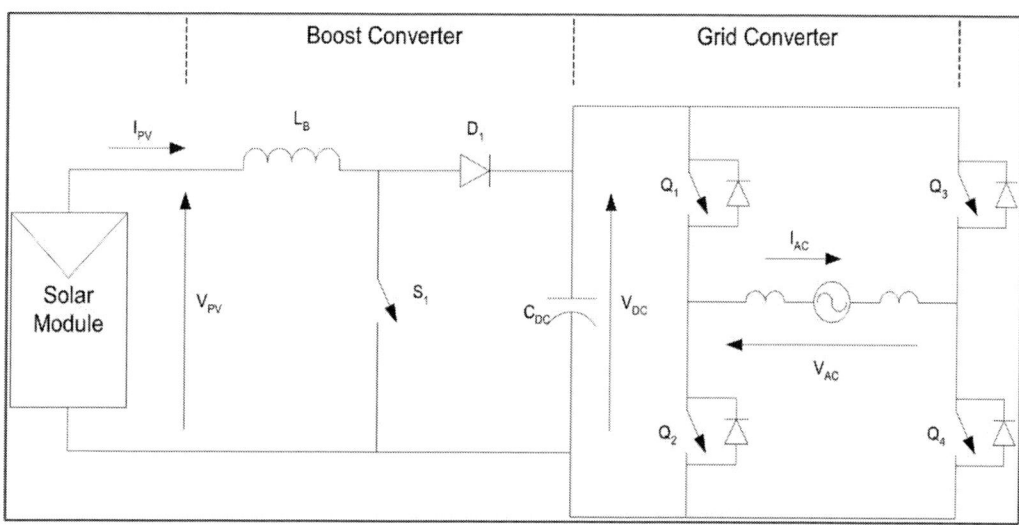

Figure 8. An illustrative PV inverter topology.

4.3.1. Common Characteristics of PV Inverters

PV inverters typically consist of two distinct stages, the PV module interface, called the boost converter, and the grid interface, called the grid converter.[8] An illustrative PV inverter topology identifying the boost and grid converters is shown in Figure 8.

The DC capacitor C_{DC} shown at the boundary of two converters is shared by the converters and it provides energy storage that functions as the buffer for energy transfer between the two converters.

The role of the boost converter is to continuously extract energy from the solar module and to transfer it into C_{DC}. The boost converter modulates the switch S_1, it continuously adjusts its duty cycle to control V_{PV} relative to V_{DC}. Adjusting V_{PV} determines the current from the solar module (labeled I_{PV}). Solar modules have nonlinear voltage current characteristics and adjusting V_{PV} is important to achieve maximum power extraction. Example voltage-current and voltage-power characteristics are shown in Figure 9. By selecting V_{PV} to correspond to point A in Figure 9, power extraction from the PV panel is maximized. The role of the boost converter is to track this operating point for changing insolation.

At the same time, the grid converter takes the energy from C_{DC} and supplies it to the grid, represented here as an AC source of voltage labeled V_{AC}. The voltage across C_{DC} (V_{DC}) always must be greater than or equal to the peak of V_{AC} to enable operation without significant AC current distortion (AC current is labeled I_{AC} in Figure 8). The grid converter controls the magnitude of I_{AC} to supply desired amount of power to V_{AC}. In a steady state, power supplied to the grid matches the power extraction from the solar module, and the voltage V_{DC} across C_{DC} is maintained at the constant value.

The grid converter also can control the phase angle of I_{AC} relative to V_{AC} to exchange reactive power with the grid. Exchanging reactive power does not require energy, and it is limited only by the current capacity of the switches Q_1 through Q_4.

4.3.2. PV Inverters' Behavior During Grid Faults

Currently, PV inverters are required to disconnect from the grid during grid faults. Experience gained from wind industry suggests that staying connected during the fault and helping to restore the voltage after the fault is cleared[9] aids system stability. It therefore is reasonable to expect that the PV industry will face similar requirements as the penetration levels increase.

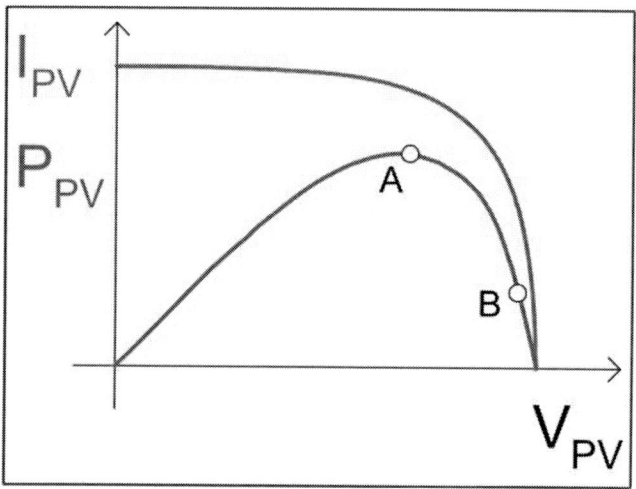

Figure 9. An example voltage current and voltage power characteristic of solar module.

Suppose, for example, that the grid experiences the fault and that due to this fault V_{AC} falls to some low value (it is "shorted" by the fault). During the fault, the grid inverter no longer can supply power to V_{AC} (power is the product of voltage and current, and if the voltage is brought to zero due to a fault, the power delivered to the AC circuit also becomes zero). This makes it impossible to remove energy from C_{DC}, and if the boost converter continues to transfer energy from the solar module to C_{DC}, V_{DC} will start to rise. The rise of V_{DC} is a signal to the boost converter that something is wrong and so it moves the operating point from point A to point B as shown in Figure 9. (Point B is shown at a power higher than zero to indicate that some active power is used internal to the inverter, and that some can be fed to the grid even during faults; generally, fault current travels through the conductors to reach the short circuit and this has associated power losses.) Operating point B can be assumed in a matter of milliseconds and it then is maintained for the duration of the fault. After the fault is cleared (many are cleared within 100 ms), the boost inverter has to transition back to point A; generally no information is provided by the manufacturers on the speed of this reverse transition.

4.3.3. Modeling PV Inverters for Transmission Planning

As discussed above, transmission planning is based on studying system recovery after various contingencies. Commonly a fault is applied to the system model and the response of the system then is studied during the fault and immediately after it is released.

As the penetration level of PV sources rises during the coming decades, these sources will displace traditional generation more and more, and become significant participants in system dynamics. To enable proper evaluation of these dynamics, PV inverters must be represented with the amount of details analogous to the representation of conventional generators. Models of conventional generators commonly include representation of generators' and turbines' inertia, detailed excitation models, models of governor controls, and models of power system stabilizers (where applicable). In contrast, PV inverters are in most cases represented simply by a fixed negative load, and no dynamic behavior is modeled.

This practice will have to change as the penetration of distributed PV increases. Dynamics of maximum power point tracking will have to be represented explicitly to enable study of the behavior of PV inverters due to changes in insolation, or during fault recovery. Similarly, the dynamics of anti-islanding detection must be modeled to enable quantifying the interaction between PV inverters and grid frequency control. Lastly, the ability of PV inverters to deliver VAR support through the distribution system to the bulk grid will have to be evaluated based on the future practices in distribution system design. This novel modeling style is used in representing PV inverters in the associated RSI report entitled *Transmission System Performance Analysis for High-Penetration PV*.

4.4. Implications for Distribution Planning and Engineering

Distributed PV generation affects distribution system planning and engineering in three essential ways.

- It affects feeder voltage regulation.
- It makes contributions to fault currents.
- It can provide an ungrounded source of voltage.

Each of these effects is discussed in the following subsections.

4.4.1. Feeder Voltage Regulation

If the penetration level of the PV inverters is sufficiently high, reverse power flow through the distribution system might start to occur during some periods of the day. This can create unanticipated conditions and cause misoperation of the utility voltage control equipment.

Generally, reverse power flow through the feeder causes a voltage gradient from the distributed PV towards the substation, and this voltage rise might push the feeder and service voltage beyond the limits suggested by ANSI C84.1.

In many situations, this voltage rise can be brought to within limits by adjusting the on-load tap-changer (OLTC) in the substation, but the control of OLTC somehow must be made aware of PV generation. A properly configured line compensation can be sufficient in most cases, but more-sophisticated control schemes based on communications of remote voltage points also are possible.

Voltage control on an example distribution feeder and options for reactive power delivery from distributed PV to the grid are analyzed and presented in the associated RSI report entitled *Distribution System Performance Analysis for High-Penetration PV*.

4.4.2. Contributions to Fault Currents and Protection Desensitization

Most modern PV inverters employ self-commutated inverters that operate in current control mode. These results in the exceptionally fast short-circuit protection and limiting of fault currents to less than 2-pu peak value that is removed within 1 ms. Compared with fault currents supplied by conventional generators, inverter fault currents are negligible and unlikely to cause significant damage.

Inverter-coupled PV sources have a more significant impact on protective relaying; transformer-connected PV inverters can provide a ground path and affect the magnitude of zero-sequence currents. This could cause protection desensitization and must be carefully evaluated on a case-by-case basis.

4.4.3. Ungrounded Source of Voltage

PV inverters can be coupled to the distribution system via transformers, and based on the transformer connection they can provide an ungrounded source of voltage to the distribution system island that is formed after the substation breaker is opened. This can cause high line-to-ground voltages on the unfaulted phases during single line-to-ground fault.

These problems can be avoided by selecting an appropriate transformer connection. Generally, good grounding arrangements also create sources for ground current that could interfere with circuit protection. A more detailed discussion and a comprehensive list of references are provided in Short (2004).

4.4.4. Software Tools Used in Distribution Engineering

Another RSI report, entitled *Utility Models, Analysis and Simulation Tools*, discusses existing software tools and outlines the strategy for future software development that will improve handling of distributed generation in the distribution engineering software. Specific recommendations are provided with regard to load flow analysis, short-circuit studies, and protection coordination.

CONCLUSIONS AND RECOMMENDATIONS

This report explores the impact of high-penetration renewable generation on the system planning methodologies, and outlines how these methodologies are evolving to enable effective integration of renewable generation sources. All three areas of system planning—generation, transmission, and distribution—are considered, and the impact of high penetration of solar PV is analyzed relative to each. The key concepts are summarized below.

5.1. Generation Planning

Generation planning is concerned with providing sufficient generation capacity to serve the load with the desired reliability, and with ensuring that the resulting generation portfolio has sufficient flexibility to cope with the variability of the load.

5.1.1. Capacity

- The emerging practice in capacity planning is to plan for the system's energy demand, not its peak load. This allows for full integration of renewable generation into the planning process. The key is in viewing intermittent renewable generation as a part of the load. The planning process is based on the net load, the system load is reduced to account for contribution from renewable generation. The amount of applicable load reduction is estimated based on historical renewable resource data scaled to predict renewable generation from existing and planned new installations. The generation capacity planning then can be handled using standard tools. Reliability calculations (such as LOLP) should consider thermal plants only.

5.1.2. Flexibility

- The flexibility of the generation portfolio is characterized by its load-following and regulation capabilities, and both should be evaluated relative to net load. Load following refers to economic re-dispatch of the entire generation portfolio every five minutes, and regulation refers to system frequency control by automatic generation control. Setpoint updates are sent every four seconds to the units participating in AGC.
- Explicit evaluation of load-following requirements in the presence of intermittent renewable generation is a new dimension in generation planning. The emerging practice is to determine load-following requirements based on a statistical analysis of net load data. Net load variability must be evaluated carefully at various levels of system load and then compared with the flexibility of the committed balance of generation portfolio. Future research should focus on defining required flexibility from the standpoint of net load, and on matching that with available flexibility from the balance of generation portfolio. Developing a methodology to accurately quantify the available flexibility also is required.
- As the penetration of renewable generation increases, net load variability eventually could become higher than the available load-following flexibility of the balance of generation portfolio. Forcing the burden of variability management on other generation then might become uneconomical, and other options to manage net load variability should be considered. Options such as load control, load shifting, use of energy storage, and curtailment of renewable generation sources all can be used to reduce net load variability at the time scale of load following. Evaluating these mitigation options (based on their cost-to-benefit ratio relative to the flexibility they provide to the system) will be an important area of future study.
- Design of efficient markets dedicated to load following is another example of needed future work.
- As in the case of load-following requirements, regulation requirements are determined based on statistical analysis of net load data, but at the finer time scale (load following relates to five-minute updates and regulation relates to four-second updates). After the regulation requirements are determined, appropriate regulation capabilities are established based on past operating experience. Appropriate

regulation for known variability is extrapolated to determine required regulation for expected variability. Present deregulated power systems in the United States operate markets for regulation service and it is reasonable to expect that these markets will respond correctly to increased needs for regulation associated with increased penetration of intermittent renewable generation. As with load following, regulation services need not be provided only by the balance of generation portfolio; other technology options are available. Emergence of new technologies and their participation in regulation service markets also will be interesting future developments.

- Accurate day-ahead renewable resource forecasting enables more-accurate unit commitment, which results in significant operating cost reductions. Improvements in renewable resource forecasting, both long term (multi-day) and short term (hours and minutes ahead), will lead to substantial benefits in system operation. Resource forecasting is an important area of future work. At the same time, resource data collection and extraction of relevant statistics are important for evaluation of future load-following and regulation requirements.

5.2. Transmission Planning

Transmission planning ensures that the transmission infrastructure can deliver power from the generators to the loads, and that all the equipment will remain within its operating limits both in normal operation and during system contingencies.

Evaluation of contingencies consists of studying a system's dynamic behavior that is precipitated by the contingencies. The dynamic model of the system is carefully maintained (in load-flow software), and time domain simulations are performed for many contingencies and then evaluated to determine which are critical. Transmission expansion and reinforcement plans are evaluated relative to their impact on critical contingencies.

- Integration of renewable generation does not require strategic changes to this process, but renewable generation has to be accurately modeled, and this accuracy becomes critically important as the penetration levels increase. Performance characteristics of inverters have to be evaluated for their impact on transmission system dynamics. The impact of the inverter's maximum power point tracking on system dynamics, and the impact of anti-islanding schemes on transmission system frequency control should be considered.

5.3. Distribution System Planning

Distribution system planning is concerned with the infrastructure that serves the load. Distribution systems are very diverse and operate under a variety of weather conditions that affect loads. They serve different types of loads with widely varying spatial topologies, and use different voltage levels and diverse system equipment. It therefore is challenging to draw generalized conclusions about distribution systems, but several observations were made that are uniformly applicable relative to high solar PV penetration levels.

- Distribution systems currently are designed under the assumption of power flow from a substation to end-use loads. Depending on the penetration level, solar PV can cause a reversal of power flow through the distribution system, and this is the likely source of problems. Correlated load and insolation data are needed to predict the maximum amount of reverse power flow for a feeder under consideration.
- Distribution systems rely on a coincidence factor of loads for sizing all of the equipment. Installed loads are not likely to operate simultaneously, and the designers take advantage of this. Equipment is sized for the expected coincident load rather than the maximum load. The probability of coincident operation of solar PV is much higher, because an entire distribution service area can easily be subject to the same insolation. This places the upper boundary on solar PV penetration; the installed peak capacity must be lower than the coincident load.
- Problems caused by high penetration of distributed solar PV can be reliably predicted by careful analysis (simulation), but it should be noted that the software tools used in distribution planning require assiduous harmonization with respect to modeling PV inverters. To enable this, a set of guidelines for modeling solar PV inverters should be created and perhaps included in the standards. This will eliminate a lot of confusion and unwarranted concerns that inevitably result from inconsistent and possibly inaccurate modeling.
- Solutions for most problems caused by high penetration of solar PV can be found in existing technology. It is reasonable, however, to expect that installing substantial distributed generation of any type will require modifications to the existing distribution systems, and that these modifications require some capital investment. To minimize this investment and build up the confidence of the industry, it will be helpful to provide plenty of examples of best practices and to showcase simple and effective solutions.
- Currently, solar PV installations are handled like installations of grid-connected distributed generation, using processes that often are slow and ineffective. As the penetration levels increase, the industry will shift towards handling PV installations like installations of appliances; the utilities will not need to be notified. On one hand, this is desirable because it reduces installation costs. On the other hand, the practice is not sustainable if the solar PVs are destined to become a significant source of energy for the power system. To avoid unexpected problems, utilities will have to include characteristics of solar PV in all areas of system planning. One way to address this is to create a set of automated tools for screening prospective solar PV installations. Any prospective installation then would be conveniently reported, screened by the system, and either approved or floated up to the utility for a more detailed evaluation. Once approved, it would get be recorded and included in all relevant databases and further used in all areas of system planning. Such screening also will enable utilities to track and report integration costs, and later recover them either through tariffs or by other means.

5.4. General Recommendations

- As the penetration levels rise, distributed PV installations might be required to provide performance characteristics similar to those of traditional generators. Such requirements could include inertial response, frequency droop characteristics, reactive power injection, and the ability to curtail production. Current inverter technology is able to support these grid-friendly features, albeit at the small penalty of inverter efficiency and slightly higher capital cost. However, many inverters currently sold are highly optimized for efficiency and have no ability to provide grid-friendly features. To change this, a careful evolution of standards, policies, and incentives is required.
- Furthermore, enabling inverters to provide grid support only completes part of the job. The other necessary part is the communication link from the system operator to every installed inverter. This communication link can take many forms; a simple broadcast on FM radio might be sufficient to accomplish the required minimal functionality, but high-bandwidth, full-duplex communication (if deployed) could be leveraged in many different ways and it would create opportunities for more-effective distribution system management, more-flexible system configurations, faster restoration, more-selective protection, efficient deployment of demand response, efficient implementation of real-time metering, introduction of creative electricity tariffs, and more. Pilot programs, sharing the existing high-speed Internet infrastructure, could help evaluate the possible benefits and guide design decisions for the dedicated infrastructure.

REFERENCES

Grigsby, L. L., ed. (2007). *Power System Stability and Control*. Boca Raton, FL: CRC Press.
Pindyck, R. S. & Rubinfeld, D.L. (2000). *Econometric Models and Economic Forecasts*. New York: McGraw-Hill.
Short, T. A. (2004). *Electric Power Distribution Handbook*. Boca Raton, FL: CRC Press.
Stoll, H. G. (1989). *Least-Cost Electric Utility Planning*. New York: Wiley-Interscience.
Willis, H. L. (2004). *Power Distribution Planning Reference Book*. 2nd edition. New York: Marcel Dekker.

REPORT DOCUMENTATION PAGE			Form Approved OMB No. 0704-0188		
	The public reporting burden for this collection of information is estimated to average 1 hour per response, including the time for reviewing instructions, searching existing data sources, gathering and maintaining the data needed, and completing and reviewing the collection of information. Send comments regarding this burden estimate or any other aspect of this collection of information, including suggestions for reducing the burden, to Department of Defense, Executive Services and Communications Directorate (0704-0188). Respondents should be aware that notwithstanding any other provision of law, no person shall be subject to any penalty for failing to comply with a collection of information if it does not display a currently valid OMB control number. **PLEASE DO NOT RETURN YOUR FORM TO THE ABOVE ORGANIZATION.**				
1. REPORT DATE (DD-MM-YYYY) February 2008		2. REPORT TYPE Subcontract Report		3. DATES COVERED (From - To)	
4. TITLE AND SUBTITLE Power System Planning: Emerging Practices Suitable for Evaluating the Impact of High-Penetration Photovoltaics				5a. CONTRACT NUMBER DE-AC36-99-GO10337	
				5b. GRANT NUMBER	
				5c. PROGRAM ELEMENT NUMBER	
6. AUTHOR(S) J. Bebic				5d. PROJECT NUMBER NREL/SR-581-42297	
				5e. TASK NUMBER PVB7.6401	
				5f. WORK UNIT NUMBER	
7. PERFORMING ORGANIZATION NAME(S) AND ADDRESS(ES) GE Global Research 1 Research Circle Niskayuna, NY 12309				8. PERFORMING ORGANIZATION REPORT NUMBER ADC-7-77032-01	
9. SPONSORING/MONITORING AGENCY NAME(S) AND ADDRESS(ES) National Renewable Energy Laboratory 1617 Cole Blvd. Golden, CO 80401-3393				10. SPONSOR/MONITOR'S ACRONYM(S) NREL	
				11. SPONSORING/MONITORING AGENCY REPORT NUMBER NREL/SR-581-42297	
12. DISTRIBUTION AVAILABILITY STATEMENT National Technical Information Service U.S. Department of Commerce 5285 Port Royal Road Springfield, VA 22161					
13. SUPPLEMENTARY NOTES NREL Technical Monitor: Ben Kroposki					
14. ABSTRACT (Maximum 200 Words) This report explores the impact of high-penetration renewable generation on electric power system planning methodologies and outlines how these methodologies are evolving to enable effective integration of variable-output renewable generation sources.					
15. SUBJECT TERMS power system planning, capacity planning, distributed generation, solar PV, photovoltaics; GE Global Research; National Renewable Energy Laboratory; NREL					
16. SECURITY CLASSIFICATION OF:			17. LIMITATION OF ABSTRACT UL	18. NUMBER OF PAGES	19a. NAME OF RESPONSIBLE PERSON
a. REPORT Unclassified	b. ABSTRACT Unclassified	c. THIS PAGE Unclassified			19b. TELEPHONE NUMBER (Include area code)

Standard Form 298 (Rev. 8/98)
Prescribed by ANSI Std. Z39.18

End Notes

[1] IEEE 1547 Standard for Interconnecting Distributed Resources with Electric Power Systems, 2003.
[2] UL 1741 Standard for Inverters, Converters, and Controllers for Use in Independent Power Systems.
[3] IEEE Standard 1366, Guide for Electric Power Distribution Reliability Indices, 2003.

[4] Load duration curves normally are plotted for an entire year. Because this is a high-level discussion, this report uses a simpler setup and the analysis is limited to one month of data.

[5] The stacking order of hydro is somewhat arbitrary and it can be used effectively to manage variability or minimum demand.

[6] Strictly speaking, wind generation should be incorporated into net load, but to simplify the discussion this was avoided.

[7] Power systems are operated as aggregates of smaller entities called control areas. This helps manage the system more effectively and facilitates trade of power between the areas. Area boundaries are largely administrative, but they are carefully monitored to facilitate billing for exchanged power.

[8] Single-stage topologies exist but they rely on high voltage (>1000 VDC) from a solar module. The NEC limits the voltage of solar modules to less than 600 VDC, therefore boost stage commonly is employed in the United States.

[9] This feature is known as low-voltage ride through (LVRT) or zero-voltage ride through (ZVRT).

Chapter 5

ENHANCED RELIABILITY OF PHOTOVOLTAIC SYSTEMS WITH ENERGY STORAGE AND CONTROLS

D. Manz, O. Schelenz, R. Chandra, S. Bose, M. de Rooij, and J. Beb

ACKNOWLEDGMENTS

The authors would like to acknowledge Ray George from NREL for providing insolation and temperature data for regions of the United States. The authors would also like to acknowledge the contribution of Tom Short from EPRI. Tom provided the GE team with detailed reliability data for two utilities in the United States. The project team would also like to thank the following individuals for their contributions to this study:

1. Jim McMahon of LBL for appliance data models, start/end time estimation.
2. Craig Cornelius for the interaction needed to obtain load modeling data and information.
3. Robert Delmerico of GE Global Research for providing data from studies on load modeling in residential communities.
4. Lawrence Berkley Labs for providing load shapes for several appliances, distinguishing week and weekend days.
5. Patrick G. McElhaney (GE C&I) for providing data on average per-state home power consumption.

EXECUTIVE SUMMARY

We observed enhanced customer reliability through the management of energy storage systems and photovoltaics (PV) during a utility outage. In an outage, the affected community would intentionally island and meet only its critical load. The timing, duration, and number of customers affected by each outage event were obtained for a single utility in 2005. These data were used to simulate outage events for a community on a distribution feeder. Overall, this technology resulted in a community experiencing fewer outages and outages of shorter duration.

The parameters considered in this analysis include three geographic regions (Golden, Colo., Hanford, Calif., and Sterling, Va.), three community sizes (10 homes, 100 homes, and 1,000 homes), and various combinations of battery capacity (0 to 10 kWh) and solar PV penetration (0%, 5%, 10%, 30%, 50%). The distribution reliability indices presented in IEEE 1366[1] were adapted to account for the management of energy storage and PV in meeting only the critical loads in each home or community. The enhancement in reliability was quantified in terms of modified reliability indices, which are pertinent to these types of communities:

- Critical SAIDI —average duration of critical load interruptions.
- Critical SAIFI—average number of interruptions per customer.
- Unserved Critical Load (UCL)—annual unserved critical load (kWh) on a circuit.

We observed a significant improvement in these three indices when PV and battery energy storage were deployed at each home within a community. The presence of more than ~5 kWh of battery capacity per home reduced each index to nearly zero (a 100% reduction). The contribution of PV to the improvement in reliability indices was less significant, contributing to ~25% reduction in each index at 50% PV penetration. The community size and geographic location had a small impact on the overall results.

The following four areas provide a vision for future research:

Load reconfiguration technology – We assumed that the loads in a home/community could reconfigure at the onset on an interruption. Identifying the technology and controls needed to perform this function is critical to achieving enhanced reliability through the use of PV systems.

Customer reliability statistics – Statistics and specific information about which customers are affected during an outage event would enhance the accuracy of the simulation. This information will be needed for both utilities and customers in order to quantify the economic benefit of enhanced reliability due to PV systems.

Load profile breakdown – A breakdown of the contributions of various appliances to the overall aggregate load of a home (and community) through a year will be needed before customers and utilities invest in home load controls.

Critical loads in communities – Residential customer surveys of appliance usage or appliance feedback to the utility are needed in order to quantify the contribution of each appliance to the total load in a home/community.

1. INTRODUCTION

In this study, photovoltaics (PV), load control, and battery energy storage systems were managed in order to enhance energy reliability for the customer. During an interruption, customers within a community are able to intentionally island, reconfigure total loads to only the critical loads, and meet the critical loads by managing PV and energy storage. The objective of this study is to evaluate the reliability improvement associated with this capability. The results of this study can be used to identify the technologies and regulations needed to enable load and energy management in a home or community. Multiple case studies were considered, including three geographic regions, three community sizes, and various combinations of energy storage and PV. In this study, PV penetration is defined as the ratio of the nameplate solar array capacity divided by the maximum 1 5-min average peak load, within the community, in one year.

Residential load modeling was performed for three regions of the United States and these results were used as an input to the reliability model. We considered adapting the reliability indices presented in IEEE 1366 in order to account for the management of energy storage and PV to meet the critical loads in each home within the community. The enhancement in reliability was quantified in terms of proposed new reliability indices that are pertinent to residential communities containing PV systems with energy storage and controls.

Three community sizes were chosen in order to evaluate the influence of coincidence factors (i.e., a larger community would exhibit a smoother load profile). Three regions were considered because each region has unique annual weather/temperature profiles, and therefore a unique heating, ventilation and air conditioning (HVAC) load profile. Additionally, appliance penetration levels vary from region to region. Fifteen-minute time steps were chosen because no data were available for shorter intervals and additional fidelity would provide computation management challenges.

In order to quantify the enhancement in reliability, the following tasks were performed:

- Reliability Modeling – Random outages were generated based on representative utility data.
- Residential Load Modeling – Total and critical load curves were constructed for individual homes and residential communities.

The reliability model was used to simulate random outages throughout the year. During the outages, the residential-load model reconfigured HVAC and appliance loads to only the critical loads. The architecture of a system that could enable load reconfiguration was not the focus of this study. As a first step, quantifying reliability enhancement will enable the development of a value proposition for such a capability, and may ultimately lead to further technology development. In order to perform this study, we made some key assumptions: (1) a residential home could intentionally island load on PV inverters and breakaway from the grid during an outage, and would be capable of managing starting inrushes, and (2) the circuit protection within the island could be resolved.

2. CURRENT STATUS OF EXISTING RESEARCH

The review of existing research is broken into two sections: distribution reliability and residential load modeling. These two distinct topics provided the basis for quantifying the reliability enhancement.

2.1. Distribution Reliability Indices

The IEEE Guide for Electric Power Distribution Reliability Indices (IEEE 1366) was developed in order to summarize relevant distribution reliability indices, outline the methodology for calculating these indices, and highlight the factors that affect the calculation of indices.

In order to understand the details of IEEE 1366, definitions and additional information are needed. In the standard, a long interruption is defined as an event whereby the voltage at the customer's connection drops to zero and does not re-establish automatically. Typically, interruptions in excess of three minutes are referred to as long interruptions, while interruptions of less than three minutes are called short interruptions,[2] but this definition varies among utilities. Additionally, the term "sustained interruption" refers to a longer interruption, ranging from three seconds in IEEE 1159 to two minutes in IEEE 1250.[3]

The primary distribution reliability indices used for sustained interruptions (outages in excess of five minutes and excluding major event days) are:

- System average interruption frequency index (SAIFI),
- System average interruption duration index (SAIDI), and
- Customer average interruption duration index (CAIDI).

SAIFI describes how often an average customer will experience a sustained interruption (greater than five minutes). Mathematically, this index is defined as:

$$SAIFI = \frac{CI}{N_T},$$

where CI is the number of customers interrupted and N_T is the total number of customers served for the area. SAIDI is defined as the total duration of interruption for an average customer over a specific period of time. Mathematically, this index is defined as:

$$SAIDI = \frac{CMI}{N_T},$$

where CMI is the customer minutes interrupted. In terms of load-based indices, the average system interruption frequency index (ASIFI) is often used to measure performance in areas with few consumers and concentrated loads.[4] Mathematically, ASIFI is defined as:

$$ASIFI = \frac{\sum L_i}{L_T}.$$

where, ASIFI is the ratio of total connected kVA of load interrupted and the total connected kVA served.

SAIDI and SAIFI are two of the most common distribution reliability indices used in the industry (see Figure 1).

The distribution reliability indices described above are used to quantify sustained interruptions. Short duration outages for some customers, such as hospitals and large industrial customers, can result in complex systems shutting down. The startup of these systems can be costly. In many cases, these customers have installed backup generation or other means of addressing short-duration outages. In particular, it is these types of outages that would benefit from the presence of distributed generation and energy storage in islanded operation. Therefore, a reliability index must not only quantify enhanced reliability for sustained interruptions, but must also quantify enhanced reliability for short-duration outages.

2.2. Residential Load Modeling

Taking into account customer data for variable numbers of customers is essential for studying the time evolution of the load in the distribution system feeders. In fact, the electricity consumption of the single residential customer is too variable in time to allow us to obtain a sound estimate of its individual load pattern. The residential load aggregation can be obtained by either working directly at the distribution system level (if the results of measurements carried out on several feeders are available), or resorting to a bottom-up approach in which the aggregated load patterns of single-house customers are computed on the basis of information obtained from real-case investigations on customer behavior, lifestyle, and appliance use. In particular, it is important to assess not only the average value of the aggregated load, but also how its probability distribution varies during the day and as a function of the number of residential customers. Previous studies [17,19,20,21] have shown that the time evolution of the average power, normalized with respect to the total contract power of the customers, has a predictable behavior, especially when the number of customers is relatively high (e.g., over 100). Yet, when the number of customers is low, the possible variations of the load power at any given time are significantly higher and strongly depend on the randomness of the customer composition and lifestyle.

We reviewed a number of articles describing the distribution load profiling techniques. A selected few are discussed in this section. Two main categories of approaches, top-down and bottom-up, were found to exist in most of these articles.

Top-down methods require data collection at the distribution substation. Even when such data are available, this approach does not easily accommodate modeling the impact of demand-response technologies. Bottom-up methods attempt to quantify the drivers for consumption and use Monte Carlo sampling to synthesize an aggregated load profile. Critical inputs are resident demographics, home size, geographic location, season, day of week, time of day, and the number of homes supplied by the feeder.

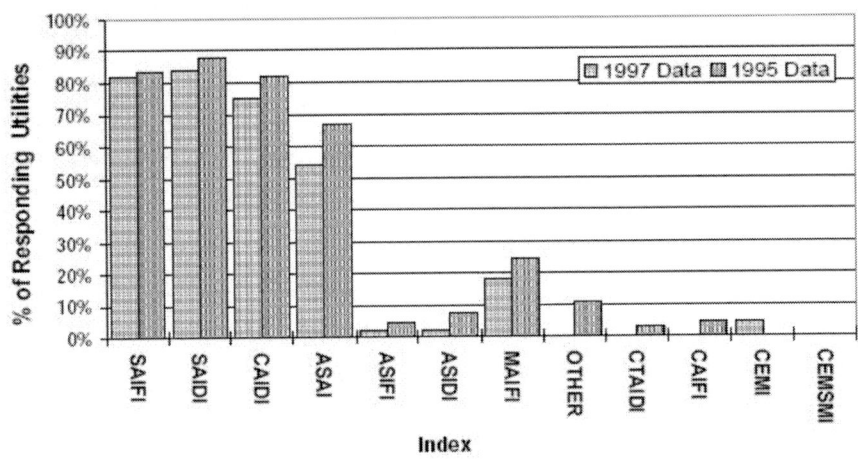

Figure 1. Percentage of responding companies using indices reporting in 1995 and 1997[5].

In Schneider & Hoad (1992),[6] researchers applied Monte Carlo sampling to data from a survey of household usage patterns. A diversity function was fit to the data, relating the diversity factor to the number of homes on the feeder. In Lee & Etezadi-Amoli (1993),[7] the authors made some general notes about coincident and diversity factors. Coincident factors differ by user type. Industrial and commercial users have higher coincident factors during the day while residential coincident factors have evening peaks. In addition, coincident factors tend to be higher during peak load. Finally, diversity factors were found to be similar for homes with electric and non-electric heating.

While focused on methods to produce short-term residential forecasts, elements of Noureddine et al. (1992)[8] apply to longer-term modeling. Loads are split between weather- and non-weather-dependent categories. Weather-dependent loads are modeled with a physics-based model while non-weather-dependent loads are quantified with an autoregressive model. Limitations of models like DOE2 are discussed—such models are house specific and ignore random usage patterns.

In Capasso & Grattieri (1994),[9] bottom-up modeling was conducted using cross-sectional (time of day) and longitudinal (day of week) use surveys as well as census data describing regional proclivities. In Carpaneto & Chicco (2006)[10] researchers found that the load profile for feeders with more than 100 homes has a predictable behavior. However, as the number of homes decreases, the profile is significantly impacted by the number of customers and the randomness of customer use patterns. A number of probability functions were fitted to the aggregated load profile. The gamma function was found to be most appropriate for extra-urban homes.

We reviewed many other articles to decide the appropriate tools for the modeling work. While the appliance load data can be built upon some existing data sources [11, 12] using simple spreadsheet-based calculations, a more detailed thermal model of the home is necessary to understand the HVAC loads. Drury et al. (2005)[13] provided an overview of a report that provides up-to-date comparisons of the features and capabilities of twenty major building energy simulation programs. The comparison is based on information provided by the program developers in the following categories:

- General modeling features,
- Zone loads,
- Building envelope and day-lighting and solar,
- Infiltration, ventilation and multi-zone airflow, renewable energy systems,
- Electrical systems and equipment,
- HVAC systems,
- HVAC equipment,
- Environmental emissions,
- Economic evaluation,
- Climate data availability, results reporting; validation, and
- User interface, links to other programs, and availability.

After careful investigation of all these tools, and based on prior experience of available resources, DOE2.2 was chosen to analyze the heating and cooling loads of a home. As DOE2.2 requires extensive inputs to perform a comprehensive simulation of all the loads, the loads, which do not have a significant impact on heating and cooling, are excluded from this model and were handled in the spreadsheet-based model. This approach is described in Section 3.4 in more detail.

3. PROJECT APPROACH

In order to quantify the enhancement in reliability we needed to identify relevant distribution reliability indices that would accurately reflect reliability improvements and offer a benchmark to reliability improvements. Once identified, distribution reliability data (duration, timing, and the number of customers affected in each outage) were used to construct a *reliability model* that could simulate random outages within a community. Finally, a *residential load model* was developed. This model consists of total and critical HVAC and appliance load data, in 15-minute intervals, for three regions and three community sizes. During an outage, total load (for a given community in a given region) is reconfigured to form the critical load, where management of PV and energy storage meets this load. Outage data were generated by the *reliability model*, while load data were generated by the *residential load model*.

3.1. Proposed Reliability Indices

We anticipate that managing loads, PV, and energy storage will reduce (or perhaps eliminate) the number of outages a customer experiences and will reduce the duration of each outage a customer experiences. Since load control will also allow customers to differentiate between critical and non-critical loads in their homes, the reliability indices should be revised to account for a customer's willingness to shed non-critical loads during a system interruption. Therefore, the following revisions to SAIDI and SAIFI are used in this study:

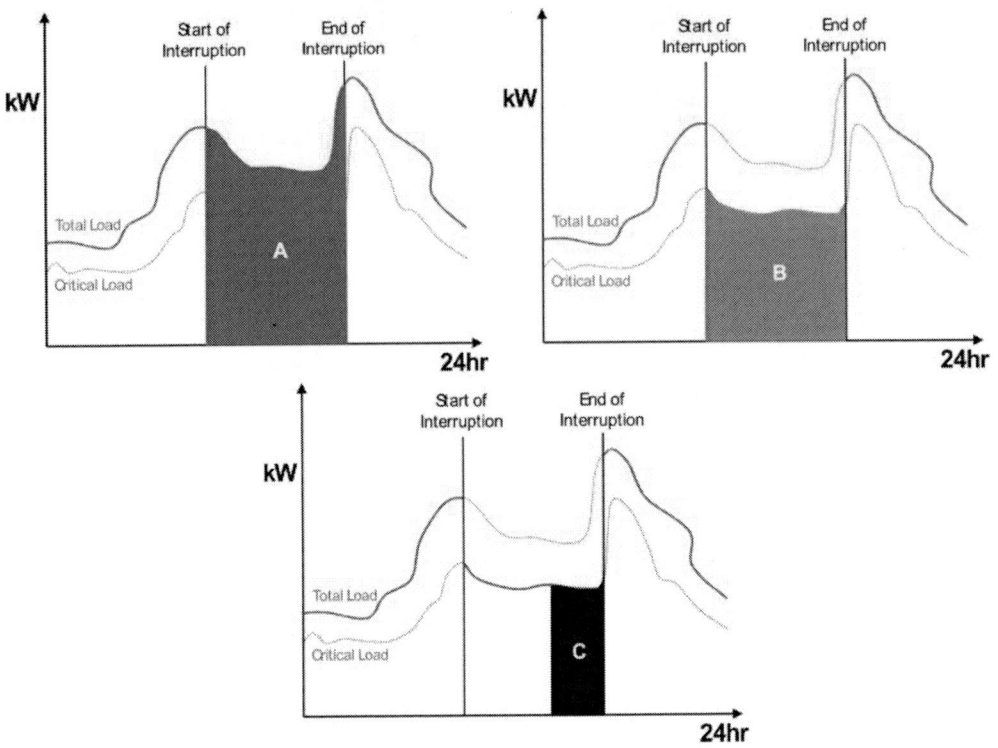

Figure 2. Graphical representation of unserved critical kWh associated with an outage.

- *Critical SAIFI* is defined as the average number of critical load interruptions experienced on a circuit.
- *Critical SAIDI* is defined as the average duration of critical load interruptions experienced on a circuit.
- *Unserved Critical Load (UCL)* is a proxy for ASIFI, defined as the annual unserved critical load (kWh) on a circuit.

In Figure 2, area "A" represents the *total* kWh unserved during an outage and area "B" represents the *critical* kWh unserved duration an outage. A PV system with energy storage and load controls will reduce the unserved critical kWh, shown as area "C". This study will quantify the enhanced reliability that PV systems with energy storage and load controls offer to residential communities based on these three indices.

3.2. Distribution Reliability Data

Reliability data for two utilities were provided by the Electric Power Research Institute (EPRI).[14] Utility A is a small utility (fewer than 200,000 customers) located in the northeastern United States. Utility B is a larger utility (greater than 800,000 customers) located in the southeastern United States. Generally, utilities collect data for outages in excess of five minutes. The data provided includes all outages in excess of five minutes (including major event days). Outage events were determined by supervisory control and data acquisition ,SCADA) measurement or by customer reporting. The data for utility A were provided with hourly time stamps for each event from 1994 to 2003, and daily time stamps for each event from 2004 to the end of 2006. Data were provided for utility B (with daily time stamps for each event) for the year 2005 only.

The reliability statistics chosen for this study were based on data from utility A. With more than 20,000 events in 10 years, utility A offered a significant database, containing 10 years of outage event data (outage duration, circuit affected, number of customers affected, weather condition, number of customers per circuit, etc.) with hourly time stamps for nine years of outage events. Although focusing on the distribution reliability performance of a single utility does not ensure that an "average" utility was considered in this study, the vast database of outage information was thought to provide the study with the best statistical data from a single data source. This ensures that outage durations obtained from one study are not intermixed with frequency data from another study. The model has been developed to incorporate distributions of outage duration (per event) and customers affected (per outage duration) as an input. Therefore, if additional data is available, additional model runs can easily be performed.

3.2.1. Outage Duration

Based on the number of outage events presented in the data for utility A, an outage duration histogram was generated for two consecutives year (see Figure 3).

Most outage durations in 2005 and 2006 were shorter than 100 minutes, though some outages were much longer. Note that each outage event affected a different number of customers.

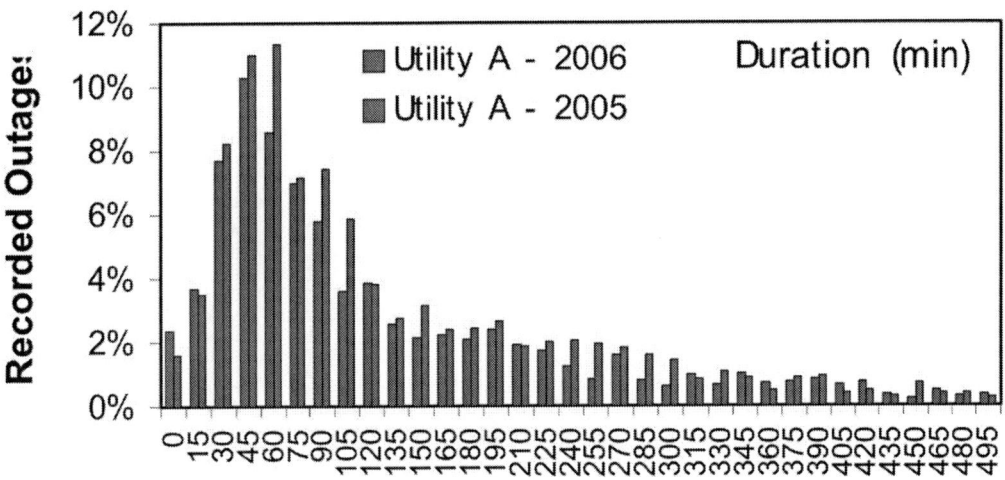

Figure 3. Historical outage duration per outage event for utility A.

3.2.2. Outage Timing

Based on the data for utility A, outages were not evenly distributed throughout the year. This trend was observable over various years of operation. The number of outage events per month in 2005 is shown for utility A and utility B in the top of Figure 4. Additionally, the number of outage events for four consecutive years at utility A is shown in Figure 4. The correlation of outage frequency to time of year can be easily observed. Based on the data provided for utility A, we examined the number of outage events over a 24-hour period (see Figure 5). These data are considered on an outage event basis, which treats each outage event as equal, independent of the number of customers affected.

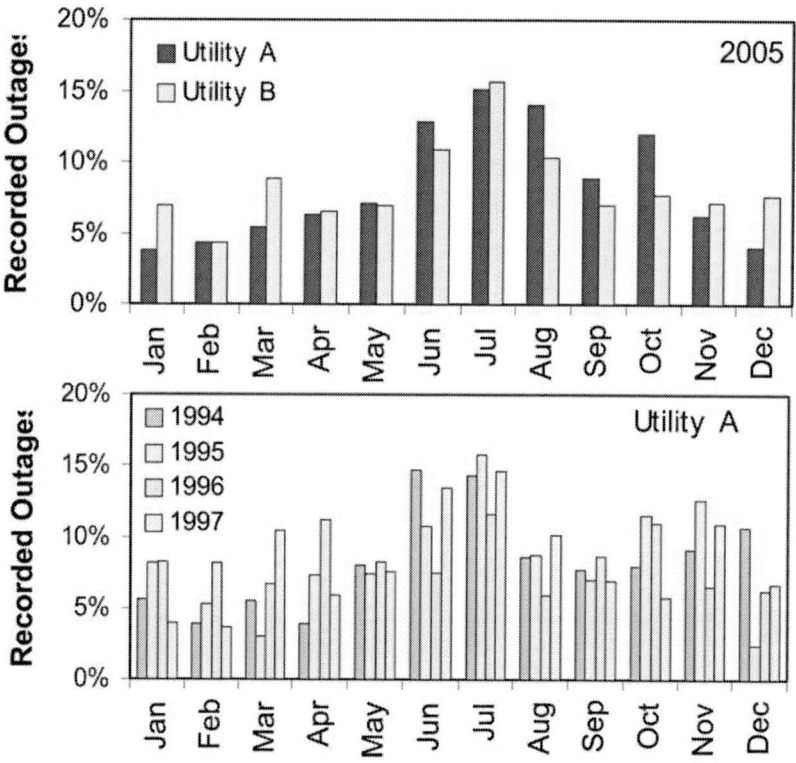

Figure 4. Number of outage events per month for utility A and utility B.

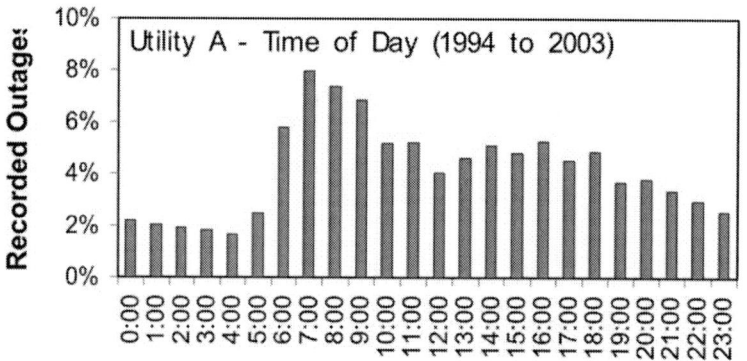

Figure 5. Hourly distribution of outage events for utility A from 1994 to 2003.

3.2.3. Outage Frequency

The SAIFI data is widely available for many utilities (see Figure 6), but SAIFI (number of outages experienced per customer) is an aggregate of outage frequency data and does not provide enough information to determine the distribution of outages experienced by each customer in a utility. Since there is a correlation between the number of customers affected by an outage event and the duration of the same outage event (see Figure 7) it was necessary to account for this relationship in order to validate the model's results against the SAIDI and SAIFI for utility A.

Based on the relationship between the customers affected and the outage duration, the energy systems model chose the outage duration and chose the number of customers affected based on the statistical distribution of customers affected for each outage (Figure 7). This was done to replicate the number of outage events experienced by utility A rather than replicate individual customer reliability throughout the year. Therefore, SAIDI and SAIFI for utility A were used to validate the simulation. Based on the data for utility A and utility B, the number of customers affected by each outage event highlights a similar relationship (see Figure 8).

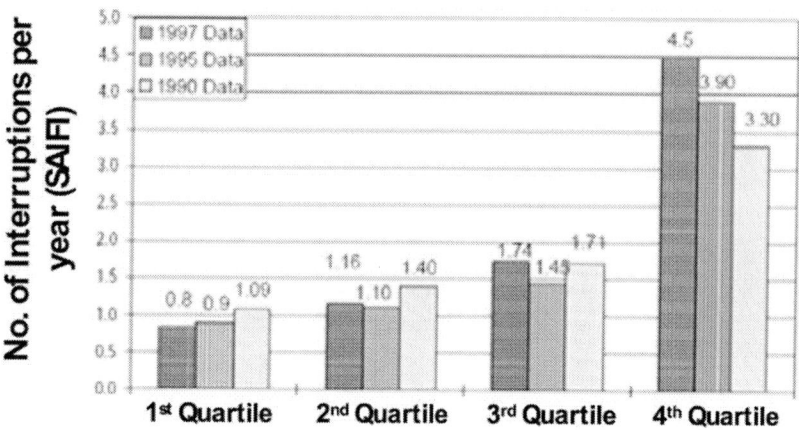

Figure 6. Utility SAIFI reported in 1990, 1995, and 1997 and sorted by quartile[15].

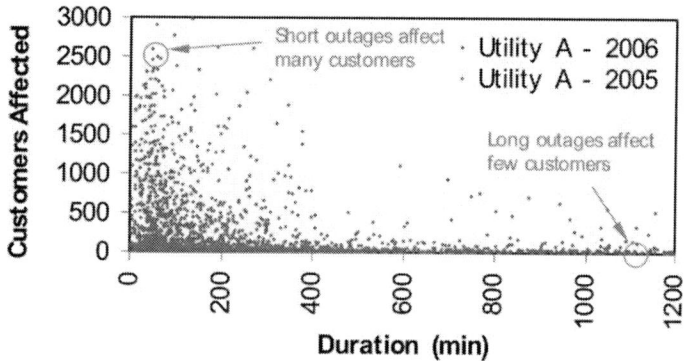

Figure 7. Number of customers affected based on various outage durations in 2005 and 2006 for utility A.

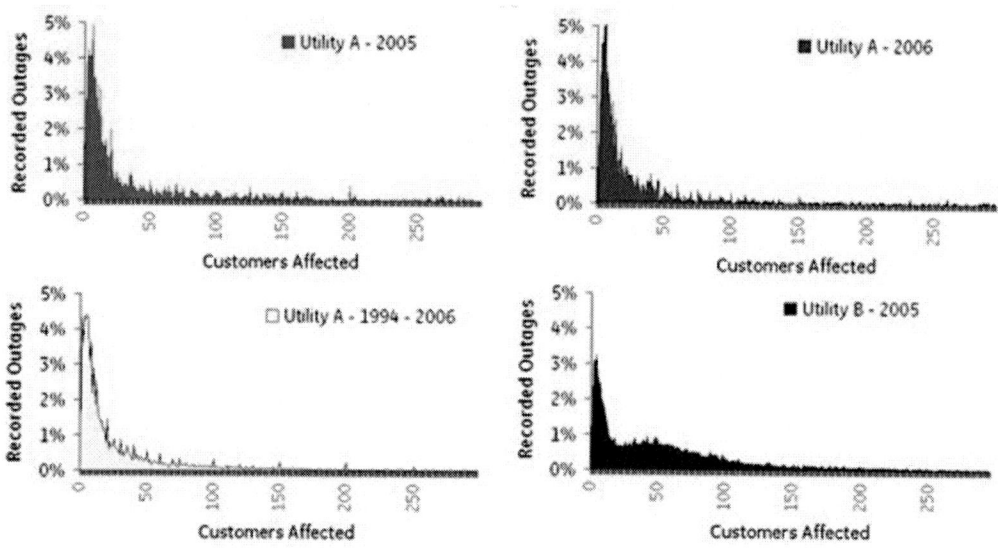

Figure 8. Customers affected during each outage event.

3.2.4. Assumptions and Factors Affecting Reliability

Many assumptions were made in order to simulate outage timing and duration. The lack of data on short-duration outages (less than five minutes) and the need to limit the size of the model, made it difficult to model short-duration outages. Since the model is discretized into 15-minute intervals, outages are rounded to the nearest 15-minute interval. This study will be unable to account for some factors, such as the correlation between lower than average solar insolation levels and poor weather conditions due to a storm. This factor presumably increases the probability of an outage. Similarly, the correlation between higher ambient temperatures (which leads to lower PV efficiency) on a "hot" day and increased outages due to hot weather will not be captured in this study. The weather observed during each outage event is shown for utility B in 2005 and utility A in 2005 and 2006 (see Figure 9).

The reporting of weather conditions is a subjective measure. Utility A and B used similar, though not identical, labels of weather condition.

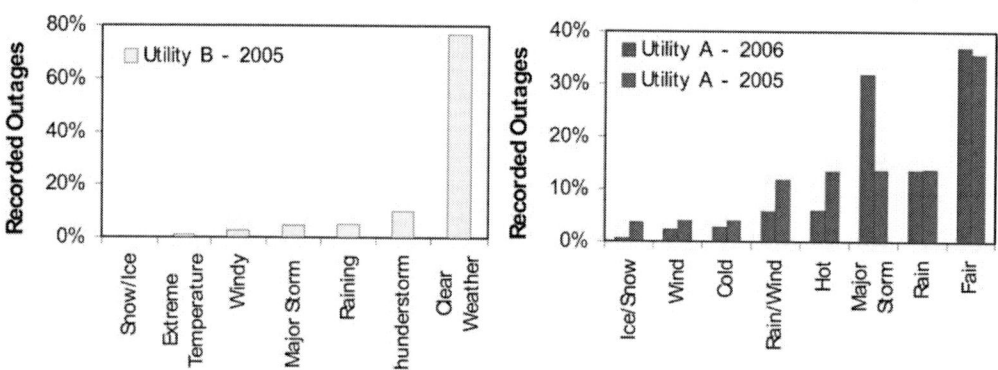

Figure 9. Weather conditions for each outage event for utility A and utility B.

3.3. Reliability Modeling Approach and Validation

In order to generate outage statistics for an outage event, we used statistical duration and timing data from utility A. The following list highlights the modeling approach:

1. For a single event, the outage duration was chosen from a bin of outage durations (15 minutes per bin) based on the discrete distribution of outage events in utility A in 2005 (see Figure 3).
2. The outage duration was chosen randomly from within the selected bin and rounded to a multiple of 15 minutes.
3. The month in which the event occurred was selected based on the discrete monthly distribution of outage events for utility A in 2005 (see Figure 4).
4. The day in which the event occurred was chosen randomly from within the month.
5. The hour in which the event occurred was selected based on the discrete hourly distribution of outage events for utility A from 1995 to 2003 (see Figure 5).
6. The event was placed on the first 15-minute interval of that hour.
7. The number of customers affected is chosen based on the relationship between the outage duration and the number of customers affected in utility A (see Figure 7).
8. A single simulation consists of an outage duration, timing, and number of customers affected. This is performed 1,800 times to replicate the number of events experienced by utility A in 2005.

Based on this approach, 1,000 simulations of a single outage event were performed. The results are summarized in Figure 10. A good match between the results shown below and the reliability data from utility A was observed.

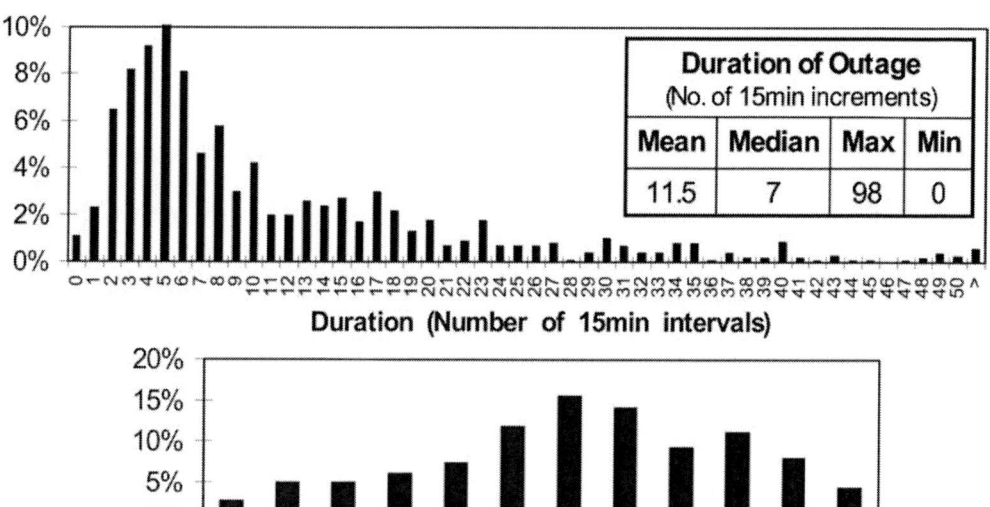

Figure 10. Outage duration and timing distribution based on 1,000 simulations.

Each outage event was input into the energy systems model and the reliability indices (described earlier) were quantified with and without the PV, energy storage, and load control system.

Based on this approach, the average of more than 100 simulations reveals that the simulation SAIDI replicates the actual SAIDI, within 5%, for utility A in 2005. Additionally, the simulation SAIFI compares within 1% of the actual SAIFI for utility A in 2005. Any discrepancy between the SAIDI/SAIFI for utility A and the simulation is associated with the statistical sampling of the number of customers affected by an event (equally probable) based on the outage duration for the event.

3.4. Residential Load Modeling

As described earlier, load shaping of residential end-users is a highly complex task; a household's energy usage is intimately linked to lifestyle-related factors that are extremely subjective and not easily defined. Furthermore census studies attempting to document this behavior don't fully resolve the problem, because they fail to consider the random variability of the demand.

The residential model is divided into two sections:

1. Matlab-based model for the top rated appliances and lighting
2. DOE 2.2 model for the heating and cooling loads.

The following two subsections describe these two approaches in more detail.

3.4.1. Appliance Load Modeling

A bottom-up approach is used to model electrical appliance loads in the home. This method models each appliance's behavior individually, allowing for a level of granularity that provides insight into the behavior of individual homes and power consumption per appliance. Furthermore, being aware of what each appliance is doing at all times enables load-shedding, allowing for continued operation of critical loads while shedding all noncritical loads, thus lowering the overall power consumption of each home. For the purposes of this report, the critical loads were lighting, refrigeration, and baseline power consumption, a definition of which will follow.

The model is composed of five appliances and lighting. Appliances were chosen by their popularity within the community,[16] frequency of use, and power consumption.[17] Once these values were understood, a statistics-based model assigned appliances, power values, and runtimes to each home in the community and dispersed the use of the assigned appliances throughout a day. The dispersion of appliance use throughout a day is modeled with help from Lawrence Berkley National Labs (LBNL) data,[19] which provide an individualized appliance usage profile over a twenty-four hour period. Furthermore, since this model is limited to five appliances, the baseline power consumption is added to each home in the community to account for most of the residential electrical loads, which are not modeled. The baseline power curve has a bimodal distribution with its peak load located during the hours of greatest electrical activity. The baseline power curve also degrades during the early hours of the morning when most of the community is expected to be sleeping.

For validation purposes, power consumption per appliance was cross checked against yearly expected consumption rates published by the Department of Energy (DOE) on Energy Guide labels and other similar sources. Furthermore, total power consumption throughout the year (including HVAC) was cross-checked against U.S. state average power consumption provided by an internal GE report. The weekly power consumption per appliance and yearly overall power consumption per home data were used to validate the model.

The model is implemented in the computational mathematics program MATLAB. Inputs to the model are read from a Microsoft Excel file and from direct input from the user (Figure 11). The Excel file contains appliance values for penetration into the community, electrical load range, use time, and variability in use start time (Table 1) where "penetration" refers to the percentage of customers in the community who own that appliance. This model only takes into account appliances that are predominantly electric; therefore appliances such as gas dryers were not considered. Furthermore, there are cases in which an appliance will have a greater than 100% penetration, most commonly observed in refrigerators. Typically 10% of homes will have a second refrigerator or standalone freezer. However, to reduce the complexity of the model, a maximum of 100% penetration was set per appliance. The impact of multiples of an appliance remains to be evaluated.

Also labeled in Table 1 are Low-/High-End Loads, which indicate the low- and high-end bounds of power consumption, in Watts, for each appliance being modeled. Section 3.4.2 further elaborates how each appliance is assigned a specific value. "Run Time Low/High" indicates low- and high-end bounds of appliance operation time; these values indicate a fraction or multiples of 15-minute intervals for which the appliance is in continuous operation. Section 3.4.3 further explains how these values are determined. Lastly, "Start Time Variability" is the standard deviation, in minutes, of the time when the appliance comes on, allowing community usage to be spread about a time continuum; this concept is further explained in section 3.4.4.

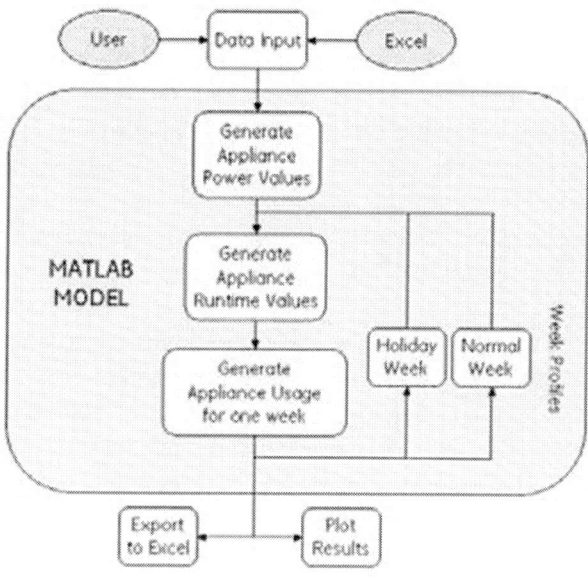

Figure 11. MATLAB appliance load modeling flowchart.

Table 1. Appliance Data Input to the MATLAB Model

Appliance	Hanford, California					
	Penetration (%)	Low End Load (Watts)	High End Load (Watts)	Run Time Low (15 min)	Run Time High (15 min)	Start Time Variability (Minutes)
Refrigerator	100	400	600	0.10	0.50	60
Dishwasher	55	1200	1400	3.00	5.00	60
Washer	68	400	600	4.00	6.00	840
Dryer	35	4500	5500	4.00	8.00	840
Range	37	2400	5000	1.00	4.00	120
Lights	100	400	1000	1.00	10.00	60

The direct input from the user consists of the community size, which is a number representing the number of homes to be simulated. As the community size grows, the total power consumption for the community becomes increasingly smooth. If the community size is small (e.g., 10 homes) the total power consumption for the community is more jagged as individual appliances can be seen turning on and off.

Based on these model inputs, a virtual community of electric loads is generated. The simulation outputs the temporal usage of appliances and lighting. The following sections further describe the generation of the community and parameters for operation.

3.4.2. Generate Appliance Power Values

Once a community size has been chosen, each home is provided with electric appliances. It is important to note that not all appliances of the same type are equal in their power consumption. Therefore, to accommodate this factor, a published range of possible power consumptions for each appliance is uniformly sampled to achieve variation within the community. Furthermore, not every home in the community is assigned every electrical appliance. Penetration of an electrical appliance into a particular community is determined by regional census data provided by the DOE Energy Information Administration (EIA). This helps account for homes that do not, for example, have a dishwasher and use a gas range and/or dryer. Therefore, some homes will have every appliance whereas other homes may only have a refrigerator and lights in addition to the baseline power consumption. Figure 12 represents the five appliances and lighting for a community of 100 homes. We can see that the refrigerator and lights are present in every home throughout the community whereas the other appliances have varying degrees of penetration; for example, in this community 67% of the homes have an electric range. Once the algorithm has finished assigning appliances to every home in the community, these values are then saved and remain fixed for the remainder of the simulation.

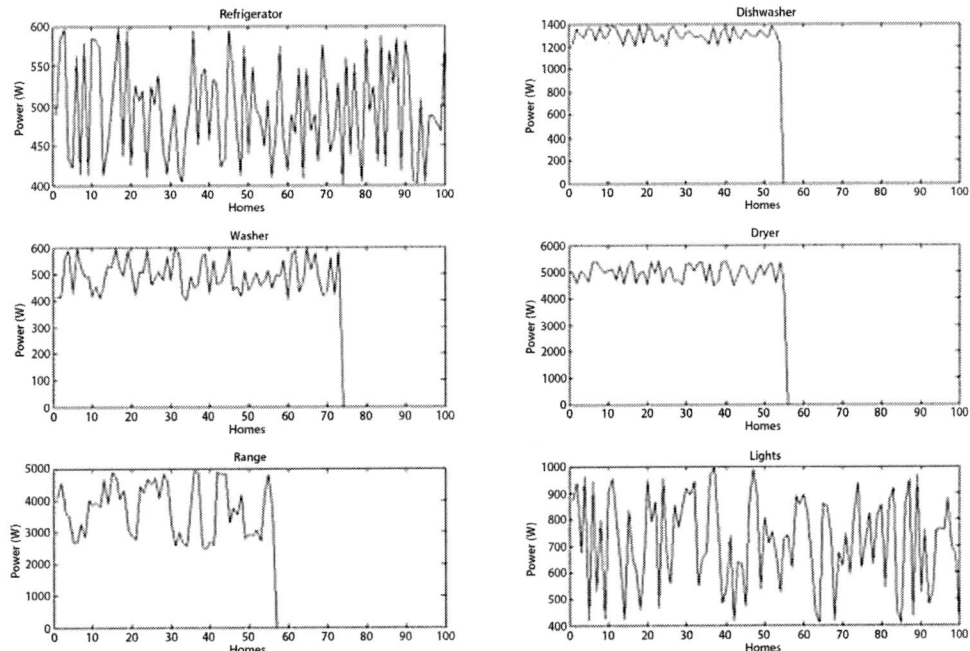

Figure 12. Appliance power and penetration for a community of 100 homes.

To further improve the accuracy of the model, some parameters should be further researched and optimized. One of these parameters is the purchasing habits of consumers; who may prefer a particular brand and/or cost. This results in a normal rather than uniform distribution of power consumption. Furthermore, this model does not take into account consumer preference variation across the different regions modeled; only appliance penetration per region is accounted for. Finally, the variation of power consumption during an appliance's operation is neglected, and is replaced by a constant rate of power consumption. After completing this step the model moves to generating runtimes for each appliance in every home of the community.

3.4.3. Generate Appliance Runtime Values

Before a typical utilization time for an appliance can be calculated, it should be confirmed that the census data, which provides appliance power and total usage time over one month, agrees with the Energy Guide yearly totals. For example, a dishwasher is reported to be in operation 8 to 40 hr per month[17]. The following can be calculated:

$$Avg = \frac{1200W + 1400W}{2} * \frac{8\frac{hrs}{mo.} + 40\frac{hrs}{mo.}}{2} * \frac{12\ months}{year} = \frac{374.4\ kWh}{year}$$

Where the first factor is the average dishwasher power consumption, the second factor is the average use in hours per month. Lastly, these factors are multiplied by months in a year to obtain the yearly kWh usage. The results compare well with an average 323.3 kWh per year expected consumption.[18] Using the 8 to 40 hrs of operation per month, a daily usage time is calculated, assuming a thirty-day month. This last operation results in a range of daily "ON" times for each appliance. It is important to consider that some appliances operate one or multiple times an hour and others operate once to multiple times a week, such as the washer and dryer. At this point, the model uniformly samples these ranges, assigning each appliance in the community an individualized runtime. Furthermore, all runtime values are regenerated every week for further variability. As an example, Figure 13 is a histogram graph illustrating binned runtimes and the corresponding number of homes for which that runtime applies.

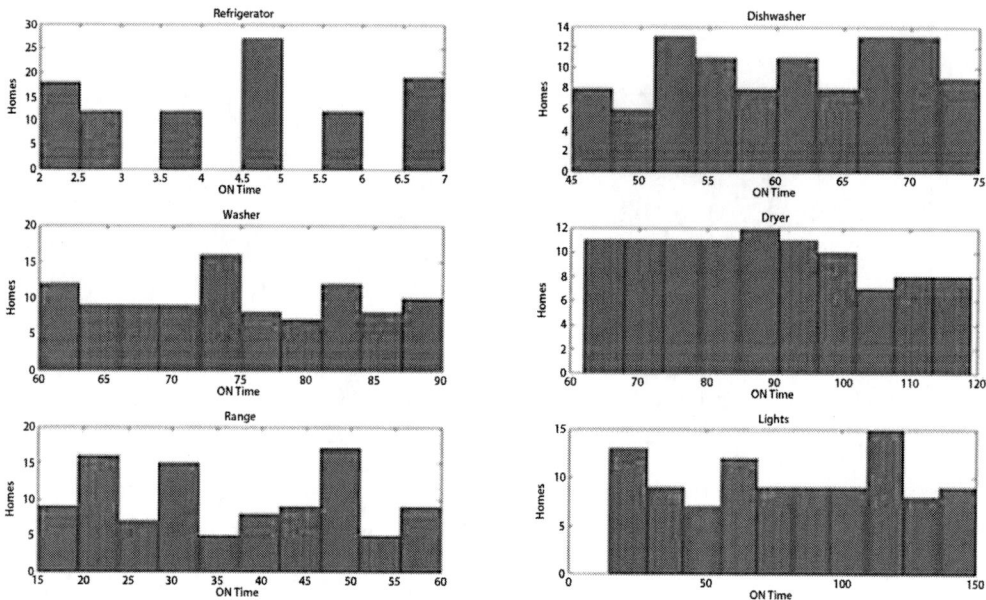

Figure 13. Appliance runtime (in minutes) for a community of 100 homes.

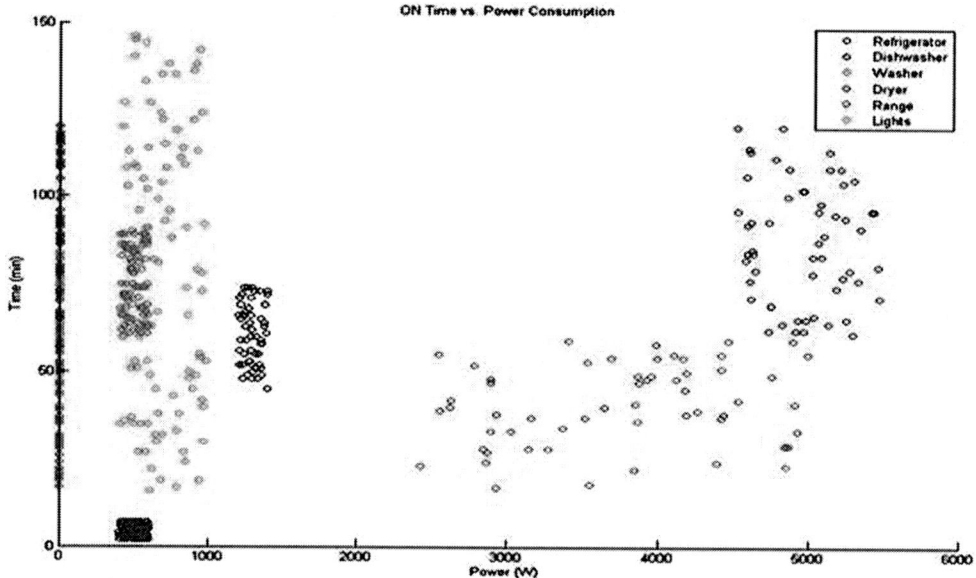

Figure 14. Appliance power and runtime for a community of 100 homes.

Enhanced Reliability of Photovoltaic Systems with Energy Storage and Controls   189

The model does not take into consideration how appliance runtimes change throughout the year. Furthermore, there is no correlation between appliance power usage and runtime, so a 5.5 kW clothing dryer could run for up to two hours and a 4.5 kW dryer could run for only one hour. Lastly, each selection of runtimes is uniformly chosen from the range of runtimes. This is an assumed behavior and it may not be an accurate representation of consumer habits.

Once appliance power consumption and runtimes are set, a visual check is performed to avoid any outliers (see Figure 14). Furthermore, the figure illustrates many runtimes at zero power. This was done to simplify the model; every home is given runtimes for every electrical appliance within it, whether it exists or not, but the product of these two will not contribute to the overall kWh consumption for the home if the appliance power value is 0 W.

At this point the MATLAB model has generated a community with electrical appliances and runtimes for each. The next step is to describe the procedure used to schedule appliances throughout the week based on the community's behavior.

3.4.4. Generate Appliance Usage for One Week

Accurately scheduling the residential loads within a community is a critical next step. This proves to be a challenging task and requires us to make many assumptions in order to keep the complexity of this model relatively low.

The model uses a pre-populated spreadsheet as shown in Figure 15. Each row represents a 15-minute interval, while columns represent appliances and lighting. The array of zeros and ones represent whether an appliance is turned on or off. There is one master spreadsheet for the entire community, and two individual day profiles—a weekday and a weekend profile. Holidays are assumed to be most similar to weekend days, and are therefore overwritten with weekend day profiles.

	Refrigerator	Dishwasher	Washer	Dryer	Range	Lights	
6:00 AM	1	0	0	0	0	1	
6:15 AM	0	0	0	0	0	0	
6:30 AM	1	0	0	0	0	0	
6:45 AM	0	0	0	0	0	0	
7:00 AM	1	0	0	0	0	0	
7:15 AM	0	0	0	0	0	0	
7:30 AM	1	0	0	0	0	0	
7:45 AM	0	0	0	0	0	1	
8:00 AM	1	0	0	0	1	0	
8:15 AM	0	0	0	0	0	0	Sunday
8:30 AM	1	0	0	0	0	0	
8:45 AM	1	0	0	0	0	0	
9:00 AM	1	0	1	0	0	0	
9:15 AM	1	0	1	0	0	0	
9:30 AM	1	0	0	0	0	0	
9:45 AM	0	0	0	1	0	0	
10:00 AM	1	1	0	1	0	0	
10:15 AM	0	0	0	0	0	0	
10:30 AM	1	0	0	0	0	0	

Figure 15. Appliance scheduling spreadsheet.

These spreadsheets were populated using LBNL graphs[19] as a reference for peak appliance demand throughout weekday and weekend days. The graphs were translated to ones and zeros within the spreadsheet. However, to avoid every dishwasher turning on at 10:00 AM, another factor, called start time variability (STV), was introduced to distribute the start time of the appliance around the 10:00 AM timestamp. Based on data from LBNL[19], the spread of start times was uniformly or normally distributed about the ones in the spreadsheet. Furthermore, as can be seen in Figure 15, both the washer and the dryer are being operated twice on Sunday, but the spreadsheet may be misleading as to their actual times of operation. Table 1 shows that both the washer and the dryer have a STV of 840 minutes or 14 hours, and their appliance demand distribution, Figure 16, shows a relatively uniform probability of dryer operation from 8 AM to 10 PM for a typical weekday in February. This simulates a little more normal behavior on the weekend but still has a mostly uniform distribution. Hence this MATLAB model will uniformly distribute the operation of the washer and dryer over this timeframe. However, the operation of a range has a more normal distribution in the morning and evening hours on the weekend and weekdays respectively, as seen in Figure 16. Therefore, the algorithm will normally distribute the range operation in the community between about 8 AM Sunday morning and at 6 PM Sunday evening.

One of the shortcomings of this model is that it does not account for the chronological order of events; it may start a dryer before a washer. However, in a large community these events would represent a very small percentage of all events. Furthermore, as the model runs throughout the year, holidays are accounted for by exchanging week profiles, as seen in Figure 11, and accommodating holidays to the appropriate day of the week by substituting a weekend profile for a weekday holiday.

The implementation of the table in Figure 15 is highly subjective and may change radically from one community to another. Appliance usage patterns drift slightly as suggested by LBNL data,[19] but to minimize model complexity these relatively minor drifts were ignored. Adding the LBNL data[19] into this model should be improved by implementing the actual equations that produced the figures shown in Figure 16. Using piecewise continuous equations could more accurately distribute the use of appliances throughout a day without confining them to uniform or normal distributions. In addition, the existence of an appliance in the home does not necessarily imply its use. The current model assumes that all appliances existing in the community are used when the model encounters a "1" in the appliance scheduling sheet; this is probably not the case in an actual community. Data available at the EIA[20] correspond to regionally dependent appliance usage patterns[21]. Further implementation of these studies would complement LBNL-provided data, thus increasing model accuracy.

The following section will address the interpretation of the simulation results.

3.4.5. Modeling Results

Once the simulation has generated a year's worth of data, interpreting and using these data is an important part of the validation exercise. One effect observed in reality is this notion of coincidence. Some utility customers could transiently demand upwards of 15-20 kW at any point in time, but the likelihood that every other home in the community demands a similar magnitude of power is statistically improbable. This concept is well understood when it comes to transformer sizing for a finite size community.[22]

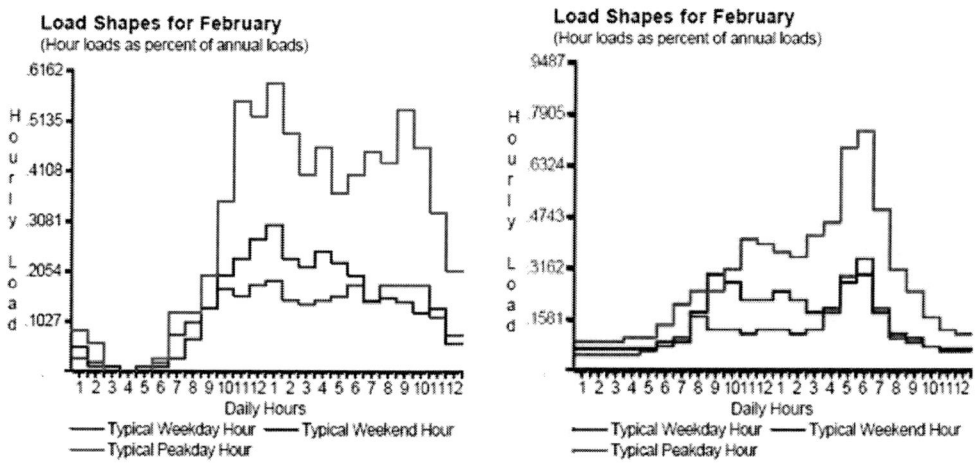

Figure 16. LBNL-provided chart[19] corresponding to residential drying (left) and residential cooking (right).

We would expect the virtual community to exhibit a similar behavior. Figure 17 represents the power drawn by a community of 100 homes on a per-home basis, that is to say, that all individual home loads have been added together and divided by the number of homes in the community. The result is an average home consumption for that community. This averaging effect, as can be seen in Figure 17, results in an average home power profile that is smooth and continuous throughout the day. However, given the bottoms-up approach of our model, we are able inspect home load profiles individually and notice higher power demand peaks as individual loads are turned on and off within the home. Figure 18 illustrates a single home where peak demand reaches approximately 12.5 kW for a few minutes. It is important to note that this peak is constituted only of domestic appliances; it does not include HVAC loads. The reader may notice that Figure 17 does not seem to be representing any load corresponding to a clothes washer. Therefore, in Table 2, the total kWh consumed per appliance during the 24-hour period is displayed, and shows that, in fact, the washer does contribute, but in a small way, and the reasons for this behavior are explained.

The first item to point out in Table 2 is that indeed the washer contributes 0.068 kWh to the overall profile in Figure 17. The main reason this value is so low is that the average home power consumption profile, as illustrated in Figure 17, represents a clothes washer usage averaged out over a community where only a fraction of the homes have that appliance; this appliance penetration factor was discussed in section 3.4.1 and more specifically in Table 1. Therefore, averaging the use of the clothes washer over the entire community lowers the average due to the homes that do not own a clothes washer, hence making it difficult to see in Figure 17. Instead, if the kWh consumption for the clothes washer in Figure 18 (a profile for an individual home with a clothes washer) were calculated, the result would be 0.85 kWh for that day, which, if projected over a year, yields approximately 200 kWh annually. This value is within the Energy Star annual kWh consumption range for clothes washers[23] of approximately 100 kWh to 400 kWh. Furthermore, it is apparent that the dryer has a ten-fold impact over the washer as shown in Table 1. This is not completely unexpected since a washer usually consists of a drum or agitator motor, a water pump and in some models, a heating element, and only the motor operates during most of the cycle. On the other hand, a clothes dryer consists of a drum motor, a blower to circulate air and a substantial heating element, all operating simultaneously.

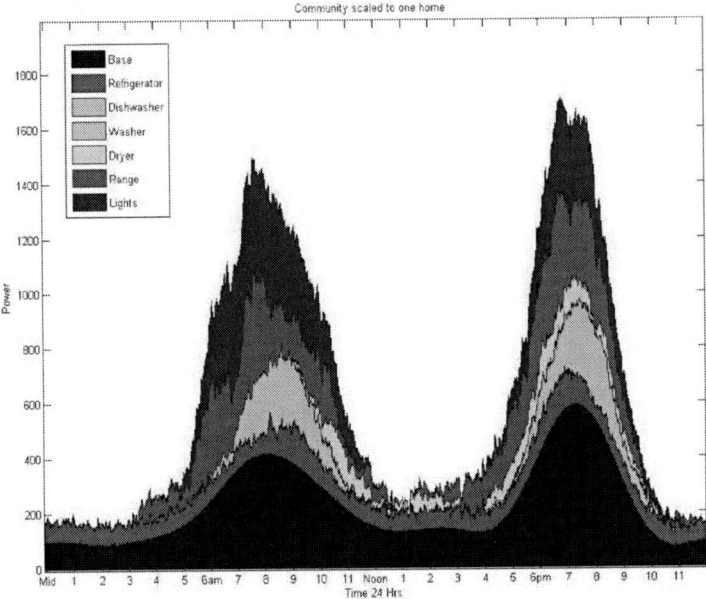

Figure 17. Power consumption by appliance (excluding HVAC) for a community of 100 homes scaled to 1 home.

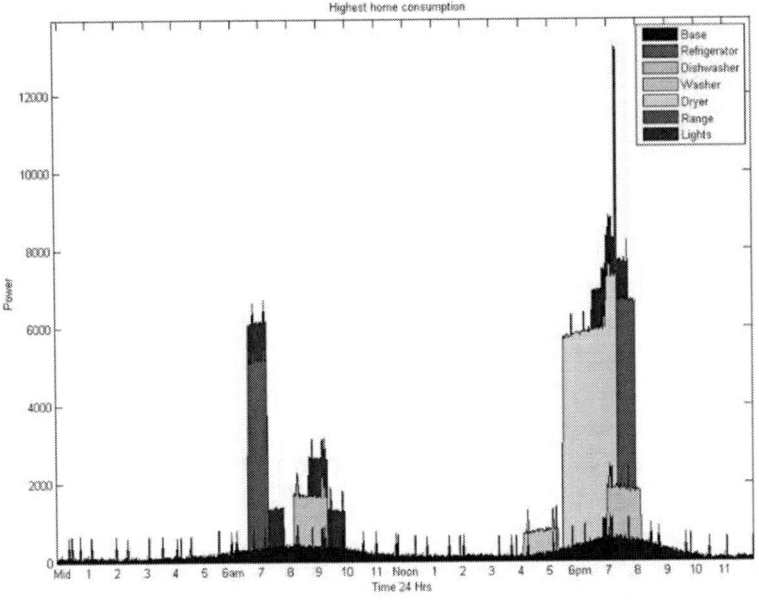

Figure 18. Power consumption per appliance (excluding HVAC) for a single home over a 24hour period.

Table 2. kWh Consumption per Appliance Represented in Figure 17

One Day of Consumption per Appliance							
Appliance	Base	Refrigerator	Dishwasher	Washer	Dryer	Range	Lights
kWh/day	1.138	1.859	1.547	0.068	0.677	2.682	4.306

Another important point to remember is that the day profile shown in Figure 17 cannot simply be repeated 365 times to represent a year's worth of appliance power consumption. This is because weekends and holidays are simulated with different user behaviors while the model also varies statistically on a daily basis.

3.4.6. Heating and Cooling Load Modeling

In this part of the study we adopted a dynamic building energy simulation program (DOE-2.2). This program was developed by LBNL as a tool for HVAC load simulation to determine the adequate size of heating and/or cooling systems. The program is able to simulate energy performance and HVAC loads of a building hour by hour for each of the 8760 hours in a year. The DOE-2.2 program is composed of four modules that are executed sequentially. The LOADS module calculates the hourly cooling and heating loads using algorithms suggested by the American Society of Heating, Refrigerating, and Air-Conditioning Engineers (ASHRAE). The SYSTEMS module simulates the performance of secondary HVAC equipment under conditions of maintaining indoor comfort within the building. The PLANT module simulates the energy performance of primary HVAC equipment, such as chiller, boiler, and cooling tower, on the basis of operating conditions and part load performance characteristics. The maximum chiller capacity can be acquired from the PLANT module. The fourth module, ECONOMICS, tackles economical benefit analysis, and was not included in the current study.

eQUEST is a graphical front end to the widely used freeware building energy analysis program DOE-2 (version 2.2). DOE-2 is a powerful building simulation program written in Fortran. However, setting up a building simulation using the DOE-2 engine is quite complicated. The user is required to provide a text file containing all building parameters such as geometry, construction materials, weather data, etc., in DOE-2's building description language (BDL). This is very cumbersome for reasonable sized buildings. eQUEST provides a layer of graphics and wizards on top of DOE-2 and contains a parser to convert the graphical inputs into BDL that can then be processed by DOE-2. For example, eQUEST allows the user to draw the building footprint shape using a sequence of keyboard and mouse commands rather than specifying each vertex in a text file. In addition, it contains features such as the ability to import building plans directly in as AutoCAD files. Another feature is that the building model can be specified via wizards, which are pre-populated with sensible default values for unknown building parameters. In addition, there is a detailed data edit mode, which exposes most of the functionality of DOE-2 while still retaining a user-friendly graphical interface.

A standard four-bedroom 2500-ft^2 colonial-style home was chosen for the simulation. Since the size of the PV array was chosen to achieve a specific PV penetration (in nameplate array size divided by the maximum 15 minute average total load), the size of the home does not impact the quantification of enhanced reliability due to the presence of PV (in % penetration). Also, without details of the potential market for enhanced reliability with batteries, PV, and load management, it is difficult to posit the size of the average home in such a community. The layout from a builder's Web site in the northeastern United States is used for this purpose. The DOE-2.2 uses meteorological data to acquire accurate and local climatically responsive HVAC loads. NREL's Resource Analysis group provided data for three locations: California, Colorado, and New Jersey. The HVAC meteorological simulation used data containing 15-minute weather information of outdoor dry and wet bulb temperatures, relative humidity, total horizontal solar radiation, normal solar radiation, and wind-speed for the whole year.

Electric Consumption (kWh x000)

	Jan	Feb	Mar	Apr	May	Jun	Jul	Aug	Sep	Oct	Nov	Dec	Total
Space Cool	-	-	0.00	0.02	0.15	0.40	0.77	0.82	0.28	0.07	0.00	-	2.53
Heat Reject.	-	-	-	-	-	-	-	-	-	-	-	-	-
Refrigeration	-	-	-	-	-	-	-	-	-	-	-	-	-
Space Heat	-	-	-	-	-	-	-	-	-	-	-	-	-
HP Supp.	-	-	-	-	-	-	-	-	-	-	-	-	-
Hot Water	-	-	-	-	-	-	-	-	-	-	-	-	-
Vent. Fans	0.10	0.08	0.07	0.05	0.03	0.04	0.08	0.08	0.03	0.03	0.05	0.09	0.75
Pumps & Aux	0.04	0.03	0.03	0.02	0.00	-	-	-	0.00	0.01	0.02	0.03	0.19
Ext. Usage	-	-	-	-	-	-	-	-	-	-	-	-	-
Misc. Equip.	0.35	0.32	0.36	0.37	0.41	0.35	0.35	0.37	0.32	0.36	0.33	0.34	4.23
Task Lights	-	-	-	-	-	-	-	-	-	-	-	-	-
Area Lights	0.34	0.31	0.35	0.34	0.34	0.34	0.34	0.35	0.33	0.34	0.32	0.34	4.04
Total	0.83	0.74	0.82	0.80	0.94	1.13	1.54	1.63	0.97	0.81	0.74	0.80	11.74

Gas Consumption (Btu x000,000)

	Jan	Feb	Mar	Apr	May	Jun	Jul	Aug	Sep	Oct	Nov	Dec	Total
Space Cool	-	-	-	-	-	-	-	-	-	-	-	-	-
Heat Reject.	-	-	-	-	-	-	-	-	-	-	-	-	-
Refrigeration	-	-	-	-	-	-	-	-	-	-	-	-	-
Space Heat	36.31	29.64	26.08	16.58	5.74	0.92	0.26	0.04	1.73	8.14	19.34	31.16	175.95
HP Supp.	-	-	-	-	-	-	-	-	-	-	-	-	-
Hot Water	0.17	0.16	0.18	0.18	0.19	0.15	0.14	0.14	0.12	0.14	0.14	0.15	1.84
Vent. Fans	-	-	-	-	-	-	-	-	-	-	-	-	-
Pumps & Aux	-	-	-	-	-	-	-	-	-	-	-	-	-
Ext. Usage	-	-	-	-	-	-	-	-	-	-	-	-	-
Misc. Equip.	-	-	-	-	-	-	-	-	-	-	-	-	-
Task Lights	-	-	-	-	-	-	-	-	-	-	-	-	-
Area Lights	-	-	-	-	-	-	-	-	-	-	-	-	-
Total	36.48	29.79	26.26	16.77	5.93	1.07	0.39	0.18	1.85	8.28	19.47	31.31	177.79

Figure 19. Detailed electric and natural gas consumption over a year.

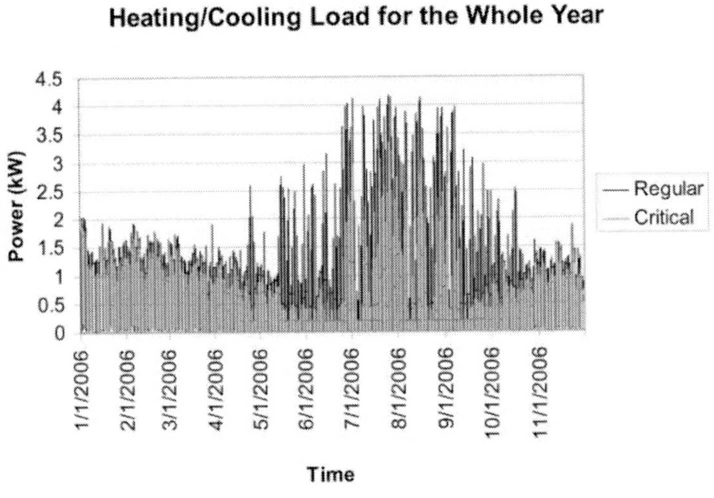

Figure 20. HVAC power consumption of a Colorado home over one year.

The other key data that were assumed are occupancy levels, HVAC set points, and lighting profiles (used only to calculate the heat gains). Many existing literature sources were consulted to obtain this data. Sample results for a whole year of simulation are shown in the figure below. The results are validated by the EIA Web site.

During outages, we assumed that the homeowners would be able to relax the constraints on the heating and cooling set points. A new table of set-points was generated with this relaxed constraint and the DOE2.2 simulations were repeated to get a new load profile, which was used for reliability calculations during outage conditions. A plot of regular (non-outage conditions) and critical loads for one whole year in Colorado is shown in Figure 20.

Furthermore, to account for different types of heating and cooling topologies, we repeated the DOE2.2 model, and the results were then weighted and combined to provide average power consumption for the region. Two methods for heating and cooling were considered. The first uses direct expansion cooling and a gas furnace, and the second method uses a heat pump for both heating and cooling needs. Data gathered for the Northeast shows a 66% use of direct expansion cooling with a gas furnace and 33% use of a heat pump; this ratio was assumed to be constant for the other two regions. Since the DOE2.2 model represents energy consumption for one home, the simulation was repeated twice for each region, once with each method. The results were then weighted and combined to generate an average home consumption for that particular region.

3.5. Energy Modeling

In order to determine the enhanced reliability associated with PV and energy storage, an energy balance model was developed. This model incorporates outage duration, timing, and the number of customers affected (based on the reliability statistics described earlier), and simulates PV output and battery output during an interruption for an individual home. Another key input to the model is the load profile (unique for each region and community size). Developing a model capable of simulating the critical and total load profiles, in 15-minute intervals, comprised a substantial component of the effort in this study.

The modeling approach is described below:

1. Loads for a community of size n, in geographic location m, have been classified as critical or non-critical for each 15-minute period of the day, for 365 consecutive days.
2. The incident solar insolation and temperature is provided for location m.
3. The PV output is calculated for location m.
4. If there is no interruption, electricity is consumed from the PV array and the grid charges the battery
5. If the battery is charged, PV is used to reduce grid consumption
6. If there is an interruption, non-critical loads are shed and available PV energy is used to meet critical loads
7. The battery is discharged to meet the remaining critical loads
8. When battery is completely discharged, critical loads are shed and an interruption occurs.
9. After the interruption, the grid meets all loads. Restart at Step 2.

This approach is highlighted in Figure 21.

Figure 21 shows an interruption occurring at different times of the day. Based on the modeling approach described above, energy from the battery and PV array was used to meet the critical loads for each home. The top left figure shows the battery meeting the entire critical load in the early morning hours (prior to sunrise). The top right figure shows the battery meeting the critical load prior to sunrise and meeting a portion of the critical load after the PV is available. The bottom left figure shows significant contribution from the PV array. The bottom right figure shows the battery discharging all of its energy prior to the end of the interruption. The customer would experience an interruption.

3.5.1. Battery Energy Storage System

Since the purpose of this study is to quantify the enhancement in reliability, a specific battery was not chosen in order to ensure technology neutrality. However, it was necessary to specify some battery parameters. The battery system was assumed to operate within a 10% to 90% depth of discharge (DOD) range for the maximum rated capacity in kWh. In this analysis we assumed that the battery (and power electronics) has a 90% charging and discharging efficiency and appropriately sized power electronics to meet the maximum critical load the home will experience during the year. It should be noted that based on statistical outage data, a battery in an average home in the United States could expect to be used, on average, once or twice per year (see above discussion on reliability statistics). We assumed that the energy storage system was located at each home in the community and represents a stand-alone battery or perhaps, in the future, the battery of a plug-in hybrid electric vehicle. For the purposes of this analysis, the relatively high efficiency may slightly overstate the impact of the battery on enhancing reliability. For details about battery specifications, see Wiegman & Lorenz[24] and Stevens & Corey[25].

3.5.2. Photovoltaics

The global insolation data were provided for three locations in the United States (Golden, Colo., Hanford, Calif., and Sterling, Va.) for the year 2003. The global insolation was translated into the actual insolation incident on the PV array. The incident insolation was then used to determine the maximum electrical power that the PV array can deliver. The following equation was used to model the PV output for a south-facing array, based on a PV array developed using the GE 200-W module. The equation was derived from Messenger & Ventre[26].

$$P_{array} = (-0.006825 T_{ambient} + 1.171) P_{incl} P_{fc} \frac{P_{global}}{P_{isoref}} P_{maxref}$$

where,

P_{global} is the global solar insolation (W/m^2) data, provided by NREL, for the three regions described above. The data varies from zero to the maximum daily solar insolation (in 15-minute intervals for an entire year).

P_{incl} is an inclination correction that varies, during the daylight hours, between zero and unity, to account for the orientation of a south-facing array with respect to ground level.

P_{fc} is a longitudinal correction that varies, during the daylight hours, between zero and unity, to account for the location of the site on the earth, and within the time zone. There are corrections for daylight savings time.

P_{maxref} is the maximum output power of the PV array at 1,000 W/m^2 and 25°C.

P_{isoref} is the reference solar insolation (1000 W/m^2).

$T_{ambient}$ is air temperature near the PV array (°C).

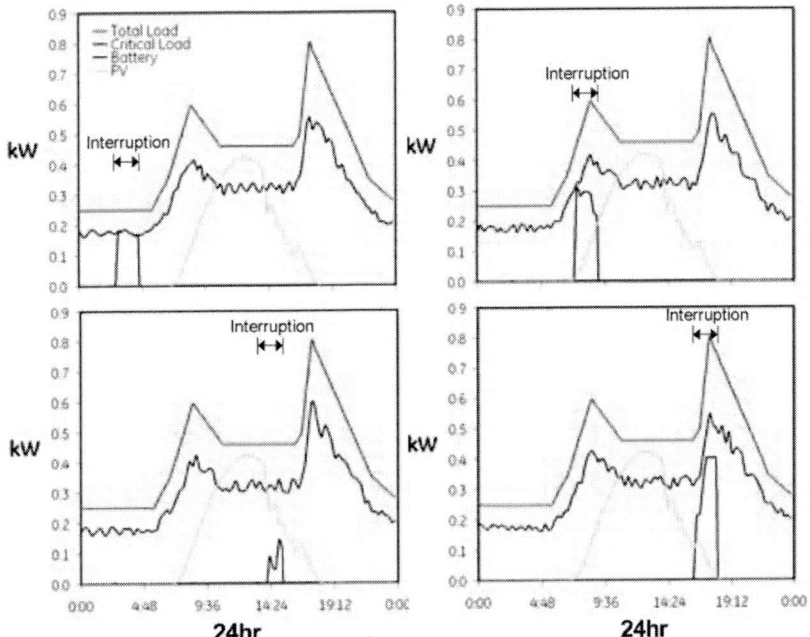

Figure 21. Energy modeling simulation validation.

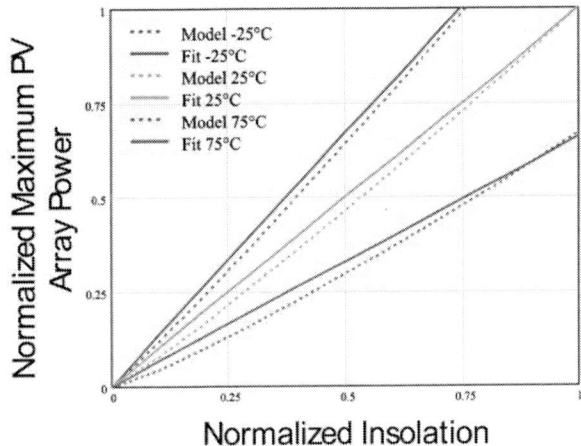

Figure 22. Normalized PV model equation fit.

The parameters related to the array performance were obtained from the specifications of a GE 200-W solar array[27] that was fitted to the Shockley Solar cell equation model and scaled using the electrical output power for a particular site. The Shockley model was used to determine the maximum electrical power of the PV array as a function of insolation level and normalized to the power level data given for a site. Since this model is complex, a linear fit equation was derived. This fitted equation introduces minimal error to the results as shown in Figure 22

The following assumptions were made:

- The temperature of the PV array is uniform over the entire array.
- The wind and ambient temperature effects on PV array were neglected.
- The effect of albedo was neglected.
- PV inverter efficiency was assumed to be 95% CEC and constant over the entire operating range of inverter operation.
- The insolation is uniform across the entire PV array.
- Effects of shading are neglected.
- The PV array is south-facing only.
- The PV array design was electrical-power driven and not based on a specific PV array design of x modules in series.

The equation used to model the PV output incorporated information and equations from multiple sources [28, 29].

Hourly PV output (kW AC) from a 1-kW south-facing array with a 25° tilt located in Golden, Colorado, was obtained from NREL based on simulations using PVWatts, a performance calculator for grid-connected PV systems. The insolation data were obtained from the NREL SPRL site in Golden and the weather data were taken from a nearby airport. The PVWatts output data were compared with the model results for five GE 200-W arrays with the same orientation. As shown in Figure 23, the hourly PV output from the model compares extremely well with the NREL data from 2003 during the middle part of the year. However, during the beginning and the end of the year the discrepancy between the model's output and the NREL data is observable. As shown in the bottom right of Figure 23, the cumulative error increases during the beginning and end of the year, but remains nearly constant during the middle part of the year. The difference between the annual kWh produced by the model and observed output from the 1-kW array in Colorado is 14%. This was deemed to be adequate for quantifying the enhancement in reliability within a distribution feeder. The model was deemed valid for a study focused on the impact of PV, energy storage, and load control on enhanced reliability in a distribution feeder.

4. PROJECT RESULTS

A design of experiments was conducted in order to identify the parameters of interest (Table 3). These parameters were combined as inputs to the model.

The community size, geographic region, PV array size (chosen to meet specific PV penetration levels), and battery capacity were taken as input parameters to the model (see

Table 4). The output table is also shown in Table 2. For the simulation presented here, the outage is 90 minutes in duration (six 15-minute intervals) and affected 872 customers. The outage started at the 7884th 15-minute time stamp of the year (March 24 at 2:45 PM). Based on the input parameters for this simulation, both the unserved critical kWh and the outage duration decreased due to the load support provided by the battery (1 kWh) and PV (2.1 kW).

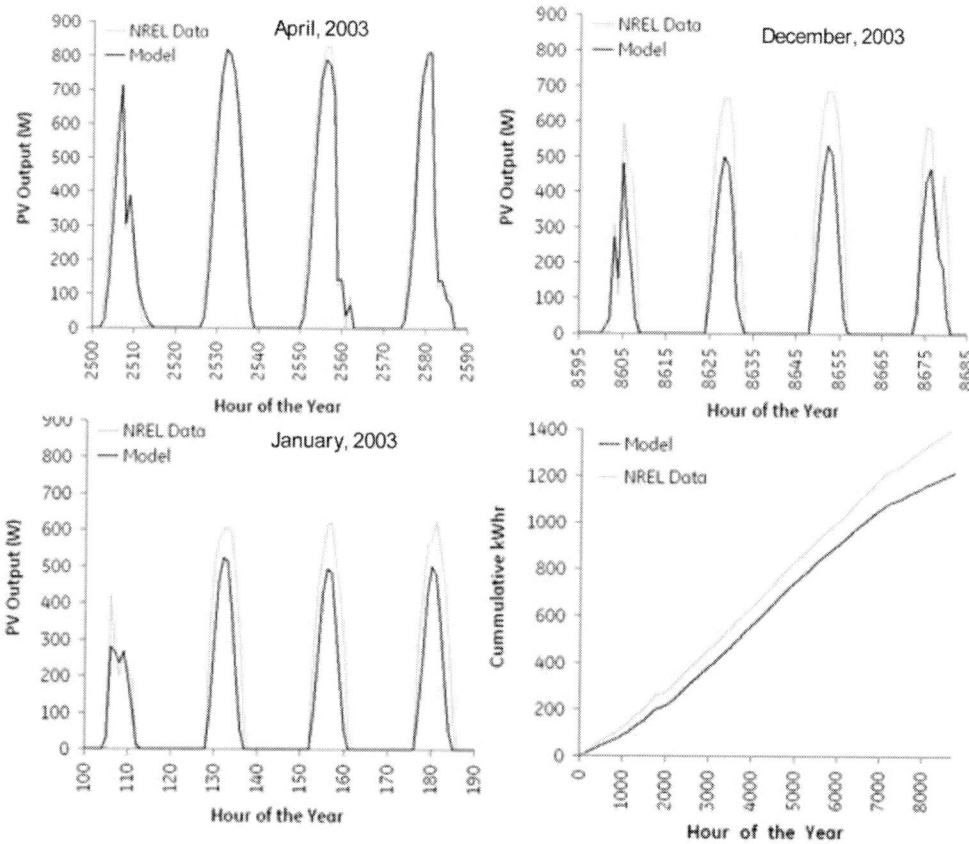

Figure 23. Model validation for Golden, Colo. Data obtained from NREL.

Table 3. Design of Experiments for Reliability Enhancement Modeling

Community Size	10, 100, 1000 homes	More homes translate into more load profile smoothing.
Geographic Region	Sterling, VA, Hanford, CA, Golden, CO	Insolation and temperature data were used for PV modeling and load modeling in three regions of the United States.
Energy Storage Capacity	0, 0.5, 1.0, 2.0, 5.0 kWh	Various battery sizes were chosen to augment the enhancement in reliability provided by solar power
Photovoltaics Penetration	0, 5, 10, 30, 50% total maximum peak 15min load.	Various penetrations of PV energy were chosen to enhance the reliability in a distribution feeder.

Table 4. Model Input and Output Tables

INPUT		OUTPUT		
Distributed Generation - Photovoltaics				
Location	CO			
Community Size	100	**Reliability Simulation**		
Photovoltaics	2.1 kW	Duration (15min intervals)	6	
PV Penetration	50.0%	Timing (15min interval)	7884	
Energy Storage - Battery		No. Customers Affected	872	
Size of Battery	1 kWh	Performance	Before	After
Round Trip Efficiency	90.0%	Critical kWh unserved	1.011	0.339
Limits from Max/Min	0.1 kWh	Outage Duration	6	2

The entire combinatorial space was considered in the study. The following cases are presented here: (1) The impact of community size and geographic region on customer reliability, and (2) The impact of PV penetration and battery storage capacity on customer reliability.

4.1. Community Size and Geographic Region

In this study, PV penetration was defined as the ratio of the maximum solar array output per home (in a specific region of the country) to the maximum total load for the home in a given year (on a 15-minute interval basis). Based on this definition, the number of arrays on the roof a home was selected to achieve a specific PV penetration level. The number of panels on a roof varies depending on the region and the total load of the home. For example, an average home in California consumes less electricity than an average home in Colorado, so a smaller array on a home in California provides the same PV penetration level as a larger array on a home in Colorado. Such scaling is important to allow for consistent comparison between regions.

The range of PV penetration levels and energy storage capacities were chosen and simulated in the energy systems model based on the reliability statistics for utility A in 2005. The many combinations of PV penetration, battery size, community size and geographic region were taken as inputs to the model. The results of the study present the fraction of each reliability index as compared with the same simulation without the presence of PV, energy storage, or critical load controls. This was done to allow for a simple comparison of relative improvement in each reliability index for each simulation. To summarize, the following reliability indices were considered:

- *Critical SAIDI* – average duration of critical load interruptions.
- *Critical SAIFI* – average number of interruptions per customer.
- *Unserved Critical Load (UCL)* – annual unserved critical load (kWh) on a circuit.

The indices are based on common distribution reliability indices (IEEE 1366), but on a critical load basis. By using reliability indices familiar to the electric utility industry, the impact of PV, energy storage, and load controls on enhancing reliability can be quantified using metrics that are meaningful to a broader audience.

The results for three PV penetration levels (10%, 30%, and 50%) for each region and community size (without the presence of a battery) are presented in Figure 24. The PV penetration levels considered cover a wide range of PV output for a given home. The results reveal a definite trend in reliability enhancement as PV penetration levels increase. These results indicate that if the PV inverter can function during an outage, the presence of PV within a community will completely eliminate some outages (improve critical SAIFI) or allow consumers to reduce their outage duration (improve critical SAIDI).

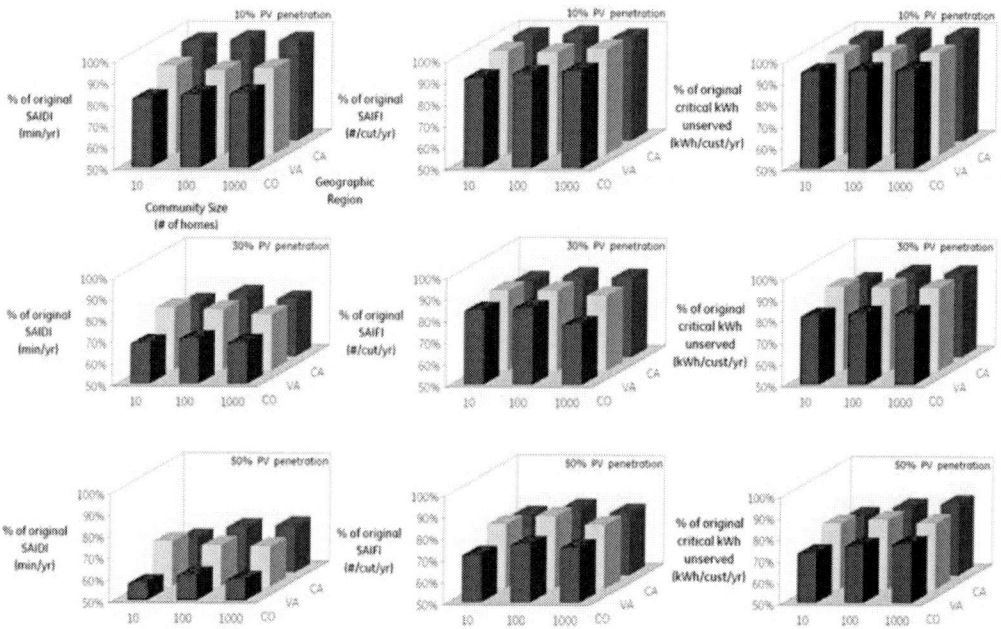

Figure 24. Customer reliability for 10%, 30%, and 50% PV penetration, with no battery, for three community sizes in three regions of the United States.

The same simulation was performed with a 1-kWh battery in each home within the community (see Figure 25). The battery provided a substantial reduction in outage duration and frequency, and largely overshadowed the improvement in reliability resulting from the increase in PV penetration levels. The relative impact of PV penetration, as compared battery size, on the overall reduction in critical SAIDI, critical SAIFI, and unserved critical kWh will be discussed in the next section.

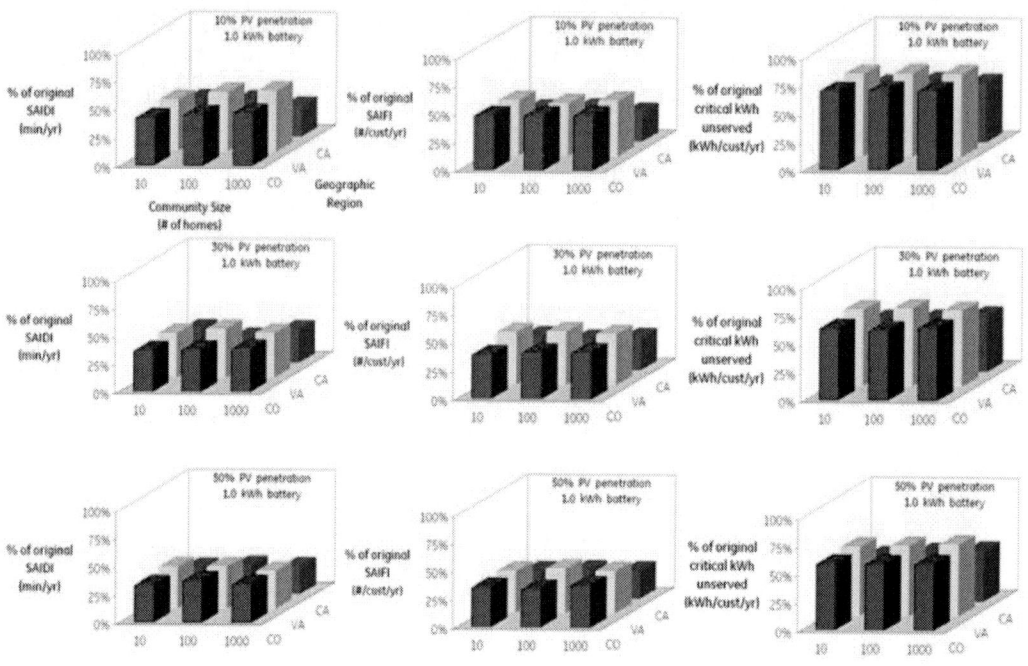

Figure 25. Customer reliability for 10%, 30% and 50% PV penetration, and a 1-kWh battery, for three community sizes in three regions of the United States.

As we mentioned earlier, the community size was intended to represent the impact of coincidence factor (a smoother load profile represents a community that was able to distribute the load more evenly during an outage). A 1000-home community should provide a smoother load profile than a community consisting of 10 homes. We therefore expected that reliability indices would show improvement for larger communities. However, based on these results, there seems to be no common trend between reliability enhancement (reduction in the proposed reliability indices) and community size. Since the electric load modeling is broken into appliances and HVAC, and the methodology for accounting for community size varies between appliances and HVAC, it is difficult to postulate that the community size is truly a measure of energy management efficacy. For the HVAC simulation there is no randomness associated with the number of homes (e.g., the value for 100 homes is obtained by multiplying the HVAC load curve for a single home by 100). Conversely, for the appliance simulation, the number of homes affects the coincidence factor. Since the HVAC results are highly sensitive to some parameters, such as the thermostat settings in each region of the country, the results rely heavily on the assumptions for thermostat settings. Additional research could further refine assumptions. Therefore, residential load modeling is cited as a significant future research activity that will greatly enhance quantification of reliability enhancement.

In the case of geographic region, each region experiences an enhancement in reliability as the penetration of PV increases. For the simulation presented in Figure 24, the reliability enhancement in Colorado is generally more pronounced than the other two regions. Since the size of the PV array is chosen based on the maximum total load in a region, and Colorado has a smaller ratio of critical load to maximum total load than the other two regions, Colorado should experience a more significant enhancement in reliability than the other two regions because ample solar power is available to meet a lower critical load during an outage. Since the classification of loads as being critical or non-critical significantly affects the results for each region, additional research on residential load modeling would allow for a more definitive conclusion on how reliability enhancement varies from region to region. Fifteen-minute appliance and HVAC load data, as well as customer preferences regarding critical load classification, may reveal that a region may have a unique preference in classifying loads as critical and deferrable. Data on consumer preferences in each region will greatly enhance this analysis.

For the simulation with the 1-kWh battery, California experiences substantial reliability enhancement, even with 10% PV penetration. Since annual electricity consumption in California is lower than that of Virginia and Colorado, a 1-kWh battery has a more significant impact on reliability enhancement in California. The presence of a 1-kWh battery in each home contributes to reliability enhancement by more than the increases in PV penetration discussed above. This aligns with the commercial building back-up power market, where batteries are sometimes used to meet critical loads during an outage.

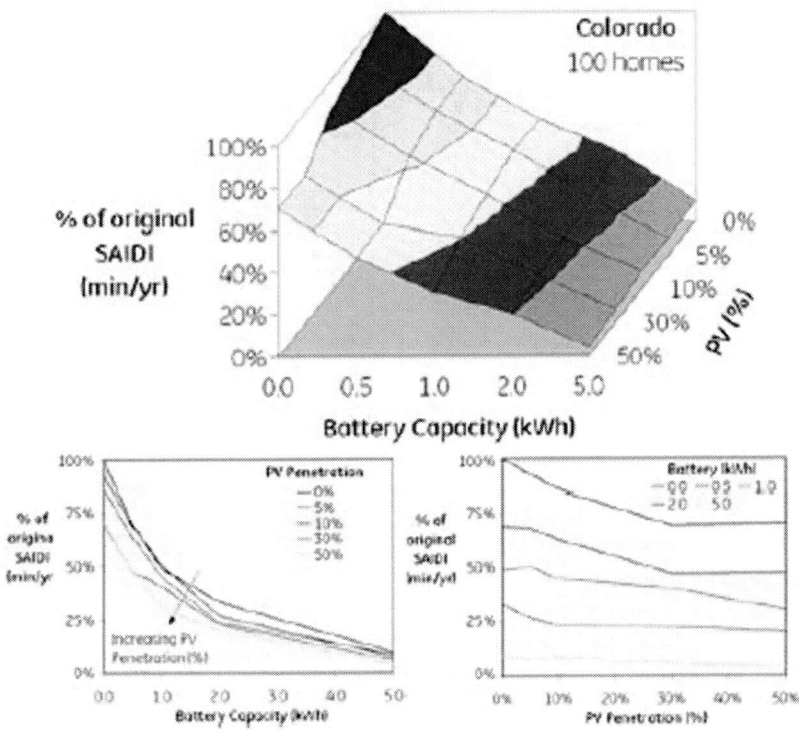

Figure 26. Critical SAIDI for a 100-home community in Golden, Colo.

4.2. Battery Size and PV Penetration

If an energy storage medium, such as a battery, is also incorporated into a community, there is an opportunity to further enhance reliability. Early analyses revealed that the presence of a 10-kWh battery would completely eliminate all outages. As a point of reference, a 10-

kWh battery is approximately the size of battery one can expect in future plug-in hybrid vehicles.

The enhancement in reliability is presented for a 100-home community in Golden Colo. (see Figure 26, Figure 27, and Figure 28). A range of battery sizes and PV penetration levels was considered.

Based on these results, a community with no energy storage (0 kWh) and no PV (0%) will experience no enhancement in reliability (100% of original index). As a validation exercise, this was observed for the indices presented in Figure 24. As PV increases, reliability improves. This is suggested by a reduction in the reliability index (decrease in average outage duration and frequency is an improvement in reliability and a reduction in reliability index). Similarly, as the battery capacity increases, reliability improves. As compared with PV penetration level, the enhancement in reliability is more substantial when the size of the battery increases. This is a logical conclusion since a charged battery is available to meet the critical load of a home during an outage at all hours of the day, independent of the presence of sun. Similar trends in reliability enhancement were observed for the other two community sizes and geographic regions.

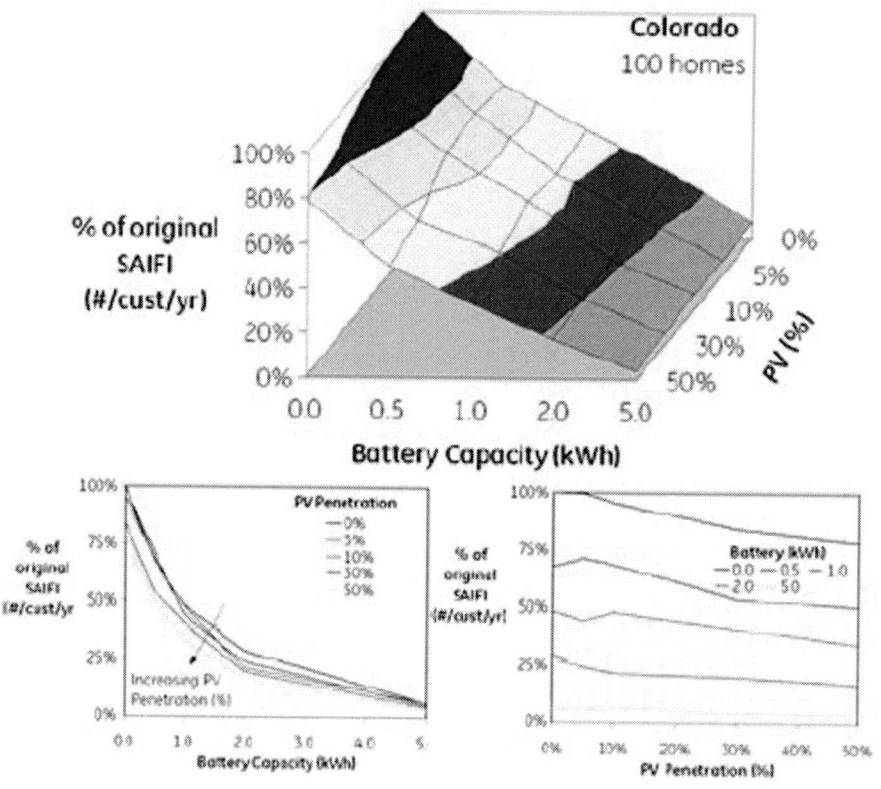

Figure 27. Critical SAIFI for a 100-home community in Golden, Colo.

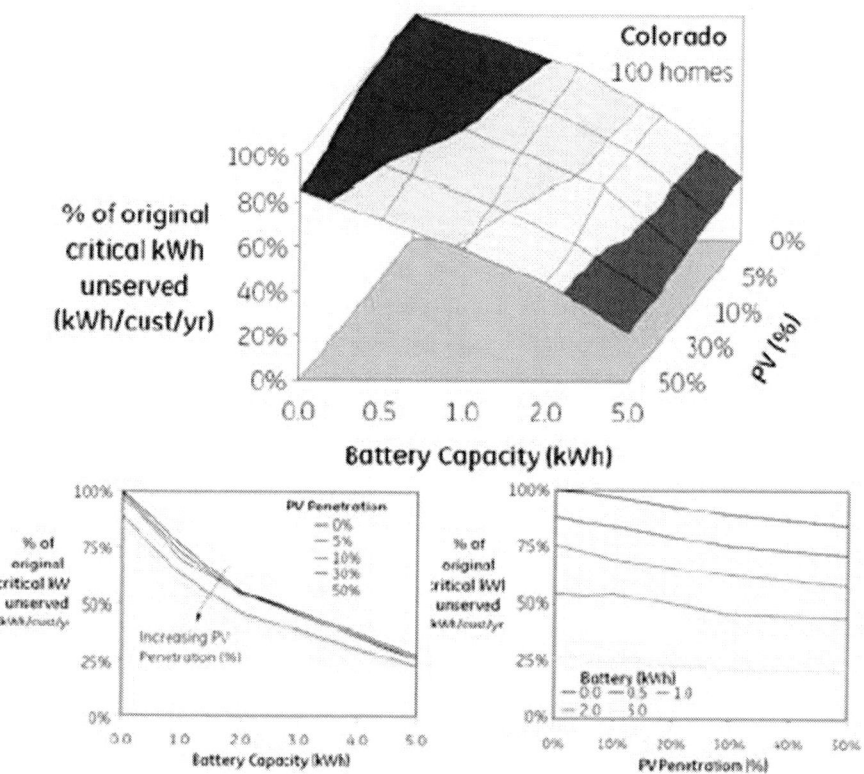

Figure 28. Critical kWh unserved for a 100-home community in Golden, Colo.

Based on this study, a significant improvement in these three reliability indices was observed when PV and battery energy storage were deployed at each home within a community. The presence of more than ~5 kWh of battery capacity per home reduced each index to nearly zero (almost a 100% reduction). The contribution of PV to the improvement in reliability indices was less significant than a battery system, though reliability enhancement was still observed for significant penetration levels. Geographic region had a small impact on the overall results, and the impact of community size could not be fully evaluated due to limitations of the HVAC modeling.

5. Gap Analysis

As in any model, some external effects cannot be captured due to a lack of information or data. For instance, if an interruption is correlated with adverse weather conditions (e.g., clouds, rain, and lightning), insolation may be low and PV may be unavailable. Unless reliability data and insolation data are sourced for the same region and timeframe, the model will not capture this nuance. No data were available that correlated insolation, temperature, and outage statistics. Such data would eliminate the need to assume outages were independent of weather conditions.

The model assumes the immediate availability of the battery and PV to meet critical loads during both short-term and sustained interruptions. The model also assumes that this "alternative supply" of electricity from PV and battery is not interrupted during grid interruptions. These assumptions were necessary in order to model the system.

Distribution reliability statistics were obtained from a single utility in the northeastern United States (Utility A). The statistics for the entire utility were used to represent the outage statistics for the communities considered in this study. Data from additional utilities would facilitate more representative analyses.

The reliability data include a description of the weather condition during each outage event. However, for the same reliability data, there is no data for temperature, wind speed, and insolation. Annual weather and reliability data from a single source for a single region would provide the necessary correlation between weather conditions and PV performance.

The model timeline is broken into 15-minute intervals for both load modeling and reliability modeling for an entire year. This decision was based on the computational power required to model shorter intervals as well as the lack of data available in shorter intervals (i.e. insolation data). The choice of 15-minute intervals for modeling implicitly assumes that any shorter-duration outages or electricity loads cannot be captured. For instance, the model does not account for transient loads due to startup, nor does the model account for the variation in load associated with restarting the grid after an interruption. We anticipate that the loads could substantially exceed the 15-minute rolling load average.

During an outage, the loads deemed non-critical during an outage will eventually become critical. As such, over the duration of an outage, critical loads will tend towards the total load. Essentially, an individual can go without certain appliances for a period of time, but at some point the individual will deem the appliance critical. This has not been accounted for in the model.

The model represents the effectiveness of energy management via the community size parameter. Load averaging over many homes will result in smoother load profiles than load averaging over fewer homes. The model assigns the number of customers affected based on the outage duration chosen from a discrete distribution provided by Utility A. The load profile is not scaled based on the number of customers affected. Instead, we assumed that the community size impacts only the smoothing of the load. Therefore, the community size parameter is a suitable proxy for energy management efficacy. One of the key assumptions of this study is the fact that for each outage event, the affected number of customers can reconfigure their loads into critical and non-critical loads, as well as operate as an islanded community, sustaining its critical loads through the use of PV and energy storage. Future research should be performed to identify the technology required to create such functionality within a community.

Further data and analysis could generate a metric of the percentage increase in reliability per kWh of battery size per kW of average critical load. This could provide a residential customer with guidance on sizing a battery for enhanced reliability. In addition, enhancing the battery model to include durability and charge/discharge characteristics, and revising its current optimization function (providing energy during outages) could provide a better representation of a battery in a residential community.

6. RECOMMENDATIONS FOR FUTURE RESEARCH

The following headings outline specific future research directions that would enable a more detailed modeling analysis based on the fields of residential load modeling, distribution reliability statistics, and energy systems modeling.

Load reconfiguration technology—We assumed in this study that the loads in a home/community could reconfigure (critical and non-critical) at the onset of an interruption. Identifying the technology and controls needed to perform this function is critical to achieving enhanced reliability using PV systems.

Customer reliability statistics—A single utility's reliability statistics were used in this analysis. Further data and specific information about which customer was affected during an outage event would enhance the accuracy of the simulation. This information will be needed for both utilities and customers in order to quantify the economic benefit of enhanced reliability due to PV systems.

Load profile breakdown—A breakdown of the contributions of various appliances to the overall aggregate load of a home (and community) through a year will be needed before customers and utilities can evaluate the effectiveness of home load control technology for either improving economics or enhancing reliability. This could be done by instrumenting many homes across the country with load data acquisition systems.

Critical loads in communities—Residential customer surveys of appliance usage or appliance feedback to the utility are needed in order to quantify the contribution of each appliance to the total load in a home/community. Identifying the critical loads is paramount to sizing PV arrays and battery storage systems.

Value of lost load—At the cornerstone of quantifying the enhancement in reliability is the understanding of the value of deferring/eliminating an outage. By understanding the economic value of avoiding an outage, utilities and customers will be in a position to invest in residential load controls, PV systems, and energy storage systems.

Availability of load data—A central data repository that consists of all non-proprietary data from this and similar studies, would enable future studies to be performed. Additionally, ease of data acquisition could enable many other, related studies.

Improve DOE 2.2—This DOE tool was used to simulate HVAC loads in a given home for three regions. By enabling 15-minute incremental outputs, higher fidelity models could use DOE 2.2 as a tool. As was mentioned earlier, no coincidence factor was available from DOE 2.2. By aggregating the HVAC loads from multiple homes (i.e., smoothing the HVAC load profile), DOE 2.2 could be used in more and varied applications.

CONCLUSIONS AND RECOMMENDATIONS

The existence of power generation on the "customer side of the meter" creates an opportunity for residential customers to not only enhance the economics of electricity consumption, but also to enhance the reliability of their electric system by islanding during utility outages. As the penetration of PV rises over the coming years, customers will have the opportunity to meet some of their most critical loads during grid interruptions. By developing the technology to reconfigure loads during an outage, individual customers will be afforded the opportunity to enhance the reliability of their electric service through the management of their loads, solar PV, and energy storage devices.

The timing, duration, and number of customers affected by each outage event were obtained for a single utility in 2005. These data were used to simulate outage events for a community on a distribution feeder. Overall, this technology resulted in a community experiencing fewer outages and outages of shorter duration.

The parameters considered in this analysis include three geographic regions (Golden, Colo., Hanford, Calif., and Sterling, Va.), three community sizes (10 homes, 100 homes, and 1,000 homes), and various combinations of battery capacity (0 to 10 kWh) and solar PV penetration (0%, 5%, 10%, 30%, 50%). The distribution reliability indices presented in IEEE 13 66[30] were adapted to account for the management of energy storage and PV in meeting only the critical loads in each home or community. The enhancement in reliability was quantified in terms of modified reliability indices, pertinent to these types of communities:

- Critical SAIDI —average duration of critical load interruptions.
- Critical SAIFI—average number of interruptions per customer.
- Unserved Critical Load (UCL)—annual unserved critical load (kWh) on a circuit.

A significant improvement in these three indices was observed when PV and battery energy storage were deployed at each home within a community. The contribution of PV to the improvement in reliability indices was observable, but less substantial than that of a battery energy storage system. Geographic region had a small impact on the overall enhancement in reliability. The impact of community size has to be reevaluated.

Finally, many specific research topics were identified that will help evaluate the impact of distributed technology on customer reliability with better accuracy.

REFERENCES

[14] IEEE Std 1366, 2003, *IEEE Guide for Electric Power Distribution Reliability Indices*, IEEE Power Engineering Society.
[15] Bollen, M. *Understanding Power Quality Problems: Voltage Sags & Interruptions*. IEEE Press, 2000, p 35.
[16] Ibid. p35
[17] *IEEE Guide for Electric Power Distribution Reliability Indices* (IEEE Std 1366), page 7.
[18] IEEE, "*A Nationwide Survey of Distribution Reliability Measurement Practices*,"

IEEE/PES Working Group on System Design, Paper No. 98 WM 218, in Annex A IEEE 1366.
[19] Schneider, K.; Hoad, R. Initial Transformer Sizing for Single-Phase Residential Load. *IEEE Transactions on Power Delivery* Vol. 7, No. 4, October 1992.
[20] Lee, Y; Etezadi-Amoli, M. *An Improved Modeling Technique for Distribution Feeders with Incomplete Information.* IEEE Transactions on Power Delivery, Vol. 8, No. 4, 1993.
[21] Noureddine, A; Alouani, A; et al. A New Technique for Short-Term Residential Electric Load Forecasting Including Weather and Lifestyle Influences. *Proceedings of the 35th Midwest Symposium on Circuits and Systems*, 1992.
[22] Capasso, A; Grattieri, WA. Bottom-Up Approach to Residential Load Modeling. *IEEE Transactions on Power Systems.* Vol. 9, No. 2, May 1994.
[23] Carpaneto, E; Chicco, G. Probability Distributions of the Aggregated Residential Load, 9th *International Conference on Probabilistic Methods Applied to Power Systems*, Sweden. 2006.
[24] *The Energy Information Administration* (EIA) Web site, (a statistical agency of the U.S. Department of Energy), Housing Characteristics Data Tables, http://www.eia.doe.gov/emeu/recs/recs2001/detail tables.html
[25] McMahon, J. *Alternative Sectoral Load Shapes for NEMS*, Lawrence Berkeley Lab, http://www.onlocationinc.com/LoadShapesAlternative2001.pdf
[26] Crawley Drury B; Hand Jon W; Kummert; Michaël; and Griffith Brent T. *Contrasting the Capabilities Of Building Energy Performance Simulation Programs*, Ninth International IBPSA Conference Montréal, Canada August 15-18, 2005.
[27] Short, T. Electric Power Research Institute (EPRI)
[28] IEEE, *A Nationwide Survey of Distribution Reliability Measurement Practices,*" IEEE/PES Working Group on System Design, Paper No. 98 WM 218, and Marinello, C. A., "A Nationwide Survey of Reliability Practices," presented at EEI Transaction and Distribution Committee Meeting, Hershey, PA, Oct 20, 1993.
[29] Energy Information Administration - http://www.eia.doe.gov/
[30] http://www.miamiok.org/departmentpages/utilitypages/ut_miscpages/applianceusechart.pdf
[31] http://www.ftc.gov/bcp/conline/edcams/eande/appliances/dwasher.htm
[32] http://www.onlocationinc.com/LoadShapesAlternative2001.pdf
[33] http://www.eia.doe.gov/emeu/recs/recs2001/detail_tables.html
[34] http://www.eia.doe.gov/emeu/recs/recs2001/hc_pdf/usage/hc6-7a_4popstates2001.pdf
[35] Stoll, H. G. *Least-Cost Electric Utility Planning*, Wiley-Interscience, 1989
[36] http://www.ftc.gov/bcp/conline/edcams/eande/appliances/clwasher.htm
[37] IEEE, SAE & AIAA, 17th Dig. Avion. Sys. Con. (DASC-17), Seattle, WA, Nov. 1998
[38] Stevens J., and Corey, G., A Study of Lead-Acid Battery Efficiency Near Top-ofCharge and the Impact on PV System Design, *Sandia National Laboratories, Photovoltaic System Applications Department.*
http://photovoltaics.sandia.gov/docs/PDF/batpapsteve.pdf
[39] Messenger, R; & Ventre, J., *Photovoltaic Systems Engineering,* CRC Press, Chapter 2.
[40] http://www.gepower.com/prod_serv/products/solar/en/downloads/gepvp200_datasheet_6 00.pdf
[41] Rauschenbach, H. *Solar Cell Array Design Handbook - Principles and Technology of*

Photovoltaic Energy Conversion, Van Nostrand Reinhold, 1980, ISBN:0-442-26842-4.

[42] Markvart, T and Castañer, L. *Photovoltaics – Practical handbook of Fundamentals and Applications*, Elsevier, 2003, ISBN: 1 8 56173909.

[43] IEEE Std 1366 2003, IEEE Guide for Electric Power Distribution Reliability Indices, IEEE Power Engineering Society.

REPORT DOCUMENTATION PAGE		Form Approved OMB No. 0704-0188
\multicolumn{3}{l}{The public reporting burden for this collection of information is estimated to average 1 hour per response, including the time for reviewing instructions, searching existing data sources, gathering and maintaining the data needed, and completing and reviewing the collection of information. Send comments regarding this burden estimate or any other aspect of this collection of information, including suggestions for reducing the burden, to Department of Defense, Executive Services and Communications Directorate (0704-0188). Respondents should be aware that notwithstanding any other provision of law, no person shall be subject to any penalty for failing to comply with a collection of information if it does not display a currently valid OMB control number.}		
\multicolumn{3}{l}{PLEASE DO NOT RETURN YOUR FORM TO THE ABOVE ORGANIZATION.}		
1. REPORT DATE (DD-MM-YYYY) February 2008	2. REPORT TYPE Subcontract Report	3. DATES COVERED (From - To)
4. TITLE AND SUBTITLE Enhanced Reliability of Photovoltaic Systems with Energy Storage and Controls		5a. CONTRACT NUMBER DE-AC36-99-GO10337
		5b. GRANT NUMBER
		5c. PROGRAM ELEMENT NUMBER
6. AUTHOR(S) D. Manz, O. Schelenz, R. Chandra, S. Bose, M. de Rooij, and J. Bebic		5d. PROJECT NUMBER NREL/SR-581-42299
		5e. TASK NUMBER PVB7.6401
		5f. WORK UNIT NUMBER
7. PERFORMING ORGANIZATION NAME(S) AND ADDRESS(ES) GE Global Research 1 Research Circle Niskayuna, NY 12309		8. PERFORMING ORGANIZATION REPORT NUMBER ADC-7-77032-01
9. SPONSORING/MONITORING AGENCY NAME(S) AND ADDRESS(ES) National Renewable Energy Laboratory 1617 Cole Blvd. Golden, CO 80401-3393		10. SPONSOR/MONITOR'S ACRONYM(S) NREL
		11. SPONSORING/MONITORING AGENCY REPORT NUMBER NREL/SR-581-42299
12. DISTRIBUTION AVAILABILITY STATEMENT National Technical Information Service U.S. Department of Commerce 5285 Port Royal Road Springfield, VA 22161		
13. SUPPLEMENTARY NOTES NREL Technical Monitor: Ben Kroposki		
14. ABSTRACT (Maximum 200 Words) This report summarizes efforts to reconfigure loads during outages to allow individual customers the opportunity to enhance the reliability of their electric service through the management of their loads, photovoltaics, and energy storage devices.		
15. SUBJECT TERMS reliability indices; distribution reliability statistics; residential load modeling; energy storage; residential load controls; photovoltaics; PV; renewable systems interconnection; GE Global Research; National Renewable Energy Laboratory; NREL		

16. SECURITY CLASSIFICATION OF:			17. LIMITATION OF ABSTRACT	18. NUMBER OF PAGES	19a. NAME OF RESPONSIBLE PERSON
a. REPORT Unclassified	b. ABSTRACT Unclassified	c. THIS PAGE Unclassified	UL		
					19b. TELEPHONE NUMBER (Include area code)

Standard Form 298 (Rev. 8/98)
Prescribed by ANSI Std. Z39.18

Chapter 6

RENEWABLE SYSTEMS INTERCONNECTION STUDY: CYBER SECURITY ANALYSIS

Annie McIntyre

ABSTRACT

To facilitate more extensive adoption of renewable distributed electric generation, the U.S. Department of Energy launched the Renewable Systems Interconnection (RSI) study during the spring of 2007. The study addressed the technical and analytical challenges that must be addressed to enable high penetration levels of distributed renewable energy technologies. The RSI project includes a Security Analysis Task designed to address the cyber security issues that may be associated with increased PV penetration. This task examines the current and future architectures that facilitate this larger penetration, identifying components and critical areas that may expose cyber vulnerabilities. This introductory review also identifies areas—such as protocol, inverter, and meter designs—that require further consideration as designs evolve and penetration increases throughout the life cycle.

ACKNOWLEDGMENTS

The author wishes to thank the following project team members and colleagues at Sandia National Laboratories, who contributed to the technical analysis and content of this report.

David Brown, Sacramento Municipal Utility District (SMUD)
Jeff Goh, Pacific Gas and Electric Company (PG&E)
Mark McGranaghan, Electric Power Research Institute (EPRI)
Tom Ortmeyer, EPRI
Mike Ropp, Institute of Electrical and Electronics Engineers
Chuck Whitaker, BEW Engineering
Randall Wetherington, Oak Ridge National Laboratory (ORNL)
Ward Bower, Sandia
David Duggan, Sandia

Glenn Kuswa, Sandia
Jason Stamp, Sandia
Juan Torres, Sandia

Executive Summary

Integration of renewable energy into the U.S. critical power-generation infrastructure will likely continue to increase throughout the next decade. Incorporating PV energy into the existing bulk power distribution grid requires consideration of potential effects associated with that integration. Cyber security issues are among these potential effects. Addressing cyber security now affords the opportunity to optimize design and build a more secure, cost-efficient solution up front. Secure operations are beneficial in that they protect both the energy consumer and the provider, ensuring continued availability and creating greater economic stability.

The RSI project includes a Security Analysis Task designed to address the cyber security issues that may be associated with increased PV penetration. This task examines the current and future architectures that facilitate this larger penetration, identifying components and critical areas that may expose cyber vulnerabilities. The aspects of cyber security for critical infrastructure, although built on information security principles, are customized for operational environments. Integrated PV aligns with operational objectives, focusing on availability and accuracy. This report defines risk in terms of threat, vulnerability, and consequence. It also presents an introductory review of cyber security in PV integration designs as it applies to power levels of 15 kV and below by defining the architecture, separating it into workable security domains, and addressing critical components. Crucial points in the architecture, such as inverters, meters, and communication pathways, require a secure design focus. These points are identified here and sample mitigation strategies are discussed. This introductory review also identifies areas—such as protocol, inverter, and meter designs—that require further consideration as designs evolve and penetration increases throughout the life cycle.

Reviewing these critical areas as penetration increases constitutes a proactive approach to ensuring reliability. By applying security controls at critical points within the architecture that facilitate continued service and stability, a balance can be struck between generation and distribution objectives and secure operations. Risks can be mitigated by employing security controls that facilitate operations and result in minimal impacts to overall architecture cost and development objectives.

1.0 Introduction

Renewable energy technologies are being increasingly integrated into the nation's critical infrastructure, and this trend is expected to continue throughout the next decade. Incorporating photovoltaic (PV) energy into the existing bulk power distribution grid necessitates consideration of potential effects associated with that integration. Among these are cyber security issues, which are now included during reviews of energy and infrastructure systems. Traditionally, many critical infrastructure systems, which include telecommunications, economics, health care, transportation, and energy, were not designed with cyber security as a primary objective. Given global events, considerable emphasis has been placed on this topic in the past several years, and securing our critical infrastructure from cyber attack has become a primary objective. To meet this objective, legacy systems must be evaluated and designs for new systems must be optimized to ensure maximum operability and security. As PV penetrates further into the distribution system, aspects of its use will become attractive adversary targets. Addressing cyber security now affords the opportunity to optimize designs and build a more secure, cost-efficient solution from the ground up. Secure operations protect energy consumers and providers, and lead to continued power availability and greater economic stability.

The Renewable Systems Interconnection (RSI) study project includes a Security Analysis Task designed to tackle cyber security issues that may arise from increased PV penetration. This task examines the current and future architectures that facilitate greater penetration, identifying components and critical areas that may prove vulnerable to cyber attack. These areas might include hardware, software applications, communication paths and protocols, or data processes. Because cyber and physical security are closely coupled, physical access control is also considered. Assessing the risk to the infrastructure as a result of PV penetration should be an ongoing effort to ensure that defenses are maintained as new designs evolve.

The aspects of cyber security for critical infrastructure are built on information security principles. Here, they are customized for operational environments. For example, the objectives of a financial institution in securing their operations differ from those of an oil refinery. Integrated PV will align more with operational objectives, focusing on availability and accuracy. Examples of security concerns in that environment might include

- Communication paths and protocols
- Access control (both cyber and physical)
- Data in storage and transit
- Standards
- Areas of control
- Situational awareness

This report contains an introductory review of cyber security in PV integration designs with power levels of 15 kV and below. The review begins by defining the architecture, separating it into workable security domains, and addressing critical components. Critical areas that require a secure design focus are then identified, followed by a discussion of sample mitigations. Finally, areas that require further technical study are enumerated.

2.0 CURRENT STATUS OF EXISTING RESEARCH

Cyber security as it relates to critical infrastructure is an expanding area of study. Research results exist that detail cyber security issues within the bulk power distribution grid. Renewable energy integration, however, is an area in which limited research on cyber security has been done. Just as distribution grid designs are evaluated, the increasing connectivity associated with renewable integration requires consideration. Much of the material on the security of PV was written before the shift in characterizing threat that followed September 11, 2001. For example, Chowdhury investigates the integration of PV into power systems and studies the flow and system behavior in a journal article[1] and a paper in a proceedings,[2] but these studies were conducted in 1990. Although these studies do not address cyber security specifically, they do suggest operational impacts of integration. The dated results, however, require that consideration be given to existing and future designs.

Recent research on security in PV integration, which focuses primarily on physical security, is limited. Physical security is important, and students evaluating the performance of PV utilization in an international environment touch on the subject in a report titled *2006 Student Research Projects, Photovoltaic Security*[3] No studies of cyber security as it relates to PV integration from a critical infrastructure protection standpoint have been done. Several research initiatives have assessed bulk power generation and transport. As discussed later in this report, an Automation Systems Reference Model (ASRM) was developed under a project for the U.S. Department of Energy (DOE) Office of Electricity Delivery and Energy Reliability. That project was designed to address cyber security issues in a control systems environment.

Other projects that are exploring control system security and critical infrastructure protection include the National SCADA (Supervisory Control and Data Acquisition) Test Bed, the Institute for Infrastructure Information Protection (I3P), and the Control System Security Program. Although many current efforts are evaluating interdependencies and vulnerabilities and characterizing risks, they do not readily include PV in the analysis phase. The National SCADA Test Bed, for example, has been focused primarily on electricity transmission security instead of distribution system security. As renewable energy technologies like PV come to be a larger part of our overall critical infrastructure, a review of cyber security for PV integration—ultimately aimed at ensuring the stability and availability of the grid—is necessary.

3.0. PROJECT APPROACH

The Security Analysis Task maintains the elementary approach of defining the problem, investigating and analyzing it, and presenting a technical summary. The following steps, then, are included in the basic approach:

- Define the architecture and future PV trends
- Identify critical areas for further analysis and research
- Enumerate potential security hazards based on the design
- Suggest mitigations or recommend further study.

The following sections present the rationale for the various aspects of the cyber security analysis.

3.1. Background for Security Discussion

It is important to enumerate the foundational cyber security aspects that guide this analysis. Historically, information security is well defined, and various research initiatives and industry forums continue to enhance the applicability of those aspects to the operational world. The analysis phase of this task requires that key security aspects be defined as they relate to the energy sector. These are described in the following subsections.

3.1.1. Elements of Security

Many different terms exist within cyber security, including "computer security," "information security," and "information assurance." The *Free Dictionary* defines information security as

> The protection of information and information systems against unauthorized access or modification of information, whether in storage, processing, or transit, and against denial of service to authorized users. Information security includes those measures necessary to detect, document, and counter such threats. Information security is composed of computer security and communications security. [4]

The same source defines information assurance, a term often used by the federal government that is now coming into use in industry, as

> Information operations that protect and defend information and information systems by ensuring their availability, integrity, authentication, confidentiality, and nonrepudiation. This includes providing for restoration of information systems by incorporating protection, detection, and reaction capabilities. [5]

For the purposes of this study, all aspects of the previous definitions have been considered. Availability, integrity, authentication, confidentiality, and nonrepudiation issues are overarching topics that guide the PV integration analysis. Again, it is important to consider how these terms relate to operational and control system environments, highlighting the objectives of critical infrastructure and energy sectors.

3.1.2. Definition of Risk

Independent of sector or environment, risk is often defined the same way, in terms of threat, vulnerability, and consequence. Numerous methodologies have been created in the past several decades to assess and measure risk, both qualitatively and quantitatively. In a control system or system of systems environment, selected processes to characterize risks can be optimized to yield more accurate results. Sandia National Laboratories performed one characterization of risk under the I3P project. That process included characterizing risk from an operational standpoint that required consideration of mission and organizational priorities such as safety and reliability.[6,7] Reports from that effort are summarized here to build a foundational view of risks to control systems and critical infrastructure.

THREAT ✕ **VULNERABILITY** ✕ **CONSEQUENCE** = **RISK**
Resources   *Weaknesses*   *Effect*   *Business Impact*

Figure 1. Components of risk [6].

3.1.2.1. Threats

Threats can exist in numerous forms, and they can be anthropogenic or environmental. These are classed as normal, abnormal, or malevolent threats. [8] Malevolent cyber threats often suggest a person or attacker with some intent to harm. Even though this can be true, one must also consider unintentional threats, such as accidents caused by humans or weather events such as hurricanes that down power lines. Threat assessments are often performed to identify potential causes of harm to an architecture or system. Threats to control systems can be assessed by [6]

- Identifying known and potential adversaries
- Analyzing each adversary's motivations, goals, and capabilities
- Assessing the threat posed by each adversary to critical system assets.

Malevolent threats or adversaries can be characterized by their level of access, motivations, and capabilities. The *I3P Risk Characterization Report* describes an adversarial threat.[6]

> A threat implies that an individual or group has the ability and access to carry out a process that creates damage to a system or exploits the system for a specific gain. Threats to control systems can come from both insiders and outsiders. For example, a disgruntled employee sympathetic to a terrorist cause will have more direct access to the control systems than a corresponding outsider.
>
> Threats can vary in capability. Capability is a function of resources such as time, money, computing power, technical knowledge, and intelligence resources. Threats and their capabilities are often divided into several specific categories such as nation-state, international terrorists, domestic terrorists, organized crime, or hackers. Although individual hackers may have malicious intent and technical knowledge, organized cyber-terrorist groups may possess the resources necessary to carry out an effective, distributed attack that produces severe consequences.

Attributes that can affect a threat's success include

- Commitment
  - Intensity
  - Stealth
  - Time
- Resources
  - Technical personnel
  - Knowledge
  - Access.

Detailed threat attributes and success factors are defined in the Sandia report entitled *Categorizing Threat: Building and Using a Generic Threat Matrix.* [8]

3.1.2.2. Vulnerabilities

A vulnerability is a weakness in a system, network, application, or process that can be exploited by a threat to create an adverse effect. [6] Vulnerabilities, which can be technical or physical in nature, can be identified through assessment activities and continual situational awareness. An unpatched vulnerability presents an opportunity for exploitation by a threat. Physical vulnerabilities, such as unlocked doors and open gates, are easily spotted. Identifying cyber vulnerabilities, such as open ports, unpatched software, and the absence of intrusion detection systems, is a more difficult task. Until a vulnerability is discovered, a system is seemingly secure. A system does not degrade over time but can become insecure instantaneously once a vulnerability is found. Locating and mitigating vulnerabilities in the design phase is a critical step. Continual situational awareness and security evaluation, however, remain necessary because threats become more sophisticated every day.

In PV integration, an example of a physical vulnerability might be the ability to easily access a circuit board in an inverter. Likewise, a cyber vulnerability might be the ability to access and alter digital usage data in a smart meter. Although some vulnerabilities appear more serious than others, it is the consequence value that allows an asset owner to make important decisions about mitigation and design.

3.1.2.3. Consequences

When defining risk, it may be consequence that gains the most attention and is heavily weighted in the equation. Consequence is the loss, damage, or impact resulting from a threat successfully exploiting a vulnerability, and can include data access and alteration, service disruption, system destruction, and public safety effects. [6]

The placement of bulk power distribution systems in publicly accessible areas creates a need to consider safety and security over a wide system area. Defining the threat and mitigating vulnerabilities is important, but understanding potential consequences and using that understanding to make solid decisions about protection and design may be most critical. Potential consequences should be considered from an operational standpoint, including both cyber and physical aspects. Interdependencies within the system and within the national critical infrastructure should be evaluated. Stamp categorized potential consequences at the 2005 I3P Workshop as[9]

- Physical impacts that encompass the set of direct consequences of control system malfunction. This includes injury or loss of life, environmental damage, and loss of property.
- Economic impacts are the resulting side effects of the physical impacts ensuing from an attack. This includes equipment or site damage, personnel compensation, and other impacts that could negatively affect the local, regional, national, or possibly global economy.
- Social impacts or "quality of life," including a loss of national or public confidence in an organization or industry, which can often lead to economic impacts as well.

- Impact on national critical infrastructure, including other dependent infrastructures such as water, agriculture, telecommunications, health care, and other energy sectors.

The technical effects resulting from a threat exploiting a vulnerability lead directly to these consequences. For example, a threat that takes advantage of an unprotected port on a system can alter system data that affect billing and load analysis. This can lead to economic impacts to the utility. Although this may be a low-impact consequence, one must also consider the possibility of access to a system that alters load, resulting in a more serious consequence. Consideration of consequences associated with vulnerabilities is crucial to ensuring that the correct levels of protection are applied in the system.

3.1.3. Domains

When addressing cyber security, it is common to break apart an architecture or set of processes into manageable layers or "domains." Evaluating domains creates a manageable approach to applying security and generally ensures that no gaps exist in the architecture. Because security controls can be dependent on one another, it is important to consider all aspects of the architecture or process set, making sure that one application of security does not make another process vulnerable.

As discussed later in this report, the bulk power grid has been analyzed from a cyber security perspective. Although some issues exist throughout the entire design, this analysis focuses on power levels of 15 kV and below.

Figure 2 depicts a simplistic view of the domains identified in this effort. Each domain drills down to a more detailed view of the components. Domain 1, which is the top level of this effort, contains the distribution substation, distribution grid, transformers, distribution buses, and house tap. Domain 2 drills down at the house level and includes the transformer and devices that interface with the distribution grid. Logically, this domain considers the movement of energy to and from the house and the associated data. Domains 3 and 4 address specific devices—the meter and the inverter, respectively. This analysis includes current designs for these devices, but proposed designs and future capabilities were also considered. As these devices evolve, future assessments should include advanced capability, potential communication functions, and accessibility.

Risk assessment was performed under the context of these domains, leading to suggestions for applying security to this effort.

3.1.4. Control Systems Security Research

Under programs funded by DOE and the Department of Homeland Security (DHS), research has been conducted that addresses security in control system environments. The consequences of a cyber event to systems within the critical infrastructure have been theorized, offering insight into needed security applications in specific operational areas. This research can be leveraged, but each new energy application should be evaluated from a security perspective. An assessment early in the design phase often reduces added costs later during implementation. With evolving technologies and applications, though, it is important to continually revisit security and consider protection levels.

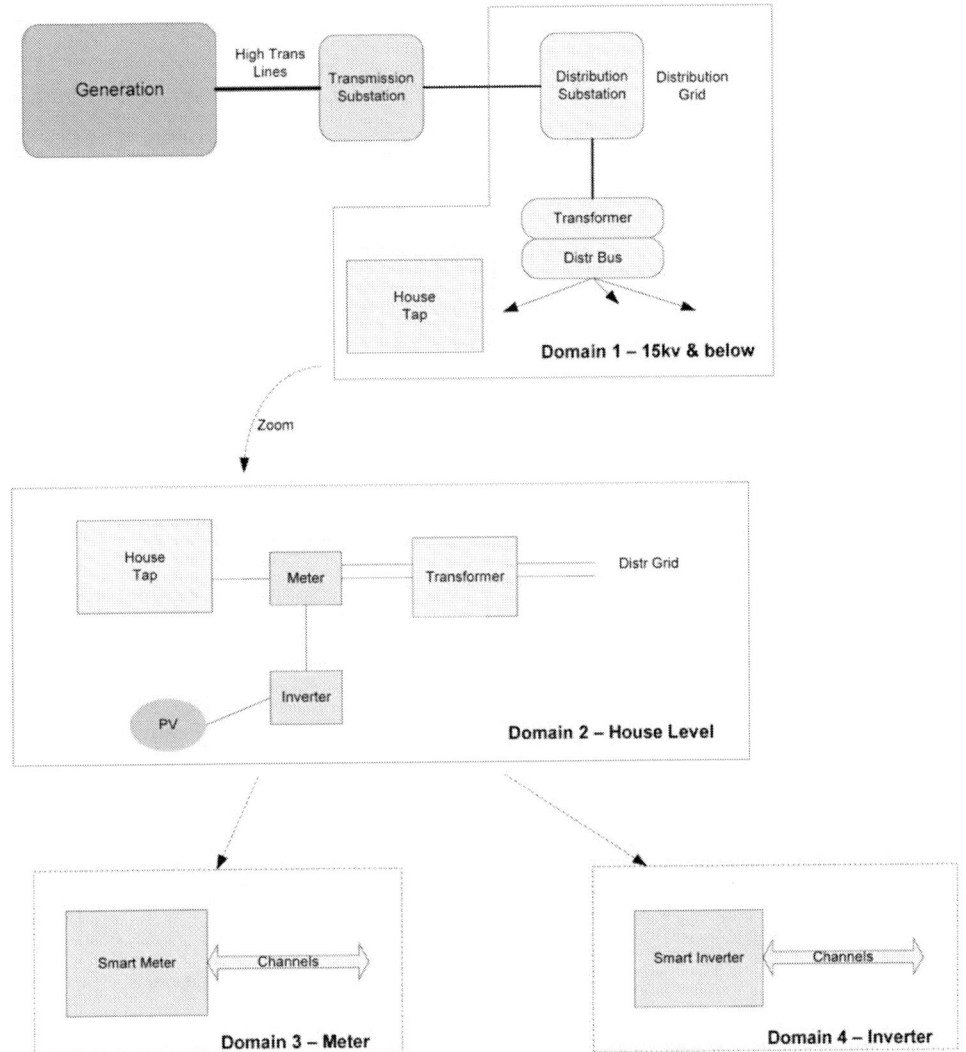

Figure 2. PV integration domains.

Previous research indicated that potential consequences can range from insignificant to major during and after a cyber event. Although total security is difficult to attain and costly, programs such as I3P are viewing survivability and recovery as part of inherent security, equally important as prevention. Although bulk power and control system issues are reflected in this analysis, a specific view of PV integration is needed to ensure that operational objectives are met without cyber security issues.

3.2. Security Model

A project conducted at Sandia National Laboratories for the DOE Office of Electricity Delivery and Energy Reliability produced the ASRM, a logical model of control systems in an operational environment.[10] The ASRM is highly detailed, but Figure 3 illustrates an adaptation. Although this model primarily considered the electric sector, it can actually be applied across control system environments such as oil, gas, and chemicals. In 2006, the model was used as a basis for applying security metrics in an operational environment under a project for the National SCADA Test Bed. [11] This effort produced guidance to assist asset owners in tailoring the model to their architecture and applying controls to needed areas to meet their security objectives.

At the utility control center and downstream, the ASRM can be used to identify areas that require security controls, such as access control and encryption. Industry guidelines exist that assist in identifying these controls. These include the North American Electric Reliability (NERC) Critical Infrastructure Protection Committee (CIPC), the National Institute of Standards and Technology (NIST) Special Publications (800 series), and the Institute of Electronics & Electrical Engineers (IEEE) standards.*

It is important to note that although this analysis focuses on power levels of 15 kV and below, the distribution grid is a complex system. Data flow and consumption at lower levels directly relate to the larger grid and affect overall utility performance. Data movement throughout the system and back to a utility, for example, is subject to the security controls identified by standards and guidelines and applied by the utility. Security at the level of 15 kV and below means that one must fully consider security at higher levels and the ultimate destination and purpose of the data in transit.

In this study, all tiers of the ASRM were considered. For example, field and infrastructure systems are included in the distribution grid. Protection mechanisms were considered for the systems and data flow at these tiers. A control center or subset of control systems that require equal consideration can also be located within the distribution grid. Automation oversight often relates to systems that link back to business systems or billing. In the case of PV, and the sale of power to and from the grid, these systems are of particular interest. As described later in this report, the future may include centralized energy management systems (EMS) that define complex connectivity between energy control and administrative systems.

4.0. PRELIMINARY ANALYSIS

Key questions in the preliminary analysis include the following:

- What devices/components in the PV integration design should be evaluated?
- What processes in the PV integration design should be considered?
- Do vulnerabilities exist and where?
- Are there any critical points of failure?
- What is the cyber risk to the PV design?
- What are the typical threats and consequences?

These questions provided the foundation for research on potential critical cyber security issues inherent in the integration of PV. Results of this analysis are outlined in the following sections.

Renewable Systems Interconnection Study: Cyber Security Analysis

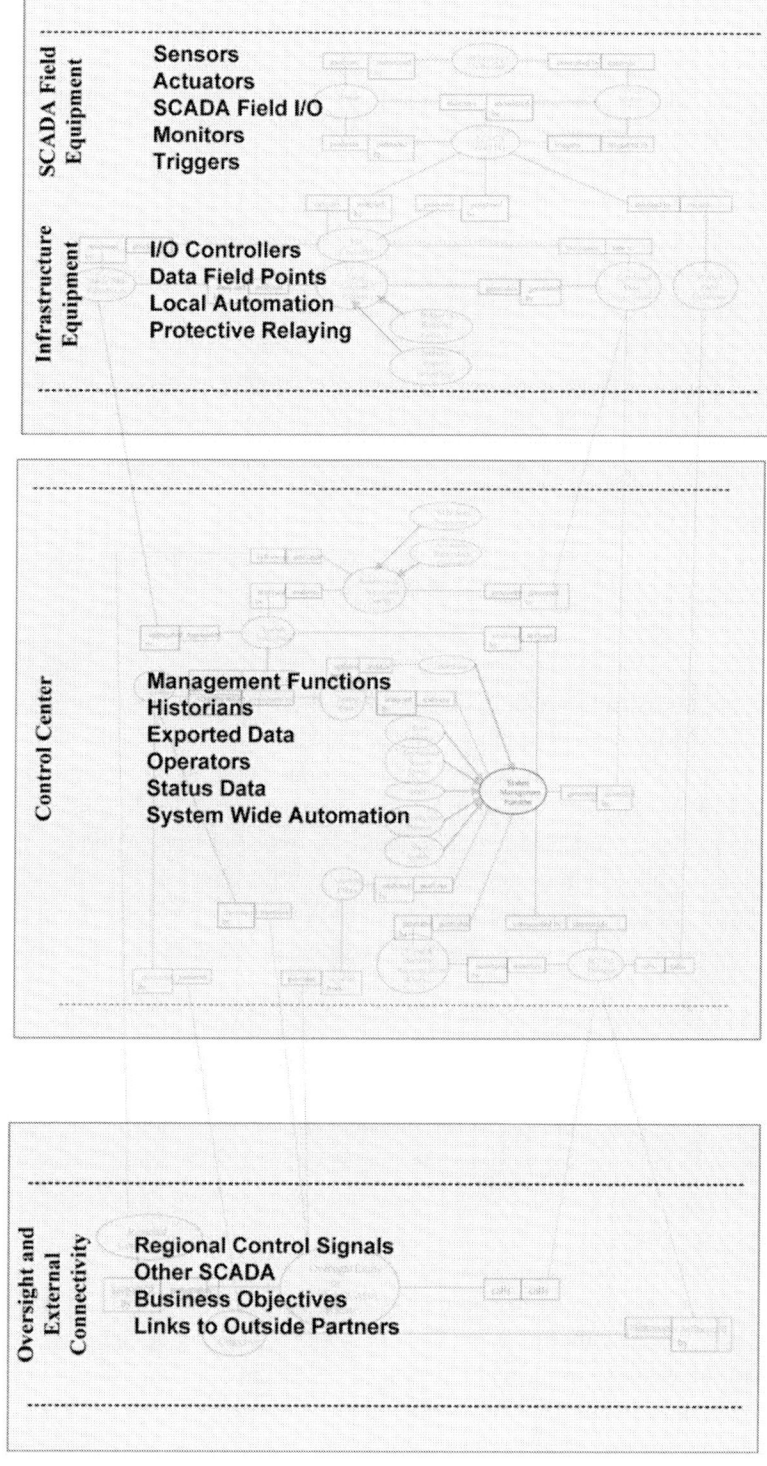

Note: I/O = input/output.

Figure 3. Automation systems reference model (adapted) [10].

4.1 Description of the Issues

Integration of PV presents new challenges in both technology and operations. Effective PV integration and an increase in penetration levels will likely be reliant on the use of advanced technology. This will result in the use of components that require two-way communication, such as inverters and meters. The flow of data and control will require that potential cyber events be considered.

Likewise, a dependence on standard and off-the-shelf technologies could present security challenges. Analyzing the design and addressing any potential security concerns now will create a more cost-effective solution for the future. Manufacturers of PV components that integrate secure solutions and processes will definitely benefit the user community. Implementing those changes early in the life cycle fosters a more secure evolution, increased stability, and eventually increased penetration.

4.2 Results of Task Research

Each concept of risk, as described in the following sections, was considered in this analysis.

4.2.1 Threats

Given the global realization that cyber incidents have become a part of daily risk to both individuals and organizations, it is important to consider the entire spectrum of threat. An infrastructure in any sector should be protected from all possible threats, normal, abnormal, and malevolent. These threats range from the inexperienced malicious hacker through organized crime to large, well-planned and -funded terrorist efforts. Although the threat can be analyzed in terms of resources and in conjunction with physical threats, it is difficult to predict an event or assume with complete certainty that one is protected from each possible threat. In terms of PV integration, it is important to look at likely threats, potential motivators, and the attractiveness of the distribution grid as a target.

Likely threats to power levels of 15-kV and below include those threats that plan to use the house or neighborhood system as an entry point to the distribution grid. A threat might find these entry points more attractive and accessible than a larger asset upstream that may have monitoring or access control. With physical and cyber controls in place at points upstream, however, it is unlikely that entering a single house could cause a large-scale incident. Several access points, such as multiple inverters in the neighborhood, may be more attractive starting points.

Details of potential consequences are outlined in upcoming sections. From this standpoint, it is important to consider a threat with physical or remote accessibility to the PV systems at the house or neighborhood level. This threat could include the curious homeowner or an individual threat performing reconnaissance to see how much control he or she could gain over the system. Although these threats do not generally have the resources or motivation to do significant harm, a poorly configured or secured system can open the door to a cyber incident.

Although it may seem unlikely that a well-funded, well-organized, and highly motivated threat would create an incident from a "shock and awe" perspective, it is still important to consider this threat from the perspective of a coordinated effort. For example, a coordinated effort might be used to produce multiple effects. Hypothetically, imagine coordinating a loss of service to strategic areas to reduce or halt communications from public media such as television and radio. This might take place to assist with a larger effort while other infrastructures are targeted. Although this simplistic example might be temporary, or easily mitigated by using a backup generator, disruptions could still occur. As a result, the potential for a coordinated attack, even if minimal, must be considered.

Whether a limited or sophisticated threat is active, the same PV components are addressed in this analysis. A detailed threat assessment should be conducted during a full vulnerability assessment of new architectures and implementations. As PV penetration evolves, standard implementations become more visible targets. Information availability on designs, especially when implemented at the house level, will increase and more data will be available to the threat. Although this cannot be stopped, vulnerabilities within components of PV can be mitigated.

4.2.2. Critical Elements and Vulnerabilities

Considering a physical perspective, domains 1, 2, 3, and 4 in Figure 2 are the primary focus of this analysis. A logical perspective includes consideration of the entire ASRM and all critical processes, along with the flow of data. Identifying the critical components requires definition of hardware, software, and processes within these views. Although focused on cyber aspects of each domain, it is important to consider physical accessibility as well. The separation between cyber and physical aspects of security is becoming increasingly less clear. The dependency between these two aspects requires the underlying consideration of physical security throughout the analysis.

It should also be noted that complex systems, such as those in electricity delivery, are evolving to facilitate optimized service. This includes delivering data to the needed recipient and fostering good decision making that leads to effective operations. This creates increased interconnectivity among systems, domains, and large-scale architectures. Interconnectivity generally creates increased risk. It is understood, however, that the trend to connect domains will likely continue and may be necessary to achieve objectives. Therefore, the result of this analysis is not to suggest disconnection, but to mitigate risks created by connection points.

Another underlying consideration in this analysis is the use of standard interfaces and a shift toward information technology (IT)-centric solutions. A movement toward standard interfaces is similar to interconnectivity. It is likely to continue to evolve to meet overall objectives, provide a cost benefit, and promote increased penetration. Standardization, however, also generally increases the availability of system details and can contribute to making the system a more attractive and accessible target. This requires forward thinking about future designs to ensure that mitigations are in place.

The following sections describe each of the critical elements, starting at the lowest level, and their potential areas of concern.

4.2.2.1. Inverter

Common inverters in use today typically have limited functionality. Advanced or "smart" inverters have been identified, however, as necessary components of future PV design. This analysis considers elements suggested for future PV designs. The main purpose of the inverter is to convert DC to AC energy, creating usable solar energy. Future PV designs are likely to include much more functionality. This could include data generation, data movement, and communication pathways. In fact, it has been stated that

> The effects of high penetration will need additional studies and inverter/controller equipment tied to the electrical distribution systems must be in communications, have intelligence to make decisions on operations, and include energy management and possible energy storage to improve the stability of the integrated grid system. [12]

Increased intelligence and smart" devices directly offer optimization, increased availability, and better decision making. [12] Considering this move forward, cyber threats can utilize data pathways as roads into the system for exploitation. Two-way communications are a valuable piece of new inverter designs, providing data for energy management. [12]

Sensors within future inverters may generate data that are useful to both the homeowner and the utility. It may be beneficial to a threat to access and alter these data, allowing the threat to change flow, alter billing, or contribute to another coordinated event. The absence of cyber and physical access controls on the data produced by the inverter may pose a vulnerability. The ability to access a physical port on the system or to obtain access through a centralized home EMS or through open wireless protocols provides an entry for a threat. Altering the load would likely only affect the home or neighborhood. A coordinated event, though, could create service availability issues across an area. Upstream changes or larger outages are unlikely but may not be completely impossible. Mitigation includes hardened physical and cyber inverter access control and upstream validity checks of the data. Monitoring for anomalies or changes in data beyond reasonable sets helps to mitigate both short- and long-term cyber events. Altered data produced at the inverter could be used to change the load, creating instability and reliance on other controls to avoid a localized trip.

Dependency on the data from a smart inverter means that altered data could affect results at various areas within the overall distribution system. Relays that prevent damage to lines may utilize this data as well as supervisory systems and EMS at various levels. It is suggested that future homes may contain a central EMS. Altered data could affect not only existing load, but could also contribute to inaccuracies in optimization, trending, and demand. Hardware and software controls to prevent unauthorized access to data production on the inverter should be considered in future designs.

4.2.2.2. Meter

Metering is one aspect that has gained notice and evolved more rapidly. Countless studies of advanced or smart metering have been performed. Many advantages to smart meters benefit both the utility and the consumer. Smart meters are being rolled out in both the United States and in parts of Europe, including rural areas. A cyber incident involving the meter may not produce a consequence that alters energy flow. But data produced at the meter can suffer economic consequences, making it an attractive target. The ability to alter usage data at the meter and hide or change consumption and production rates could result in billing inaccuracies. For example, this could give the appearance that more energy was sold back the grid than was actually returned, or that very little energy was consumed. It is assumed that a utility would detect a large anomaly. Stealthy, moderate changes to data over a long period of time, however, could result in significant economic benefit to the threat carrying out the cyber incident.

Data produced at the meter move along standard communication pathways to administrative systems at the utility. These data are used for billing as well as aggregated in trending and usage analyses. A series of cyber events that altered meter data could have the potential to significantly modify these analyses over a period of time. Although security and safety are often the primary objectives of utility companies, economic accuracy is definitely a concern as well. Like the inverter, the smart meter must have physical and cyber access control. If a centralized EMS is employed at the house level, controls on the ability to "write: or change meter data are necessary. These controls may not require significantly new technology, but considering their structure during the design phase will definitely save costs to upgrade after implementation.

4.2.2.3. Communications

At the component level, it can be concluded that control of data is most valuable. Advanced smart components clearly rely on the ability to move data to specific destinations. It can be said, then, that the communications pathways included in the PV integration system may be the most critical and valuable pieces of the implementation.

It is first necessary to consider where communications exist within the integrated architecture. Figure 4 illustrates one example of logical communications within an architecture.

It can be assumed that smart components might use two-way communications in the following functions:

- Energy production and usage data moving to and from the inverter
- Energy usage data moving from the meter.

Data can move to various places, including an EMS that could be off site or in house. The EMS would interface with smart appliances and then travel off site. The data can also move to the utility for load adjustment. This could occur at a control center, an intelligent relay at a substation, or onto a system or historian for trending analysis. Data can also move directly to the utility for billing and trending analysis, which normally occurs on the utility's administrative network.

Data movement presents challenges. One benefit to the two-way communications design is the interactive capability. Moving data directly to where a utility requires the information, when the utility requires it, creates more effective and stable operations. Increased interactive capability, though, also presents an increased number of access points and a greater opportunity for altered data. Senate Bill S.1115 suggests secure, dependable communications, and the smart grid will rely on these communications to perform critical functions.[12] Communication protocols and mechanisms that have been suggested as possible standards include[12]

- Protocols:
    - BACnet
    - TheLonTalk
- Mechanisms:
    - TCP/IP Ethernet

- Wireless such as Bluetooth and WiFi
- BPL/Broadband over power line

These protocols and mechanisms are currently in use in some architectures. Standard protocols and mechanisms are required for increased interoperability and ease of implementation, but often create easier, more visible targets. The use of TCP/IP in two-way communications for PV integration, for example, opens the architecture to many existing TCP/IP vulnerabilities. Generally, the longer a mechanism or protocol is in use, the more available configuration and implementation details become across Web sites and within documentation. If security is not evolving at that same rate of availability, the standard becomes vulnerable. An inevitable shift toward commercial IT products may also be likely, posing inherent risks by potentially introducing common vulnerabilities. Physical and cyber access control on ports, routers, and other networking equipment that facilitate communications is important and must be considered throughout the life cycle.

Once data are in transit, the information is subject to the same risks that standard mechanisms present across a large physical area. In addition to access control, an encryption mechanism may be useful in protecting the integrity of the data in transit. Likewise, signature mechanisms to ensure the authentication of the data produced may be useful. This would also assist in preventing data alteration resulting from a "man-in-the-middle" attack, which is certainly possible when utilizing a standard mechanism such as a TCP/IP network. Data moving to control centers, administrative systems, or historians should be protected at those locations as well. These sites have been analyzed in other control system research projects and the results apply across sectors.

The value of transit, sensor, load, or billing data should be protected at levels according to the priorities and values assigned by the utility. Monitoring and validity checks on the data assist in determining whether data have been altered significantly. This, as well as signed or authenticated data, can double as a static intrusion detection system that alerts the utility to an event that produced altered data.

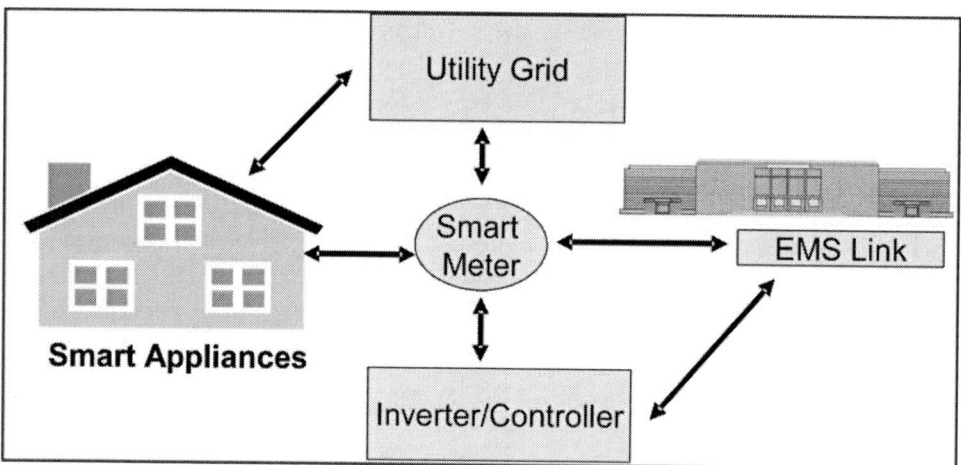

Figure 4. Smart components and logical communications[12].

As in all control system and operational environments, standard IT protective mechanisms are not easily applied. Virus protection software, off-the-shelf access controls, and real-time intrusion detection and prevention systems are not readily available or easily implemented. Unfortunately, increased PV penetration will likely increase motivations by threats that seek to exploit any existing vulnerabilities. In many instances, protocols and communication mechanisms are leased by utilities from commercial entities. Utilities may not have sufficient control over implementation of security controls within the communication backbone. If a utility has the option to own and manage its own communication infrastructure, though, it has complete control of security.

There are still many positive reasons to include a communications layer in PV architectures, such as, for example, stability and reliability of the grid. Communications provide[13]

- Monitoring capability
- Demand management and automated decisions
- Advanced metering
- Coordinated control
- Distributed generation, integration, and storage
- Real-time analysis
- Pricing controls
- Protective mechanisms.

As the trend of utilizing two-way communications continues, standardized IT functionality may become increasingly present in PV architectures. This could include interfaces with smart appliances and home-level management and monitoring systems. As the industry considers standard implementations to facilitate these functions, security must also be taken into account in the design to mitigate emerging risks. Overall reliability and efficiency can clearly be increased by using two-way communications. A recommended model for approaching security in operational layers, however, would be beneficial to ensure that mitigations are applied accordingly and that potential consequences are contained. Standard approaches such as the IEEE guidelines represent the future; currently, utilities are using disparate configurations to meet data requirements. [14] Increased communications facilitate added functionality and can increase survivability during a cyber incident through situational awareness and response. Overall operational objectives can be met securely by achieving a balance between security and functionality.

4.2.2.4. Upstream Distribution Substations

When evaluating a design of 15 kV and below, security at the distribution substation level should also be considered. Any control system, EMS, digital sensor, or data aggregation point at a substation should be protected using applicable security mechanisms. These mechanisms should include a combination of physical and cyber access controls. A layered approach to securing the perimeter and then protecting systems with cyber access control will prevent tampering, data alteration, and service disruptions resulting from cyber-based effects. Disruptions caused by physical damage are extremely difficult to prevent over widespread geographic areas. Mitigating data loss in those instances requires redundancies at data storage and aggregation points, such as backup systems and archives. Substations provide a location for those systems, but protective mechanisms are needed to ensure that access is controlled, data integrity is maintained, and monitoring is in place to detect intrusion. Conventional data management systems and networking equipment at these locations may be sufficiently protected with standard IT controls. Operationally focused security mechanisms can assist in protecting control systems at these locations. Such security mechanisms are being identified by industry forums and research projects and are becoming increasingly available in the form of guidelines and recommendations for control system environments.

4.2.3. Potential Consequences

As mentioned in the previous sections, potential consequences can be derived from scenarios developed by investigating possible designs. One scenario, altered data at the inverter, could affect load balancing and faulting, which could disrupt service on a localized level, such as a neighborhood or a microgrid service area. Local hardware damage is also possible. Large, cascading failures are unlikely, but cannot be completely disregarded. Coordinated localized outages could result in widespread effects, especially if carried out by a well-motivated threat that places emphasis on interdependent infrastructures. Altered data at the meter results in billing and trending inaccuracies. Although this may have little effect on delivery, over time, the utility could experience financial consequences. Communication backbones provide added functionality and stability to the grid. These lines of communication facilitate the following benefits to the consumer and the utility:

- Service stability and reliability
- Improved delivery
- Accurate usage measurement and prediction
- Distributed generation
- Accurate pricing.

Protective mechanisms at each layer are necessary to ensure that data integrity is preserved and data flow is maintained. An approach to security should include mechanisms that address the following aspects:

- **Prevention** entails hardening the systems and process structure to make sure defenses are in place to maintain proper function. Redundant architectures should be employed to facilitate operations.
- **Awareness and reaction** includes the ability to detect a cyber anomaly that results in an impact. This includes information management that supports decision making during a cyber event and fosters reactions that end the event with minimal impact.
- **Recovery** means that when or if a cyber event occurs, a structure is in place that supports redundant capability or backup data, mitigating the lasting effects of the event.

Hardware, software, and design solutions can serve as protective mechanisms, ensuring that these steps are accomplished. Digital data, the primary subject of protection, must be controlled to make sure that energy generation, distribution, and consumption are maintained at necessary levels. Data integrity, authentication, and encryption technologies can be used to protect these data during production and transport.

A complete denial of service to an area can be accomplished by inflicting physical damage to areas of the distribution grid. A common assumption is that this event is more likely and easier to carry out than a cyber event. Although that conclusion may be true in some cases, specific objectives may prompt a cyber-focused event launched by a threat with a specific intent. The stealth capabilities of a cyber event often foster long-lasting effects, and if detected, require time and resources by the utility to troubleshoot. Without monitoring and detection in place, cyber events may go unnoticed for long time periods, allowing a threat to carry out more reconnaissance and produce targeted, lasting consequences.

Distributed architecture and increased functionality also create the possibility of unintentional consequences. Curious homeowners or operator errors at the component level can also produce negative consequences. Hardware- and software-based access controls, in addition to well-configured house-level EMS, can prevent accidental disruption and physical damage.

To ensure that the overarching objectives of PV integration are met, protocols, data transport mechanisms, and overall implementation activities must be standardized. Because standard implementations often create more visible targets, security should be applied at each significant area of the architecture and accompanying security guidelines are recommended.

A first step toward identifying and mitigating vulnerabilities, thereby reducing consequences, is an assessment of the architecture. "Red teaming" is an assessment process that can be customized and performed on an architecture at various levels of maturity. A red team assessment or a risk assessment can be a paper-based or hands-on evaluation that identifies and analyzes critical processes, access points, potential failure points, and possible vulnerabilities. An example of red teaming in the design phase can be reviewed at http://www.idart.sandia.gov, which illustrates how even notional architectures can be assessed.

5.0. RECOMMENDATIONS FOR FUTURE RESEARCH

Considering cyber security and potential consequences early in the design life cycle is more cost effective and generally offers more comprehensive protective mechanisms than "bolt-on" solutions addressing systems that have already been implemented. As standard designs and implementations evolve, specific areas that can represent critical points of failure or significant vulnerabilities should be evaluated from a security perspective. Based on the objectives of distributed functionality and interconnected systems, each valuable component of the system should be evaluated, along with the communication pathways. System components can include

- Standard protocols and mechanisms (such as broadband over power line), any associated vulnerabilities, existing threats, and potential mitigations
- Advanced metering designs
- Advanced inverter designs
- Evolving EMS

- Digital sensors
- Secure wireless for reliability

Evaluating designs from an access control perspective is recommended. Protecting access to the component's configuration as well as the data it produces is paramount. Cost analyses of advanced components for security can also be helpful. A balance between implementation and security can be found, especially if evaluated early in the life cycle. A review of both the ramifications and benefits of standard implementations is suggested. An accompanying set of security guidelines and desired protective mechanisms during implementation assists in mitigating risks.

Finally, continued review of consequences associated with increased penetration is recommended. A specific example is the potential for any coordinated events that may create a cascading failure. This requires a review of interdependent infrastructures and potential threat scenarios.

Conclusions

Increased PV penetration and distribution capabilities can enhance grid stability. Distribution fosters evolving ways to manage energy from generation through distribution to consumption. Advanced designs of critical components facilitate the capabilities of new energy management methods. These smart components utilize digital controls and communication pathways to supply data to specific locations within the architecture. Movement and management of these data affects load balancing and fault tolerance, as well as administrative functions such as billing. The distributed architecture and cyber capabilities create potential vulnerable access points. Data generated at inverters and meters must be authentic and protected during generation and transit. Standardized implementations and common communication protocols are necessary to increase PV penetration and ensure stability. Standard implementations, however, create attractive targets and must be protected through security controls. A blend of physical and cyber access controls, monitoring, intrusion prevention, encryption, and signature technologies can help mitigate the risks of altered data. Data alteration could have consequences for load, hardware, and administrative information. Even though widespread denial of service as a result of exploiting a single component is unlikely, coordinated cyber events should be considered, including other interdependent infrastructures.

Information technology security has evolved extensively, but readily available security technologies are not easily applied in operational or control system environments. As components and critical processes evolve in future architectures, a risk assessment that includes a review of threat, vulnerabilities, and potential consequences is recommended. Implementing security controls as the system designs evolve easier and more cost effective than retrofitting existing designs.

Given the importance of the distribution grid and our national critical infrastructure, security should be considered as all new energy systems adapt to provide new capabilities. A balance between generation and distribution objectives and secure operations can be met by applying security controls at critical points within the architecture. These controls will facilitate continued service and stability. Risks can be mitigated by employing security controls that promote operations and present minimal impacts to overall architecture cost and development objectives.

REFERENCES

[1] Chowdhury, BH. "Optimizing the Integration of Photovoltaic Systems with Electric Utilities." *IEEE Transactions on Energy Convergence*; Vol. 7, No. 1, March 1992.

[2] Chowdhury, BH. "Effect of Central Station Photovoltaic Plants on Power System Security." *Proceedings of the 21st IEEE Photovoltaic Specialist Conference.* Kissimmee, FL, May 1990.

[3] Hightower, A. "2006 Student Research Projects, Photovoltaic Security." *Occidental College Renewable Energy Education Fund 2006 Annual Report.* Los Angeles, CA: Occidental College, March 2007.

[4] "Information Security." *Dictionary of Military and Associated Terms.* 2005. U.S. Department of Defense. http://www.thefreedictionary.com/information+security. Accessed August 29, 2007.

[5] "Information Assurance." *Dictionary of Military and Associated Terms.* 2005. U.S. Department of Defense. http://www.thefreedictionary.com/information+assurance. Accessed August 29, 2007.

[6] McIntyre, A.; Stamp, J.; Cook, B.K. *I3P Risk Characterization Report.* Research Report No. 9. Hanover, NH: Institute for Infrastructure Information Protection (I3P), May 2007. http://www.thei3p.org/repository/researchrepo9.pdf.

[7] McIntyre, A.; Stamp, J.; Cook, BK; Lanzone, A. "Workshops Identify Threats to Process Control Systems." *Oil and Gas Journal*; Vol. 104, Issue 38, October 9, 2006; pg. 44-50.

[8] Duggan, D; Thomas, S; Veitch, C; Woodard, L. *Categorizing Threat: Building and Using a Generic Threat Matrix.* SAND2007-5791. Albuquerque, NM: Sandia National Laboratories, September 2007.

[9] Stamp, J. "Drivers and Concerns for SCADA Security." Presented at the I3P 2005 Workshop. Houston, TX, June 2, 2005.

[10] Stamp, J; Berg, M; Baca, M. *Reference Model for Control and Automation Systems in Electric Power.* SAND2005-6286P. Albuquerque, NM: Sandia National Laboratories, November 2005.

[11] McIntyre, A; Becker, B; Halbgewachs, R; Tejani, B. "Security Metrics Taxonomy for Control Systems." Presented at the Process Control Systems Forum (PCSF) 2007 Annual Meeting. Atlanta, GA, March 6-8, 2007.

[12] U.S. Department of Energy (DOE), Office of Energy Efficiency and Renewable Energy. *Advanced Integrated Inverters/Controller and Energy Management Systems Program.* Washington, DC: DOE, July 2007.

[13] McGranaghan, M; Ortmeyer, T; Key, T; Barker, P; Smith, J; Crudele, D; Roth, T. "Study Area 1: Grid Issues and Research Needs." Presented at the RSI Status Meeting, Washington, DC, July 26, 2007.

[14] "Power System Communications Committee (PSCC)." IEEE Power Engineering Society. http://www.ewh.ieee.org/soc/pes/pscc/index.html. Accessed September 2007.

Distribution

1	MS1104	Margie Tatro, 6200
1	MS1104	Rush Robinett, 6330
1	MS1033	Charlie Hanley, 6335
1	MS1110	Jeff Nelson, 6337
1	MS1124	Jose Zayas, 6333
1	MS1108	Juan Torres, 6332
1	MS1033	Doug Blankenship, 6331
1	MS0734	Ellen Stechel, 6338
1	MS1108	Jason Stamp, 6332
1	MS1108	David Wilson, 6332
1	MS1108	Jaci Hernandez, 6332
1	MS1108	Michael Baca, 6332
1	MS0455	Shannon Spires, 6332
1	MS1108	Steven Goldsmith, 6332
1	MS1108	Jeff Carlson, 6332
1	MS 0899	Technical Library, 9536 (electronic copy)

End Note

* See http://www.nerc.com/~filez/cip.html for the NERC CIP, http://csrc.nist.gov/publications/PubsSPs.html for the NIST 800 series, and http://ieeexplore.ieee.org/Xplore/dynhome.jsp for the IEEE standards.

CHAPTER SOURCES

The following chapters have been previously published:

Chapter 1 – This is an edited reformatted and augmented version of a National Renewable Energy Laboratory publication, report NREL/SR-581-42300, dated February 2008.

Chapter 2 – This is an edited reformatted and augmented version of a Sandia National Laboratories publication, report SAND2008-0946P, dated February 2008.

Chapter 3 – This is an edited reformatted and augmented version of a National Renewable Energy Laboratory publication, report NREL/SR-581-42298 dated February 2008.

Chapter 4 – This is an edited reformatted and augmented version of a National Renewable Energy Laboratory publication, report NREL/SR-581-42297, dated February 2008.

Chapter 5 – This is an edited reformatted and augmented version of a National Renewable Energy Laboratory publication, report NREL/SR-581-42299, dated February 2008.

Chapter 6 – This is an edited reformatted and augmented version of a Sandia National Laboratories publication, report SAND2008-0947P, dated February 2008.

INDEX

A

access, 89, 150, 213, 215, 216, 217, 218, 220, 222, 224, 225, 226, 227, 229, 230
accessibility, 218, 222, 223
accounting, 202
acid, 105, 106
ACL, 1, 13, 50
active current limit, 1, 48, 50, 52
adaptation, 220
adjustment, 111, 112, 225
advancements, vii, 1, 4
adverse effects, 72
adverse weather, 205
aggregation, 99, 175, 227
agriculture, 218
air temperature, 197
algorithm, 95, 186, 190
alters, 218, 224
amplitude, 71, 79, 81, 82
assessment, 109, 137, 217, 218, 223, 229
assets, 3, 30, 216
attacker, 216
authentication, 215, 226, 229
automatic voltage regulator, 1, 7
automation, 79
Automation Systems Reference Model (ASRM), 214
awareness, 213, 217, 227

B

bandwidth, 13, 14, 31, 83, 86, 87, 88, 89, 90, 168
banks, 62, 69, 112
barriers, vii
base, 7, 9, 66, 67, 69, 78, 79, 94, 110, 115, 121, 122
batteries, 70, 71, 105, 106, 194, 203
behaviors, 4, 18, 193
benchmarking, 142
benchmarks, 33
beneficial effect, 152
benefits, 3, 57, 59, 67, 77, 88, 114, 142, 144, 146, 154, 166, 168, 228, 230
Bluetooth, 89, 91, 226
bounds, 112, 185
breakdown, 172, 207
break-even, 66, 69
bromine, 106
building code, 110
businesses, 148

C

cable television, 89
cables, 63, 68
cadmium, 104, 105
candidates, 87
carbon, 105
case studies, 108, 136, 173
case study, 127
CDC, 161, 162
CEC, 101, 198
certification, 66
challenges, vii, 57, 83, 85, 114, 143, 144, 155, 157, 173, 211, 222, 225
chemicals, 220
City, 70, 95
classification, 203
climate, vii
climate change, vii
clothing, 189
coal, 65, 146, 156, 157
cogeneration, 142
collaboration, 144
commercial, viii, 9, 64, 83, 98, 107, 108, 109, 115, 116, 117, 143, 145, 176, 203, 226, 227

communication, 32, 59, 83, 85, 87, 88, 89, 92, 98, 108, 109, 136, 137, 142, 168, 212, 213, 218, 222, 224, 225, 227, 228, 229, 230
communication systems, 87, 88
communication technologies, 89
communications channel, 68, 86
communities, 171, 172, 173, 178, 202, 206, 207, 208
community, viii, 65, 68, 171, 172, 173, 177, 184, 185, 186, 187, 188, 189, 190, 191, 192, 194, 195, 196, 198, 200, 201, 202, 203, 204, 205, 206, 207, 208, 222
compatibility, vii, 57, 58
compensation, 8, 112, 122, 125, 163, 217
competition, 89
competitiveness, 143
complement, 190
complexity, 67, 152, 185, 189, 190
compliance, 4, 72
composition, 175
computation, 173
computational mathematics, 185
computer, 65, 71, 85, 111, 149, 215
computing, 151, 216
conditioning, 84, 94, 112, 173
conductor, 88, 112, 117
conductors, 77, 88, 112, 139, 162
confidentiality, 215
configuration, 68, 70, 71, 81, 118, 119, 121, 123, 125, 127, 129, 226, 230
connectivity, 89, 214, 220
consensus, 82
constant rate, 187
construction, 148, 150, 193
consumers, 174, 187, 201, 213
consumption, 61, 73, 88, 105, 119, 120, 143, 171, 175, 184, 185, 186, 187, 189, 191, 192, 193, 194, 195, 203, 208, 220, 224, 228, 230
consumption rates, 185
contingency, 23, 148
cooking, 191
cooling, 177, 184, 193, 195
coordination, viii, 67, 84, 86, 92, 97, 99, 107, 110, 164
copper, 87, 88, 91
correlation, 145, 180, 181, 182, 189, 206
corrosion, 88
cost, vii, viii, 17, 59, 60, 61, 66, 77, 78, 79, 83, 84, 85, 87, 88, 90, 91, 95, 98, 105, 122, 139, 142, 143, 146, 147, 148, 150, 151, 154, 156, 160, 165, 166, 168, 187, 212, 213, 222, 223, 229, 230, 231
creativity, 156
critical infrastructure, 212, 213, 214, 215, 217, 218, 230

criticism, 107
crystalline, 104, 105
current limit, 1, 12, 13, 23, 48, 50, 52, 150
customer data, 175
customer preferences, 203
customer service, 115, 117
customers, viii, 79, 85, 92, 94, 96, 99, 107, 110, 111, 144, 150, 171, 172, 173, 174, 175, 176, 177, 178, 179, 180, 181, 183, 184, 185, 190, 195, 199, 206, 207, 208
Cyber security, viii, 212, 213
cycles, 79, 90

D

damping, 24
data availability, 145, 177
data collection, 166, 175
data gathering, 121
data generation, 224
database, 2, 4, 5, 7, 9, 111, 179
deficiency, 107, 147
deficit, 18, 23
degradation, 88
demonstrations, 86
denial, 215, 229, 230
Department of Defense, 231
Department of Energy, vii, 57, 185, 209, 211, 214, 231
Department of Homeland Security, 218
depth, 72, 196
deregulation, 148
desensitization, viii, 107, 164
designers, 62, 167
destruction, 217
detectable, 81
detection, 81, 82, 83, 84, 85, 92, 102, 163, 215, 217, 226, 227, 229
detection system, 217, 226
detection techniques, 84
deviation, 6
DHS, 218
differential equations, 149
digital communication, 87
discontinuity, 120
dispersion, 184
distributed load, 115
distributed renewable energy, vii, 57, 211
distribution, vii, viii, 1, 4, 5, 9, 14, 32, 57, 58, 59, 60, 61, 62, 63, 65, 66, 67, 68, 69, 70, 73, 74, 75, 77, 78, 79, 83, 87, 88, 93, 94, 95, 96, 97, 98, 99, 104, 105, 107, 108, 109, 110, 111, 112, 114, 115, 118, 121, 123, 132, 136, 137, 141, 142, 143, 144, 150,

151, 160, 163, 164, 167, 168, 171, 172, 174, 175, 177, 179, 180, 181, 183, 184, 187, 190, 198, 199, 201, 206, 207, 208, 212, 213, 214, 217, 218, 220, 222, 224, 227, 228, 229, 230
diversity, 65, 160, 176
drawing, 63, 71
drying, 191
durability, 206

E

economic competitiveness, 60
economic consequences, 224
economics, viii, 57, 58, 65, 85, 207, 208, 213
electric power system, vii, viii, 1, 4, 141, 142, 144
electric power system planning methodologies, viii, 141
electrical distribution systems, viii, 107, 224
electricity, vii, 60, 71, 79, 90, 95, 109, 142, 143, 145, 150, 156, 168, 175, 195, 200, 203, 206, 208, 214, 223
electrolyte, 106
e-mail, 87
EMS, 58, 78, 79, 85, 90, 92, 93, 94, 95, 96, 98, 99, 220, 224, 225, 227, 229
encryption, 220, 226, 229, 230
end-users, 184
energy, vii, viii, 1, 2, 3, 4, 14, 30, 31, 32, 57, 58, 59, 60, 61, 64, 70, 71, 77, 78, 79, 80, 84, 86, 90, 93, 95, 97, 98, 99, 105, 106, 109, 113, 141, 142, 143, 144, 145, 148, 149, 152, 153, 154, 156, 157, 159, 160, 161, 162, 165, 167, 171, 172, 173, 175, 176, 177, 178, 181, 184, 193, 195, 196, 198, 199, 200, 201, 202, 203, 204, 205, 206, 207, 208, 212, 213, 215, 218, 220, 223, 224, 228, 230
energy consumption, 64, 195
energy prices, 95
energy supply, 143
energy transfer, 161
engineering, 86, 110, 142, 145, 151, 163, 164
environment, 4, 5, 73, 85, 213, 214, 215, 220
equipment, viii, 62, 79, 82, 84, 88, 89, 105, 107, 108, 109, 110, 114, 115, 119, 121, 132, 136, 137, 146, 148, 152, 163, 166, 167, 177, 193, 217, 224, 226, 228
Europe, 224
evolution, 168, 175, 222
excitation, 97, 149, 150, 163
exercise, 190, 204
expenditures, 148
exploitation, 217, 224
extraction, 70, 161, 166
extracts, 10

F

fault tolerance, 230
federal government, 215
fidelity, 173, 207
field tests, 68
financial, 213, 228
first generation, 65
flexibility, 2, 3, 15, 30, 77, 100, 141, 154, 157, 159, 164, 165
fluctuations, 11, 14, 60, 64, 65, 72, 98
force, 148
forecasting, 80, 95, 97, 145, 150, 151, 153, 166
freedom, 110
frequency droop control, 1
fuel consumption, 152
full capacity, 62

G

gel, 106
General Electric, 1, 4, 33, 57, 66, 100, 114
geometry, 193
Georgia, 101
Germany, vii, 1, 4
global economy, 217
governor, 9, 149, 163
graph, 188
grounding, 107, 164
growth, vii, 60, 109, 142, 143, 150, 153, 154
growth rate, 60, 109
guidance, 3, 30, 143, 206, 220
guidelines, 3, 30, 110, 111, 118, 139, 142, 167, 220, 227, 228, 229, 230

H

harmonization, 142, 167
Hawaii, 103
hazards, 214
health, 213, 218
health care, 213, 218
histogram, 179, 188
homeowners, 195, 229
homes, 65, 88, 142, 172, 173, 175, 176, 177, 184, 185, 186, 187, 188, 191, 192, 199, 202, 206, 207, 208, 224
host, 81, 82, 85
humidity, 145, 194
hurricanes, 216
hybrid, 71, 196, 204

I

ideal, 90
identification, 86
idiosyncratic, 99
imagery, 80
imbalances, 2, 31
immunity, 83
improvements, 177
individuals, 171, 222
induction, 66, 67
inductor, 113
industries, 145
industry, 4, 109, 137, 142, 144, 145, 148, 151, 152, 153, 162, 167, 175, 201, 215, 217, 227, 228
inefficiency, 61
inertia, 2, 18, 20, 31, 76, 151, 158, 163
information technology, 223
infrastructure, vii, viii, 57, 58, 61, 78, 88, 94, 109, 136, 137, 142, 144, 148, 150, 166, 168, 212, 213, 214, 220, 222, 227
Institute of Electrical and Electronics Engineers, 1, 211
insulation, 88
integration, vii, viii, 2, 58, 59, 60, 86, 141, 142, 143, 144, 151, 152, 154, 155, 157, 164, 165, 167, 212, 213, 214, 215, 217, 219, 220, 222, 225, 226, 227, 229
integrity, 97, 215, 226, 228, 229
intelligence, 90, 216, 224
interface, 63, 100, 144, 149, 161, 177, 193, 218, 225
interference, 83, 89
interoperability, 89, 226
investment, 142, 143, 146, 148, 167
irradiation, 11, 12, 32, 47
islands, 67
isolation, 94
issues, vii, viii, 1, 4, 57, 59, 61, 62, 64, 65, 69, 73, 83, 98, 99, 104, 107, 109, 121, 137, 211, 212, 213, 214, 215, 218, 219, 220, 224
Italy, 101

J

Japan, 66, 78, 101
Jordan, 141

L

latency, 87, 88, 89, 90
Lawrence Berkley National Labs (LBNL), 184
lead, viii, 62, 67, 84, 92, 98, 105, 106, 107, 122, 145, 166, 173, 213, 217, 218
leakage, 121
life cycle, 211, 212, 222, 226, 229, 230
light, 61, 76, 111, 121, 155
light conditions, 61
linear systems, 149
lithium, 105
load imbalance, 159
local area networks, 89
low-voltage ride-through, 1, 2, 4
LTC, 58, 65

M

machinery, 105
magnitude, 64, 88, 113, 136, 161, 164, 190
man, 226
management, viii, 58, 59, 78, 85, 86, 87, 105, 109, 138, 142, 160, 165, 168, 171, 172, 173, 177, 194, 202, 206, 208, 220, 224, 227, 228, 230
mapping, 72, 146
market share, vii
mass, 72, 76, 99, 159
materials, 193
matter, 113, 121, 162
maximum power point tracking, 1, 163, 166
mean time before failure, 58, 59, 62
measurement, 13, 14, 87, 179, 228
measurements, 136, 175
media, 88, 223
melt, 97
messages, 87
meteor, 96
meter, 63, 83, 89, 108, 110, 137, 143, 150, 208, 211, 212, 217, 218, 224, 225, 228
methodology, 165, 174, 202
Mexico, 72
Microsoft, 185
mission, 215
models, 3, 7, 9, 14, 23, 30, 34, 63, 73, 95, 109, 110, 115, 116, 137, 142, 145, 149, 150, 160, 163, 171, 176, 184, 191, 207
modifications, 3, 4, 5, 13, 30, 84, 145, 167
modules, 10, 14, 32, 59, 83, 98, 104, 105, 117, 122, 161, 170, 193, 198
momentum, 89
motivation, 222
multiples, 185

N

National Renewable Energy Laboratory, 58, 66, 100, 101, 102, 233
natural gas, 156, 194
negative consequences, 229
network elements, 149
networking, 226, 228
New England, 65
nickel, 105
nodes, 119, 129
normal distribution, 93, 145, 190
North America, 220

O

oil, 213, 220
Oklahoma, 64
operating costs, 79, 88, 152, 153, 154, 155, 157
operating range, 113, 198
operations, viii, 12, 61, 107, 150, 160, 212, 213, 215, 222, 223, 224, 225, 228, 230
opportunities, 109, 138, 142, 168
optimization, 60, 95, 144, 206, 224
oscillation, 149
oversight, 220

P

Pacific, 211
parallel, 8, 66, 73, 82, 85, 92, 93
participants, 162
pathways, 97, 212, 224, 225, 229, 230
permit, 63, 70
PES, 209
petroleum, 156
phase-locked loop, 1
photovoltaics (PV), vii, viii, 1, 109, 142, 171, 173
physical phenomena, 149
physics, 176
plants, 23, 61, 72, 79, 97, 152, 157, 165
pleasure, 141
PM, 190, 199
point of common coupling voltage, 58, 69, 81
policy, 98
pollution, 97, 99
portfolio, vii, 15, 141, 144, 148, 152, 154, 156, 157, 159, 160, 164, 165, 166
Portugal, 138
positive feedback, 67, 68, 78, 82, 150
positive sequence load flow, 1, 67
postal service, 89
power generation, viii, 61, 62, 93, 105, 107, 122, 143, 145, 208, 214
power lines, 216
power plants, 79
power system stabilizer, 1, 7
present value, 148
prevention, 66, 73, 82, 90, 219, 227, 230
price signals, 89
principles, 212, 213
probability, 68, 71, 88, 146, 167, 175, 176, 182, 190
probability density function, 68
probability distribution, 68, 175
project, 65, 66, 70, 72, 171, 211, 212, 213, 214, 215, 219
proliferation, 99
propagation, 83
proposition, 173
protection, viii, 68, 72, 77, 81, 87, 88, 92, 97, 107, 142, 150, 151, 164, 168, 173, 214, 215, 217, 218, 227, 228
protective mechanisms, 227, 228, 229, 230
public safety, 217

Q

quality of life, 217
quantification, 194, 202
quartile, 181

R

radiation, 194
radio, 88, 89, 168, 223
radius, 113
ramp, 3, 30, 64, 79, 92
reactions, 228
reading, 69, 83, 89
real time, 90, 94, 96, 159
reality, 86, 107, 190
recall, 160
recommendations, 164, 228
recovery, 23, 31, 32, 162, 163, 219
redundancy, 88
regression, 145
regression analysis, 145
regulations, 60, 61, 173
regulatory agencies, 96
reinforcement, 166
relevance, 31
reliability, viii, 2, 3, 4, 23, 30, 31, 32, 60, 62, 78, 85, 87, 88, 89, 96, 99, 100, 139, 144, 145, 146, 147, 148, 150, 151, 164, 171, 172, 173, 174, 175, 177,

178, 179, 181, 183, 184, 194, 195, 196, 198, 199, 200, 201, 202, 203, 204, 205, 206, 207, 208, 212, 215, 227, 228, 230
renewable energy, vii, viii, 57, 152, 154, 177, 211, 212, 214
Renewable Systems Interconnection (RSI), vii, 57, 211, 213
requirements, 2, 3, 4, 30, 32, 66, 82, 84, 87, 89, 90, 95, 97, 108, 109, 110, 121, 129, 137, 141, 144, 146, 147, 148, 154, 157, 159, 160, 162, 165, 166, 168, 227
researchers, 78, 176
reserves, 146
resistance, 62, 86, 112
resolution, 68
resources, 3, 30, 32, 88, 93, 111, 154, 177, 216, 222, 229
response, 18, 47, 48, 49, 50, 51, 52, 53, 67, 73, 79, 81, 82, 86, 90, 92, 95, 97, 142, 149, 150, 153, 159, 162, 168, 175, 227
response time, 79, 92, 97
responsiveness, 47
restoration, 142, 168, 215
retirement, 147, 157
revenue, 62
risk, 31, 212, 213, 215, 216, 217, 220, 222, 223, 229, 230
risk assessment, 229, 230
rules, 15, 63, 70, 96
rural areas, 224

S

Sacramento Municipal Utility District, 72, 211
safety, 68, 87, 215, 217, 225
Sandia Voltage Shift, 58, 73
saturation, 65
savings, 66, 95, 139, 152, 197
scale system, 67
scaling, 200
scope, 85, 139, 148
security, viii, 211, 212, 213, 214, 215, 217, 218, 219, 220, 222, 223, 225, 226, 227, 228, 229, 230, 231
semicircle, 113
Senate, 225
senses, 71
sensitivity, 14, 145, 152
sensors, 230
services, 3, 31, 60, 84, 99, 160, 166
SFS, 58, 73, 74
shape, 71, 155, 193
shock, 222
shortage, 90, 154, 159

side effects, 217
signals, 47, 48, 52, 83, 84, 86, 87, 88, 89, 92, 93, 94, 95
silicon, 104, 105
simulation, 2, 4, 5, 10, 13, 14, 18, 19, 47, 52, 67, 74, 76, 114, 142, 146, 147, 148, 155, 157, 167, 172, 176, 177, 181, 183, 184, 186, 190, 193, 195, 197, 199, 200, 201, 202, 203, 207
simulations, 2, 4, 5, 9, 10, 13, 14, 15, 17, 18, 74, 75, 149, 150, 166, 183, 184, 195, 198
sine wave, 65
smart com, 225, 230
smoothing, 65, 199, 206, 207
sodium, 106
software, 84, 85, 109, 137, 138, 142, 164, 166, 167, 213, 217, 223, 224, 227, 228, 229
solar photovoltaic (PV) installations, vii
solution, viii, 76, 77, 78, 79, 96, 212, 213, 222
Spain, vii, 1, 4
specifications, 196, 198
speed of light, 87
spreadsheets, 190
Spring, 72
stability, vii, viii, 2, 4, 10, 23, 24, 31, 32, 57, 58, 85, 86, 109, 148, 149, 150, 162, 212, 213, 214, 222, 224, 227, 228, 230, 231
stabilizers, 163
standard deviation, 145, 185
standardization, 77
state, vii, 4, 18, 19, 23, 31, 32, 48, 65, 71, 72, 113, 118, 138, 144, 149, 161, 171, 185, 216
state-level renewable portfolio standards, vii
states, 95, 148
statistics, 166, 172, 179, 183, 184, 195, 196, 200, 205, 206, 207
storage, viii, 3, 30, 59, 60, 61, 62, 63, 70, 71, 77, 78, 79, 80, 84, 85, 86, 90, 92, 93, 94, 95, 96, 97, 99, 104, 105, 106, 113, 142, 143, 154, 160, 161, 165, 171, 172, 173, 175, 177, 178, 184, 195, 196, 198, 200, 201, 203, 204, 205, 206, 207, 208, 213, 215, 224, 227
structure, 9, 14, 225, 228
structuring, 156
style, 163, 193
sulfur, 106
supplier, 105
supply disruption, 62
support services, 61
Sweden, 209
Switzerland, 105
synchronization, 63
synthesis, 90

T

tanks, 71
target, 111, 146, 147, 222, 223, 224
team members, 211
teams, 86
techniques, 82, 85, 86, 109, 137, 175
technologies, vii, 1, 4, 59, 60, 69, 88, 89, 143, 160, 166, 173, 212, 214, 218, 222, 229, 230
technology, vii, viii, 1, 3, 4, 30, 57, 59, 60, 61, 62, 68, 86, 87, 89, 97, 105, 143, 144, 146, 160, 166, 167, 168, 172, 173, 196, 206, 207, 208, 222, 225, 230
telecommunications, 109, 213, 218
telephone, 87, 88
telephone conversations, 87
temperature, 63, 95, 112, 145, 171, 173, 195, 198, 199, 205, 206
terminals, 81, 131
territory, 150
terrorist groups, 216
terrorists, 216
testing, 72, 109, 137
testing program, 72
threats, 215, 216, 217, 220, 222, 224, 227, 229
time periods, 121, 229
time resolution, 80
topology, 109, 132, 137, 161
total costs, 148
trade, 170
transformation, 8, 160
transmission, vii, viii, 1, 2, 3, 4, 5, 12, 14, 17, 18, 19, 30, 31, 32, 86, 88, 89, 96, 97, 105, 131, 141, 142, 143, 144, 146, 148, 149, 150, 151, 153, 154, 160, 162, 164, 166, 214
transport, 68, 89, 214, 229
transportation, 213
treatment, 142

U

uniform, 187, 190, 198
uninterruptible power supply, 58
United, 68, 69, 70, 104, 145, 160, 166, 170, 171, 173, 178, 194, 196, 199, 201, 202, 206, 224
United Kingdom, 68, 69

United States (USA), 68, 70, 104, 145, 160, 166, 170, 171, 173, 178, 194, 196, 199, 201, 202, 206, 224
universities, 68
urban, 176
UV, 58, 63, 81, 82

V

validation, 7, 94, 177, 185, 190, 197, 199, 204
vanadium, 106
variables, 86
variations, 13, 14, 17, 18, 32, 61, 62, 121, 154, 175
vector, 113
vehicles, 105, 204
velocity, 158
ventilation, 173, 177
vision, 172
vulnerability, 212, 215, 217, 218, 223, 224

W

Washington, 231
waste, 61
water, 62, 71, 90, 96, 109, 191, 218
water heater, 90
weakness, 217
wealth, 145
wear, viii, 107
Western Electricity Coordinating Council, 58, 67
wholesale, 156
Wi-Fi, 89
wind farm, 80, 82, 152
Wind generation, vii, 1, 4
windows, 81
wires, 62, 88, 89
Wisconsin, 102
workers, 81
worldwide, 99

Y

yield, 215

Z

zinc, 106